WEYERHAEUSER ENVIRONMENTAL BOOKS
William Cronon, Editor

Weyerhaeuser Environmental Books explore
human relationships with natural environments
in all their variety and complexity. They seek to
cast new light on the ways that natural systems
affect human communities, the ways that people
affect the environments of which they are a part,
and the ways that different cultural conceptions
of nature profoundly shape our sense of the world
around us. A complete listing of the books in the
series appears at the end of the book.

THE COUNTRY

The Greening of the San Francisco Bay Area

IN THE CITY

RICHARD A. WALKER

Foreword by William Cronon

UNIVERSITY OF WASHINGTON PRESS

Seattle and London

The Country in the City *is published with the assistance of a grant from the Weyerhaeuser Environmental Books Endowment, established by the Weyerhaeuser Company Foundation, members of the Weyerhaeuser family, and Janet and Jack Creighton.*

© 2007 by the University of Washington Press
Printed in the United States of America
Design by Pamela Canell
12 11 10 09 08 07 5 4 3 2 1

University of Washington Press
P.O. Box 50096, Seattle, WA 98145
www.washington.edu/uwpress

Library of Congress Cataloging-in-Publication Data
Walker, Richard, 1947-
The country in the city : the greening of the San Francisco Bay Area / Richard Walker.
p. cm. — (Weyerhaeuser environmental books)
Includes bibliographical references and index.
ISBN-13: 978-0-295-98701-9 (hardback : alk. paper)
ISBN-10: 0-295-98701-4 (hardback : alk. paper)
1. Urban landscape architecture—California—San Francisco Bay Area.   2. Urban renewal—Environmental aspects—California—San Francisco Bay Area.   3. Landscape ecology—California—San Francisco Bay Area.   4. Environmental protection—California—San Francisco Bay Area.   I. Title.
SB472.7.W35 2007   712.09794'6—dc22   2006034211

To my father, Robert A. Walker,
who brought us West and
took us camping and fishing
every summer

and

to my mother, Louise C. Walker,
who taught me to love gardening
and first led me to Rachel Carson

May their enthusiasm for this piece of earth
pass to my little force of nature, Zia.

CONTENTS

MAPS

FOREWORD *Thinking Globally, Acting Locally*

William Cronon

In 1969, having just been forced to step down as executive director of the Sierra Club after a dispute with the club's board about financial management, the colorful, larger-than-life David Brower founded a new organization called Friends of the Earth that soon became a worldwide leader in environmental politics. As the motto for this new group, Brower offered a prescriptive phrase that has since gone on to become a defining theme of environmental activism: "Think globally, act locally." However broad or systemic environmental problems might be, they almost always express themselves in local places, so that we often encounter them most directly in our own homes. Moreover, Brower was suggesting that effective environmental activism also begins at home, in the seemingly small day-to-day actions within our own communities that when multiplied together by thousands and millions of other people doing similar things, ultimately produce our myriad impacts on the creatures and ecosystems of the planet. The other key insight expressed in this slogan is that people can feel immensely disempowered and dispirited trying to solve the problems of an entire planet. Rather than leading to effective activism, the global perspectives of environmentalism can all too easily yield passivity and despair. By calling for local action, then, Brower was focusing people's attention on places where they could unquestionably make a difference if only they chose to do so: in their own homes and communities and daily lives.

Given the widespread acceptance of "think globally, act locally" as a key principle of environmental politics and ethics, it is a bit surprising that relatively few environmental histories have been written in this spirit. Except when dealing with a single local campaign (for instance, the controversy surrounding Love Canal in the late 1970s), academic studies of environmental activism have typically focused on geographical scales well above the local community. More often than not, national legislation and the actions of federal agencies have received disproportionate attention. Certainly this was true of Samuel Hays's classic *Conservation and the Gospel of Efficiency*, which arguably launched the modern field of environmental political history, and it was equally true of Roderick Nash's *Wilderness and the American Mind*. Hays's magisterial study of post–World War II environmentalism, *Beauty, Health, and Permanence*, pays much more attention to local activism and conflicts, but the vast scope of the book prevents any one place from receiving more than a few pages of discussion. Although these books and their successors certainly detail political conflict in particular local places, their larger goal has generally been to attach those places and controversies to a wider national narrative.

We do have a few environmental histories of individual states, with William Robbins's two-volume study of Oregon (published by the University of Washington Press in this series) standing as the most distinguished benchmark of the genre. But below the level of the nation or the state, we have essentially no long-term histories of local environmental politics. The reasons for this are many, but can no doubt be attributed partly to environmental history's long-standing interest in the public lands and the federal policies that have shaped places like national parks and national forests. Access to federal records is often a good deal easier and fuller than is true of more local archives, and this too may help account for the national and state-level scale of many environmental histories. Even those scholars who *have* focused on smaller geographical areas to analyze local environmental politics—here, Andrew Hurley's *Environmental Inequalities* remains among the most distinguished examples—have tended to focus on a single issue or campaign.

For all these reasons, Richard Walker's *The Country in the City: The Greening of the San Francisco Bay Area* should be welcomed as a truly pathbreaking work. There is really no book quite like it. Not only does it offer the first comprehensive history of environmental activism in the communities surrounding San Francisco Bay from the nineteenth century to the present, it is in fact the first comprehensive study of local environmental politics that has ever been written for *any* American city. As such, it offers a model that we can only hope other scholars will emulate. Rarely have the benefits of "thinking globally but acting locally" been more evident than in the pages of this book—making it all the more appropriate that David Brower was himself a prime example of a San Francisco environmentalist engaging in activism internationally, nationally *and* locally, all at the same time.

Richard Walker is one of the nation's leading economic and historical geographers, but he is also a child of the Bay Area with a deep love of one of America's most stunningly beautiful metropolitan regions. Having served for more than three decades on the faculty of the University of California, Berkeley, he has been personally involved in many of the political campaigns described in this book and has had first-hand access to individuals who led and contributed to those campaigns. As a result, his narrative of San Francisco environmental politics combines careful research and long-term historical analysis with a strong sense of personal engagement that is unusual in a work of academic scholarship. Like many Bay Area activists, Walker is as deeply committed to social justice as he is to environmental protection, making him an especially passionate chronicler of historical struggles in which Left-Green alliances have probably been more common than in the nation as a whole.

Walker's title—*The Country in the City*—recalls a classic work of environmental scholarship by the British literary critic Raymond Williams entitled *The Country and the City*. In that book, Williams traced the history of pastoral tropes and conventions in English literature, paying particular attention to the ways in which artistic representations of rural nature have tended to suppress the human (and animal) labor that has gone into creating and sustaining that nature. Like Williams, Walker is acutely conscious of the role that social class plays in defining who has access to which landscapes, and what kinds of human uses of the natural world are and are not considered appropriate by different groups of people with different degrees of wealth and power.

But rather than follow Williams out into the rural countryside to recount (for instance) the familiar story of how San Franciscans like John Muir and David Brower protected natural areas like Yosemite or Grand Canyon,

Walker organizes his narrative around struggles to protect natural areas right within San Francisco's metropolitan district. By focusing on the country *within* the city, he is able to avoid the environmentalist dichotomy that too often segregates non-human nature from places of human dwelling and labor. Given Walker's long-standing interest in political economy, the result is a book that regularly juxtaposes the forces of urban development and economic growth with the natural areas that lay in the path of those forces. The heroes of his story are the growing numbers of people who sought to defend those beloved places from what Walker sees as the inevitably destructive effects of an expanding capitalist economy.

This emphasis on land conservation in the face of capitalist economic growth gives Walker's book both its narrative unity and one of its most important messages for contemporary readers far beyond the Bay Area. Although the focus of public concern might shift from decade to decade—from an emphasis on romantic nature in Olmstedian public parks during the early years to growing concerns about pollution, toxicity, and biodiversity by the late twentieth century—the central riddles of land conservation are perennial. The growth of the built environment produces essential amenities for many urban residents—and great wealth and power for a few—but it also brings a host of attendant costs and challenges. Among the most important of these have to do with protecting public space against private privilege, and sustaining natural ecosystems and human health against political economic forces that regularly undervalue their importance.

Here, Walker offers a contrarian note of hope. Although there are plenty of things to be angry about when contemplating the history of past American abuses of the environment—and this book is forthright about such anger—there is much to celebrate as well. For Walker, environmental politics in the Bay Area have been characterized by remarkably broad-based citizen coalitions—transcending the divisions of class or race or gender—of people who whatever their other political beliefs were deeply committed to the protection of urban nature as a beloved and defining feature of their communities. Having no patience for those who proclaim the so-called "death of environmentalism," Walker sees the successful struggle to build a powerful public commitment to environmental protection in San Francisco and its neighbors as a bellwether for how such work should be done elsewhere in the future. Only by reaching out at the local level to growing numbers of citizens increasingly aware of their shared interests, he says, can the forces that would otherwise destroy "the country in the city" be defeated. By acting locally against global forces that potentially threaten all landscapes and communities, we defend not just the local nature of places like San Francisco Bay, but all those other local natures that together comprise the nation and the world as well.

PREFACE

For me, this book is an attempt to go home again. As a child of the Bay Area, I grew up with the Santa Cruz Mountains as a backdrop, the open foothills behind Stanford as a playground. The orchards of Santa Clara Valley were still in bloom, the redwoods of Big Basin and Memorial Park beckoned, and the beaches of Pescadero, San Gregorio, and Capitola were practically empty. My parents took us on long family vacations every year, where we breathed the colors of California and the West on extended car trips. This was well before air-conditioning was introduced in automobiles, and the windows stayed wide open through the heat of the valleys and the cool of the big trees. I loved the lay of the land.

As a teenager, I was awakened to environmental politics by the Save the Bay movement and the Sierra Club's campaign to protect the Grand Canyon. Close to home, I heard of strange conflicts over Stanford University's land-use plans—conflicts in which my dad played a big part—while I was still young. Naturally, I ended up on the side opposite my father and the university administration he loved. But he had only himself to blame for taking me out into the mountains.

In college I lived in Europe for a time and saw a different kind of landscape than I was used to. I have been back there many times. While California was the frontier of modernity when I started traveling abroad, something grabbed me about that old land across the Atlantic: not just the ancient monuments and churches, but a landscape so lovingly maintained by the human hand. I got to see it before agricultural modernization erased the livelihood of the villages and turned them into summer homes for urbanites.

Upon entering graduate school, I went east to Baltimore, Maryland. Johns Hopkins University had one of the few environmental programs anywhere in the country in the early 1970s. Reds Wolman had put together a wonderful collection of faculty and students in the newly assembled Department of Geography and Environmental Engineering. There I learned to think more deeply about things such as water policy and forest change.

Perhaps more than anything, the experience impressed me with how extraordinarily different parts of the United States are. I became a regionalist simply by realizing how much of a *Californian* I am. I recall the shock, for instance, on discovering that the Chesapeake Bay—with a coastline five times as long as California's—has hardly a speck of public access. How could that be? The beaches back home were public. I had never thought about it, but that expectation was in my bones.

On returning to the University of California, I taught courses in environment and pollution for several years before being diverted by urban and economic geography. My career as an environmental activist was short but sweet. I produced, along with Michael Storper and Elaine Mariolle, a report on the finances of the state water project, which helped sink the project. Through that experience, I got to know such water activists as Phil Williams, Bill Davoren, and Mark Dubois.

Through Ellen Widess, my partner at the time, I was immersed in the world of occupational health, pesticides, and environmental activism. She introduced me to such stalwarts of labor and pollution struggles as Tony Mazzocchi, Nick Arguimbau, and Mike Belliveau and made me understand what it meant to be devoted to—and not just write about—a good cause. It was a great time of life.

Over the years, I have crossed paths with many other wonderful people mentioned in this narrative, including Carl Anthony, Greg Karras, and David Brower. Unfortunately, I missed out on meeting some of the key players of Bay Area environmental history, such as Dorothy Erskine, Mel Scott, and Caroline Livermore. Some of the elders—Sylvia McLaughlin, Bill Kortum, Florence LaRiviere—were still around to interview, but the ranks of those born before World War II have grown thin. On the other hand, I am happy to have made the acquaintance of a number of younger people now involved in environmental work, such as Mateo Nube, Robin Grossinger, and Vivian Chang.

My early political education was as a New Deal Democrat, thanks to my parents. But it has been good to discover all the old liberal Republicans lurking in the greenbelt story, because I still remember a day when Republican did not equate to George W. Bush and his Horsemen of the Apocalypse. By the end of the sixties, I had become much more radical than my folks, much to my father's chagrin, but I am still full of that good old-time religion of the New Deal. I have never made a very good revolutionary.

Nevertheless, I discovered Karl Marx by reading *Capital* with David Harvey at Johns Hopkins. I still do Marxist economics in my day job, but it has never altered what I came to believe deeply in my youth: that protecting the earth—the land, the beasts of the field, the fish of sea, and, along with them, the health and good sense of the human species—is an absolute. I absorbed the lessons of Rachel Carson, Barry Commoner, and John Muir before I ever saw the light of political economy, and it has stuck with me. It was from environmental activists, not Leninists, that I first learned the virtue of a radical militancy in defense of things precious. It feels good to get back on the green bandwagon after years in the wilds of global political economy.

It's not as though everyone got the message that to be red, one should be green, and vice versa. There have been too many leftists who missed the boat on the environment and too many environmentalists who refused to grasp the nettle of capitalism. But there were many great thinkers around Berkeley and the Bay Area who did, and I could safely mingle in this swarm of fellow travelers. In the vast mainstream of American ideology, Berkeley-bred ideas are anathema, but I will defend our local political culture to the end. We have a bad habit of being right in the end, if a little crazy along the way. And this is why I champion so fervently the special place that is the Bay Area.

Most of all, this is a book for everyone in the Bay Area. It is meant to help those who live in the San Francisco Bay Area appreciate what they have and what they have done. They can use this book as an exercise in collective memory. Anyone who has walked the trails of Mount Diablo, picnicked in Tilden Park, surfed the Sonoma coast, or simply taken in the open vistas of the Golden Gate might want to know how the wonders they enjoy came to be. It is a civic duty to pay tribute to all those who made the greensward possible, and it should be a pleasure for those who enjoy our parks, streams, and bay to celebrate what we have.

The tale I have to tell is by no means dispassionate, but it is no less objective and worth reading for that. I have tried to tell it as honestly and thoroughly as possible. My red side tells me I should have been more critical of everything and everyone, but my green side wants this to be an upbeat lesson in the art of the possible (which we sorely need in these dire times). I have tried to be comprehensive, but it is impossible to be exhaustive within the covers of one book—especially if it is to be readable for the wider public. There are scores of green groups, hundreds of key fights, thousands of activists, and a million acres of open space and water in the Green City—this is only some of their stories. I have expressly downplayed the national and international sides of Bay Area environmentalism, which are better known, in order to show that even the big fish, including Muir and Brower, had to swim in the local pond.

I am happy to acknowledge some terrific help on this project. A trio of friends who keep the red-green flame burning are Gray Brechin, Matt Williams, and Marty Bennett, and all provided inspiration and assistance, including reading various chapters. I want to thank those busy activists who gave of their valuable time to be interviewed, and special thanks to those who read individual chapters and answered further questions: Ted Smith, Bill Kortum, Seth Adams, Caryl Hart, Larry Orman, John Woodbury, and Robin Grossinger. Among academic colleagues, Hal Rothman very kindly read chapter 4; Terry Young caught several gaffs in chapter 3; and Louise Mozingo made me rething the history of parks. Bill Cronon read the entire manuscript and provided a gentle guiding hand, as well as the kind of encouragement that keeps one going when book production becomes a slog.

I have learned so much from students over the years of teaching and advising that it is hard to thank them all individually. Many of them know more about environmental history and practice than I, and I have made good use of this fact. Some, including Joan Cardellino, Briggs Nisbet, Peter Cohen, and Monica Moore, became practitioners. Others, such as James McCarthy, Jake Kosek, Kate Davis, Mike Heiman, Scott Prudham, Rod Neumann, David Igler, Carmen Concepcion, Rachel Frosch, and Julie

Guthman, became professors. Jen Sokolove, Greig Guthey, Louise Dyble, and Shana Cohen have recently completed their doctorates, and Sarah Thomas, Juan DeLara, and Carmen Rojas are still in graduate school.

I have been ably assisted by many people along the way to finishing this book. Peter Cohen, Sona Chilingaryan, and Juan DeLara have done great work as my research assistants. Librarians at History San Jose, San Mateo History, Marin Library's California Room, and the Bancroft Library, especially Susan Snyder, head of Public Services and Ann Lage of the Regional Oral History Office, have given fully of their time. Julidta Tarver of University of Washington Press saw the potential in my early scribblings and brought my little ship into port; Marilyn Trueblood of the press ably oversaw the production of the book; and Kris Fulsaas did an admirable job of "municipal housekeeping" on the manuscript. All of them know that I don't use the term "scribbling" in a purely metaphorical sense.

Thanks are owed, as well, to many people whose influence on this book has been indirect but nonetheless invaluable. Here's to all the environmental journalists who have made my job easier, such as John Hart, Jane Kay, Harold Gilliam, and the team around David Loeb at *Bay Nature*, not to mention those stalwarts putting out the newsletters and broadsheets of the green organizations of the Bay Area. Special credit is due to a host of marvelous photographers whose images bring to life the pleasure of the Bay Area's landscape—especially those who generously let me make use of their talents. I can't say enough about the support and inspiration of my friends in the *Retort* group in Berkeley. And a tip of the hat to George Henderson, from whom I cribbed my title.

And thanks most of all to Annie for her ineffable enthusiasm, indispensable support, and, best of all, her love.

ABBREVIATIONS

AAAS	American Association for the Advancement of Science
AAG	American Association of Geographers
ABAG	Association of Bay Area Governments
AEC	Atomic Energy Commission
APEN	Asian Pacific Islander Environmental Network
ARB	Air Resources Board
BAAQMD	Bay Area Air Quality Management District
BCDC	Bay Conservation and Development Commission
CALFED	California-Federal Water Agreement
CalTrans	California Department of Transportation
CBE	Citizens (now Communities) for a Better Environment
COAAST	Californians Organized to Acquire Access to State Tidelands
CRLA	California Rural Legal Assistance
EBMUD	East Bay Municipal Utility District
EBRPD	East Bay Regional Parks District
EDF	Environmental Defense Fund

EPA	U.S. Environmental Protection Agency
FOE	Friends of the Earth
GGNRA	Golden Gate National Recreation Area
IGS	Institute of Governmental Studies
MALT	Marin Agricultural Land Trust
MMT	Methylcyclopentadienyl manganese tricarbonyl
MROSD	Midpeninsula Regional Open Space District
MTBE	methyl tributyl ethylene
MTC	Metropolitan Transportation Commission
NRDC	Natural Resources Defense Council
OPR	Governor's Office of Planning and Research
PCL	Planning and Conservation League
PG&E	Pacific Gas and Electric Company
PODER	People Organized to Defend Economic and Environmental Rights
POS	People for Open Space

POST	Peninsula Open Space Trust
PUEBLO	People United for a Better Oakland
RCRA	Resource Conservation and Recovery Act
RFF	Resources for the Future
SARA	Superfund Amendment and Reauthorization Act
SCAMP	Sonoma County Agricultural Marketing Program
SLUG	San Francisco League of Urban Gardeners
SMART	Sonoma-Marin Regional Transit
SPUR	San Francisco Planning and Urban Renewal (now Research) Association
SVTC	Silicon Valley Toxics Coalition
TALC	Transportation and Land-use Coalition
TNC	The Nature Conservancy
TPL	Trust for Public Land
WPA	Works Progress Administration

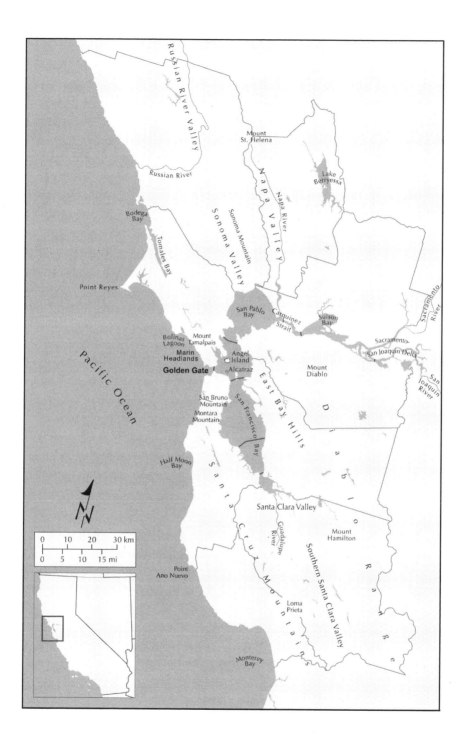

MAP 1. Major Topographic Features of the Bay Area. Drawn by Darin Jensen.

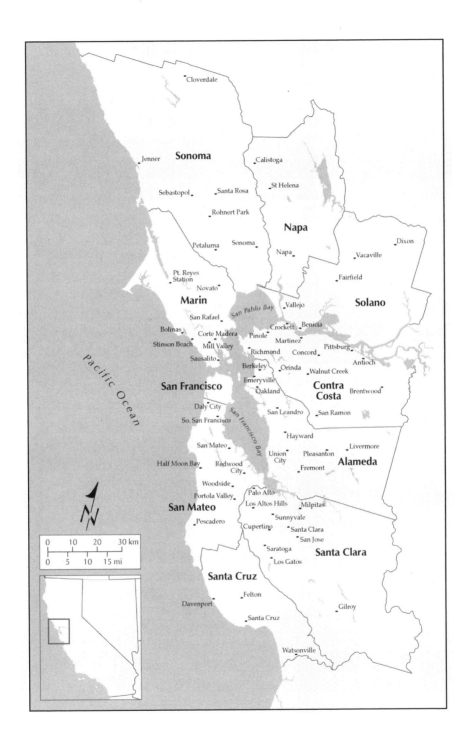

MAP 2. Bay Area Counties and Cities. Drawn by Darin Jensen.

THE COUNTRY IN THE CITY

The Greening of the San Francisco Bay Area

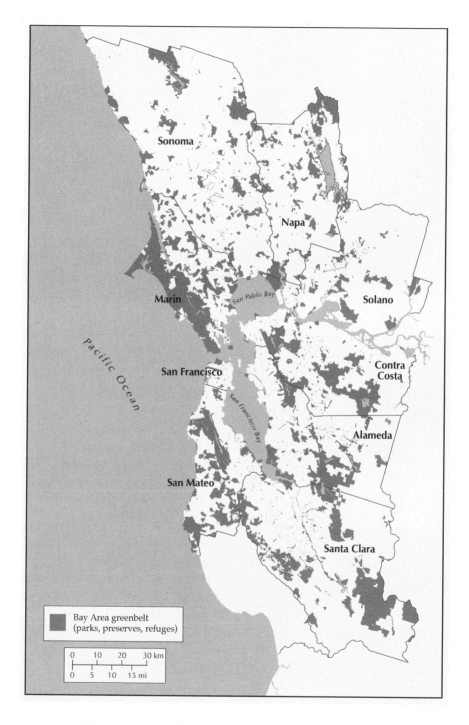

MAP 3. The Bay Area Greenbelt, 2005. Drawn by Darin Jensen. Courtesy of Bay Area Open Space Council, GreenInfo Network, Greenbelt Alliance.

INTRODUCTION *Saving Graces*

T he San Francisco Bay Area is more greensward than asphalt jungle, more open space than hardscape. Like the universe, it is mostly empty space. But what looks empty at first glance is not; it is a different kind of matter than the cityscape of bright lights and buildings. It is filled with a jumble of fields, woods, farms, and grasslands, of mines, marinas, creeks, and bays. This open space is sometimes called the urban greenbelt, but it is really a coat of many colors besides green—yellow, brown, blue, and gold—and less a belt than a quilt of countryside tucked into the folds of the metropolis.

In the nine-county San Francisco region,[1] 3.75 million of 4.5 million acres are greenbelt and open water, and less than 750,000 acres lie beneath buildings and pavements. More than 1 million acres of open space are protected, or are what advocates call secure greenbelt. The Bay Area has the most extensive such greensward in the country. It includes a patchwork of more than 200 publically owned parks and reserves within 40 miles of San Francisco covering an area larger than Yosemite National Park. At the heart of the metropolitan region is the bay itself, another 725,000 acres of water and wetlands guarded by government regulation. Agriculture occupies the largest swath of unpaved surface, 1.8 million acres,[2] and woodlands the rest, about a half-million acres.

This is the country in the city. Cities like the Bay Area are involuted throughout their length and breadth with elements, traces, and re-creations of the countryside, including farms, ranches, parks, watersheds, wildlife reserves, trails, creeks, beaches, restored wetlands, and more. These are as much as part of the geography of urbanism as offices, roads, airports, apartments, and shopping malls. Our green and pleasant lands are not something apart from the city, something left over from nature's beneficence. They are part and parcel of the city—that most profoundly unnatural of human artifacts. The greenbelt, the open bay, and the clear breezes of the Bay Area are ours to enjoy because of intense human activity. This book lays out the jigsaw geography of those places, the nonurban urban realm, and tries to arrange the pieces into a coherent narrative.

The making of the Bay Area's civic greensward has been a long and arduous process, built up over decades and fought over acre by acre. Every park, reserve, and trail required money, mobilization, lobbying, and electoral pressure. Every farm and watershed was developed through investment, hard work, and sacrifice. Every place not paved meant holding back developers and highway builders with local ordinances, regional commissions, and greenbelt alliances. Every mile of bay not reeking of sewage and garbage dumps is due to regulations, legal maneuvering, and political will. In short, the greenbelt is not a natural product of social progress and dawning enlightenment, but something that has been won through decades of effort.

This book tells the story of how the Bay Area got its green groove. It begins with the stirrings of conservation in the time of John Muir and the Sempervirens Club. It then lays out the agricultural and mining legacy of the region and their ongoing mark on the geography of the city. Next, it turns to origins of the recreational park system, "pleasure grounds" such as San Francisco's Golden Gate Park and the East Bay's Tilden Park. For the post–World War II era, it jumps north to Marin County and south to the San Francisco Peninsula to observe their astonishing success in stopping the suburban bulldozers and setting aside their beautiful backyards as nature preserves.

After that comes the all-important fight to save San Francisco Bay—the blue-green heart of the region. Subsequently, the narrative expounds on the battles all around the Bay Area to control urban growth rates, as well as the utopian movement for a regional plan. Following that, it turns to the use of open-space districts and land trusts in the wake of declining government funding. The story next moves farther north to the Wine Country in Napa and Sonoma counties and the lively fight for the greenbelt at the fast-growing urban fringe. The final two chapters look deep within the built-up city at the battle against pollution and toxic chemicals, in the neighborhoods of the working class and the dark-hued poor.

Every city produces its own antithesis: a working, living, vibrant countryside. This is true in a double sense of an *urbanized countryside* and a *ruralized city*. We need to take account of both senses of the country in the city. On the one hand, the economic process of urbanization necessarily demands a countryside to service urban needs: farms, quarries, reservoirs, dumps, and the rest. The urban periphery is, in an important sense, a huge backyard of the city, and one that is fiercely exploited. That is the urbanized countryside.

On the other hand, the first kind of countryside disappears as the city expands; it is platted, bulldozed, paved, and built upon. But there is a counterflow to urban advance, one that leads to setting aside open spaces, reserves, and other naturalized areas and that reworks nature into parks and other recreational spaces. It includes individual backyards that capture a bit of personal space and greenery. This is the ruralized city. It is born of a reaction to the bulldozers of urban development leveling the countryside and the desire to have something of the rural and the wild near at hand. It even reaches out into the working countryside to salvage the land from exploitation by the supply lines of the city.

This is not the ordinary view of the city and the country. Conventionally, the city is the built-up area and the country is a thing apart where nature reigns. As the city grows, it obliterates everything that came before. Environmentalists refer to this advance as "the wall of sprawl"; Adam Rome speaks of "the bulldozer in the countryside"; and Edward Soja calls it the "postmodern city," in which history is erased from the landscape. Such imagery is powerful and by no means altogether wrong, but it is nonetheless misleading.[3]

The city and country coevolve. One cannot say that the rural came first and the city later intruded upon it. George Henderson has rightly said of California that "city and countryside developed in tandem." There was no settled countryside that preexisted the urban; the Gold Rush of 1849 created "instant cities," and California quickly became one of the most urbanized places on earth. If we persist in seeing cities as latecomers invading pristine rural landscapes, we are complicit with the profound American amnesia regarding history and geography. The countryside has always depended on the city, coexisted within the urban realm, and left its mark on urban forms.[4]

This mingling of country and city complicates the social geography of environmental sentiment. Raymond Williams's pathbreaking work, *The Country and the City*, can help us sort this out. Williams shows how deeply intertwined the urban and rural social orders of England have been and how this gave rise to a particular upper-class "structure of feeling" about the English landscape.[5] Americans are no different. Their views about conservation and the land were born out of the encounter of the city and the county: from moving back and forth between the rural and urban, from their dreams of remembered youth in the country, and from urban sophistication and civic nightmares. Even more, these movements emerged out of struggles over the *nearby* countryside, where the city gave way to the suburbs. The impulse continues right down to environmental justice advocates wrestling with the meaning of a wholesome and livable environment in the central city.

The environment most urbanites care about is not primarily wilderness and nature far from where they live. The wilderness obsession was made the mantra of environmental history by Roderick Nash's *Wilderness and the American Mind*. Students would do better to start with Leo Marx's *The Machine in the Garden* and Henry Nash Smith's *Virgin Land*, because they portray the effort by Americans to come to grips with modernity in the countryside and the loss of the rural past as seen from the city. Urban historians such as Kenneth Jackson, in *The Crabgrass Frontier*, and Peter Schmitt, in *Back to Nature*, are also clear about the mingling of the country in the city. But environmental historians have only just begun to break with the wilderness tradition and get back to the cities.[6]

This unity of country and city is evident in the San Francisco Bay Area. People here have commonly been immersed in the city *and* in love with the country, notably Lake Tahoe, Yosemite, and Big Sur. Elites could bridge the geographic gap by owning both city homes and forest hideaways or by taking the railroad to Del Monte Hotel on the Monterey Peninsula. Even John Muir lived on the edge of the city when he wasn't tramping around the mountains. The middle classes made their forays into the countryside by car and on limited vacation time but enjoyed their campgrounds, fishing holes, and beaches. A common source of affection for the countryside is the memory of childhood excursions to the mountains or the coast.

The most cherished environments of the Bay Area have often been ones nearest the city, because they have been the most accessible, the most visible, and the most threatened. There are sainted venues like Muir Woods, the Napa Valley, and Point Reyes, but just as important are the more everyday places, glanced at out the window, such as Angel Island, Mission Peak, or the Carquinez Straits. These have been the sites of some of the fiercest and most defining battles over the fate of nature and of urban society in this region, and they are the ones that make up the story recounted here.

For a city of its size—seven million people, the fourth-largest metropolitan area in the United States—the San Francisco Bay Area is remarkably blessed with breathing room. The watery expanse of the bay is the centerpiece and the region's trademark. The surrounding rings of mountains and hills provide a topographic counterpoint to the bay and plains. The central Coast Ranges are rich with wooded canyons, open ridges, chaparral-coated hillsides, and oak-studded savanna. The place is blessed with great vistas and dramatic fog banks, thanks to the fresh breezes blowing off the Pacific Ocean, and combines an abundance of bright sunshine and rainfall in the same climatic package. It is one of the most beautiful cities on earth, many say, and no doubt nature has been kind to those of us who live here.

Nevertheless, one old myth of Bay Area exceptionalism needs to be put to rest. Its political stance did not spring from nature, despite the fact that this has been an article of faith among local greens. Dorothy Erskine, founder of People for Open Space, once declared that "our surroundings are so dramatic that I think that it has generated a certain emotional drive that led to victory in the battle for the Bay, and advanced our region farther than other parts of the United States in consciousness about the environment." Regional planner Jack Kent felt the same way: "What distinguishes the San Francisco Bay Area from its metropolitan peers is the remarkable way a dramatic geographic setting has . . . had a decisive influence in shaping a Bay Area 'way of life.'"[7]

All this would carry more weight if Los Angeles—so often disparaged by San Franciscans—were not so clearly more spectacular in its topography and more idyllic for human habitation. Looked at with some dispassion, the Bay Area is notoriously foggy, its mountains the lowest in the Coast Ranges, its hills mostly bare and dry in summer, its ocean front bitterly cold, its bay muddy, its creeks puny. What Los Angeles has lacked is not natural beauty but failure of imagination about saving its natural blessings.[8]

No, nature did not grant us all that we see. It did not give us parks and playgrounds, open space reserves and vineyards, piers and sailboats with which to enjoy the bay. And it certainly did not carry an insurance policy that assured the bay would never be filled, the mudflats fouled, and the rivers made to run uphill or that the hills would not be dug up for minerals, cut down for housing tracts, or lain to rest beneath freeways. It didn't give us trees in the right places, enough picnic tables and trails, or sufficient water for such a huge population.

It is all, in an important sense, art and artifice. A friend from New York declaimed many years ago on a visit to lush Golden Gate Park that "It's amazing to think that the Bay Area once all looked like this!" Amazing, indeed, since the previous sand dunes were pretty barren until made into a great horticultural garden by the park builders. Nor are the lovely vineyards of the Napa Valley any less a product of the human hand. Even in places as wild looking as Point Reyes, there is a profound imprint of two centuries of cattle grazing. The whole of the greensward is an act of social engineering. So let's clear our eyes about the blessings of nature, and see them through the prescription glasses of human action.

The greensward is the Bay Area's saving grace. But not because it naturalizes the metropolis. Rather, open spaces humanize the city. To say a city is made more humane by nature is no paradox. After all, the green lands are as much human- as nature-made, and therein lies their secret pleasure. This is a living landscape redolent with meaning for those who love both city and country and who realize how hard it is to reconcile the two. Open spaces close at hand are no passive backdrop to the city or leafy stage setting. No, every inch of the greenbelt, even the air we breath and the water we drink, vibrates with history and politics.

Who, then, had a hand in making the Bay Area's greensward? Who fought to humanize the city, protect its bay, hold on to its fresh air? The primary role of women in this vibrant history is striking. Women are everywhere in the story told here. They have served as leaders, innovators, and minions, and they came from every direction, as housewives, mothers, lawyers, and scientists. There's just no getting around the sheer number of women. Indeed, women such as Mary Bowerman, Caroline Livermore, Amy Meyer,

Janet Adams, and Pam Tau Lee are the real heroes of this story. They have done it all.

Yet women are all but invisible in traditional environmental history. Rachel Carson always appears, but there is no comprehensive study of women in American conservation. Robert Gottlieb tries to even up the balance sheet by including health and sanitation work done by progressives such as Alice Hamilton. And Susan Schrepter has shown that women were a vital presence in mountaineering and wilderness protection. Glenda Riley has assembled a remarkable collection of women writers and scientists coming to terms with nature in the far West, but hers is a strictly intellectual history. Environmentalism has not been just a man's world, and by getting down to the local level we get beyond the upper crust of male leadership, men's tendency to push themselves to the forefront, and bias toward charismatic figures. Too much ink has been spilt over the famously male yearnings for wilderness, a land ethic, and deep ecology in accounts of American environmental history.[9]

There is a hidden history of conservation that turns on the outlook and actions of women. In an era when women were still known by their husband's names, they were already having a significant impact on conservation projects and the spread of its principles through the network of civic and garden clubs, and a few were emerging as professionals and scientists. In the postwar era, women had an even greater public role in environmentalism; there were more college graduates than ever, and more were pursuing such callings as botany, landscape architecture, and engineering. But most were still housewives, looking for outlets for their energy and intelligence.[10]

Times have changed, and today's women environmentalists are usually in the workforce, and many are professionals. They are numerous in the nonprofit sector. And they are very present in pollution control and anti-toxics struggles, but by no means exclusively. As one veteran of the postwar generation notes, with good humor, "I was a typical 'do-gooder'; we're all gone now. Now good works are all done by executive directors."[11]

Why are women so prominent in conservation and toxics battles? Most likely it derives from women's age-old role as the prime caregivers and nurturers of children, family, and the domestic sphere. This was hardly invented by bourgeois society, though it was certainly amped up in the nineteenth century. And while women have steadily emerged into the workforce and public life, this intense involvement with domestic concerns and familial well-being has not passed away. It would be a mistake, however, to confuse the "domestic" with the private sphere; women's concerns have repeatedly been taken into public forums, from the "municipal housekeeping" reformers of Progressive-era Chicago to postwar suburban women mobilizing the political right in Orange County.[12]

In being thus engaged, women come up against problems that can only be called "environmental": Are my children safe? Where can they play? How are my friends' and family's health and well-being? How fares my community, and who gets to decide? If women develop a strong green bent, it has usually arisen from such things as the loss of nearby woods where children play, the contamination of local drinking water, or the stench arising from a nearby factory. One hears such reasons repeatedly in interviews with green activists or environmental-justice advocates, along with the more generic reflections on the local sources of their love of nature, including days at the beach, walks in the redwoods, or an affection for wildflowers.[13]

Leading male conservationists, such as John Muir and David Brower, have been more inclined to be high romantics, who slip easily into lofty abstractions about spiritual reunification between a transcendental nature and the interior self, and the like. Women environmentalists are more likely to be realists grounded in the local, communal, and ecological, who take their cue from pragmatic lessons of maintaining a home, building a support network, or gardening.[14] This is not to claim that all men are one way and all women another or that it is always in their natures to be thus; indeed, there has been considerable convergence in perspectives in recent years. But when the dust settles after we argue over statistics and essences, the difference in outlook remains too widespread to ignore.

The Bay Area environmental story bears witness to the importance of the elite in land conservation and nature protection. Greens are typically upper class and white, from the businessmen conservationists of the Progressive era to the Stanford- or Harvard-trained lawyers of today's national organizations such as Earthjustice. And the San Francisco region is wealthier than any other city in the United States, with a disproportionately large number of capitalists, managers, and professionals. But what compels the well-to-do to take the path of conservation? In the era of George W. Bush, it ought to be plain that environmentalism is not the inevitable calling of the national bourgeoisie and satraps of the American empire—quite the contrary, one might say. So why is the green band in the spectrum of elite ideas so wide in the Bay Area?

Certainly, it begins with the experience of the beautiful places around northern California, from the sea to the Sierra, and with a capacious sense of a collective "backyard" on Mount Tamalpais or along the San Mateo coast. The upper classes are prodigious consumers of space and nature, with large properties, large houses, and a large sense of command of the landscape. They are also keenly aware of the relation between a comely environment and land values. So they might be expected to try to create great swaths of desirable space near to where they live, in the name of conservation. They are, in short, great consumers of the landscape.[15]

But there's a rub. All is not class harmony when the forces of capital storm the gates of landscape consumption. Many times the redoubts of the rich have been threatened by urban expansion, and many times the privileged have had to take on developers head to head. This contradiction has been prominent in the Bay Area, from freeway revolts to stopping bay fill. It leads people to oppose their own class interest in fighting off developers, industry, and investors and amounts, in the larger sense, to biting the hand that feeds. Conservationists usually call this "opposition to *blind* progress," but the howls of protest from business against these class traitors say that it is opposition to blind *profit*. More often than not, it has been men on the side of "the city profitable," and women on the side of "the city livable," as Maureen Flanagan shows for the urban reformers of Chicago during the Progressive era.[16]

Nor is greenbelt protection reducible to elite consumption spaces and their defense against developers. Bay Area greens have held to a remarkably broad notion of public interest and public space. This has its roots in Frederick Law Olmsted's ideas about public pleasure grounds and Progressive notions of the public good; it expanded with the coming of New Deal social liberalism and public works and reached utopian heights among some Modernist regional planners after World War II; and, finally, it got a huge shot in the arm from the New Left radicalism of the 1960s and its aftershocks. A key difference between the conservative conservation of Westchester County, New York, as documented by Jim and Nancy Duncan in *Landscapes of Privilege*, and the socially liberal conservation of the Bay Area is that the vast majority of open space set aside here is public land for public use. From Point Reyes to Alcatraz, from Mount Diablo to the Bay Trail, the people—of all classes and kinds—are welcome and even encouraged to partake of their place in the greensward.[17]

Nor has environmentalism been the exclusive province of the upper classes or white people, even in the wealthy Bay Area. The bumper sticker made famous by Richard White, "Are you an environmentalist or do you work for a living?" paints a double-yellow line between classes, a line that really ought to be dotted. We should not be too quick to dismiss the mass of ordinary people as advocates of environmental protection. As Andrew Hurley shows so well in the case of Gary, Indiana, it is always a question of *the kind of environment* each class and race confronts in a city. The environmental struggles of different sides have their own integrity, even if inflected by class and race prerogatives or injustices.[18]

Environmentalism is a movement made up of a remarkable mixture of people and projects.[19] It is in no way reducible to the national conservation organizations, such as the Wilderness Society and National Wildlife Federation. This is particularly true of the Bay Area. No place in the country has witnessed more outbreaks of green activism. In fact, it was here that modern environmentalism first became a mass *political* movement in the 1960s with the sudden ballooning of the Sierra Club and the Save San Francisco Bay Association. And they are only two among dozens of local green organizations. The big groups stand out like downtown skyscrapers—monumental, symbolic, and dominating—but they are far from the sum total of the landscape of urban environmentalism.

The Bay Area's greenbelt has had considerable working-class support for decades. Workers have not been oblivious to the need for parks, recreational areas, and fishing habitat. Common folk and unions have lent support to conservationists back at least to the formation of the East Bay Regional Parks in the 1930s, and that continued through the creation of the Golden Gate National Recreation Area in the 1970s and resurfaced in such campaigns as putting urban growth boundaries around the cities of Sonoma County in the 1990s. Surveys show a rock-solid support for parks and open space among people of color, especially Latinos, both as users of public space and as voters for parklands.

In the Bay Area, environmentalism has steadily broadened its appeal, particularly in the domain of pollution control. Over the last generation, a host of environmental-justice groups dealing with brownfield issues in the inner city have raised new and troubling concerns about toxic environments. While their demands are not the same as those of mainstream greens seeking suburban open space, it is perverse to define such groups out of the environmental camp altogether. If anything, they have a deeper and more encompassing notion of what constitutes the urban environment and what it takes to make it livable. And their views have altered those of the larger conservation community.

Environmentalism is more than the sum of class, gender, and race positions or interests. It has an irreducible life of its own. The green tendency has been a pillar of Bay Area life and politics for more than a century, and it inflects every other social movement to some degree. It has been as significant a marker of the place as the Gold Rush, Silicon Valley, or the Beat movement. Raymond Williams would have called this local tradition of politics and opposition a "militant particularism."[20] We can call it, simply, a *green political culture.*[21]

Places are not just where things happen. They develop a character all their own that builds up over time, feeds upon itself, and affects all who pass through its bracing atmosphere. It is not given by nature or the mystic past; it has to be reproduced, like anything else. Harvard University has a distinctive culture; so does the Pentagon; so does as large and ill-defined a place as Hollywood. We make reference to these cultures all the time, but when it comes to whole cities or regions, we balk at making the same assertions. A city seems too big, people too flighty, the world too intrusive in this globalized era.[22] And certainly one cannot argue by assertion. This book is a sustained case for calling the Bay Area green.

Politics is not just elections and candidates, winning and losing. Politics is a system of parties, nonprofit organizations, civic groups, popular movements, business associations, and legal rules. At the same time, political life implies a public sphere, public debate, and public conflicts. These are the social cauldrons in which ideas grow, movements are born, and political hegemony is sustained. That is, politics is organized action, with an institutional fabric that must be built up and kept up to be effective. That, in turn, takes financial resources, social solidarity, and able leadership. Bay Area environmentalists have constructed an exceptionally effective network of organizations, from the Marin Conservation League to Save

Mount Diablo, which has continued to grow healthy new branches over the years.[23]

The public sphere also has a language and conventions that must be employed and appealed to in order to sway others to one's cause. This politic syntax takes on a particular slant depending on the time and the place. Consider the divergent public discourse of the United States and Europe about global warming in the early 2000s; the two hardly seemed to inhabit the same planet. At a smaller scale, in the Bay Area, ideas such as climate change, habitat loss, and disappearing species are much more common currency in the local media and classrooms than in, say, Houston. Professors, pundits, and politicians must all make reference to green causes from time to time to be taken seriously.[24]

Culture, whether we're speaking of politics, religion, or nation, is a social order, greater than any one person for the most part. It is the sea in which human beings swim, the air we breath. It helps define who we are and where we belong. To grow up, live, and work in the Bay Area is to partake of its green culture. At the same time, culture is not a structure that determines what we do and how we think. It is more than ruling ideology, repeated habits, and the rules of the game. It is an ethical universe in which we operate, a workaday grammar of good behavior that must be deployed by each and every person and can be reworked to deal with new situations as they arise. And being a system of moral sentiments, it is about "doing the right thing." In a green moral order, protecting nature is both good in itself and the road to the good life for the people of the city.[25]

In this sense of political culture, the Bay Area is not just a place full of green spaces, green projects, and green organizations. It is a place where green enthusiasms, like open space, are built into the fabric of urban life. It is a city whose identity is wrapped up in the local environment and its protection. Kenneth Olwig refers to this kind of affection for place as the "invisible fellowship of the landscape," and it is partly symbolic, partly material, and thorough communal.[26]

Of course, no place in the modern world stands alone, least of all a city at the crossroads of global technology, finance, and commerce like the San Francisco Bay Area. There is an intense traffic in people and ideas between different places. John Muir was born in Scotland, raised in Wisconsin, and toured Central America before coming to California. Frederick Law Olmsted only passed through California for a few years. Judi Bari moved to the redwood country from New York, Mike Belliveau from New England, and Caroline Livermore from Texas. None of this is unusual; many heroes of this story were born elsewhere. One can also point to reverse migrations, such as Petra Kelly and Jose Bové, both of whom spent part of their youth in Berkeley before going back to Germany and France, respectively, to raise green hell. The Bay Area's environmental culture has always been a fertile mixture of intensely local struggles informed by larger visions of nature, planning, government, law, and health.[27]

Nothing about the green cast of local political culture is inevitable. Nor is there anything preventing environmental causes from taking a rightward course. For example, population control, a perennial undercurrent of American environmentalism, has at its root eugenics and racial purification. The German Nazis claimed to have learned a great deal from California in their programs of racial cleansing, and they put great stock in environmental purity, as well. The Sierra Club still has a large contingent that would like to make border closure a plank of the club's policy. The comparison with Orange County or San Diego should remind us that the direction politics takes rests on local political culture; we are lucky that it took a more liberal as well as a greener path in the Bay Area.[28]

A staple of environmentalist faith in the Bay Area, as elsewhere, has been the belief in a beautiful, elegant, and virtuous nature. In this view, nature has its own logics of ecosystems, topography, and hydrology apart from those of humankind, even when significantly altered by human action, and those natural systems must be respected, admired, and, where degraded, restored to something like their previous vitality. Is this just a fantasy about untouched nature and the forest primeval? Not at all, though there are naive adherents to any set of beliefs. At root, the environmental ethic has a sound basis in science and a reasonable grasp of the human imprint on an urbanized countryside. Nonetheless, it has instilled in the denizens of the city a necessary critique of the fetishes of market value, anthropocentricism, and theology. For this, a sense of nature outside and beyond human affairs is not an idle dream but an essential point of leverage for rethinking the way we inhabit the earth.[29]

At the same time, environmentalism in the Bay Area took a distinctly left turn along the way. Local conservationists are much more militant and socially committed than those most anywhere else in the country. Thanks to the uncompromising stance of John Muir, the Sierra Club ended up more progressive than its contemporaries, the Audubon Society or the Boone and Crockett Club. In the mid-twentieth century, the Bay Area kept on turning left (or at least down a road less traveled) behind the likes of Dave Brower, Martin Litton, Dorothy Erskine, and Sylvia McLaughlin, as Bay Area greens turned against such mainstream ideas as nuclear power, hydroelectricity, and clear-cutting.

But is this enough? On the left, it has often been said that bourgeois environmentalists are not radical because they don't understand the grip of capitalism on the earth.[30] No doubt, it is true that the grasp of most activists on the dynamics of capital accumulation is shaky. Yet, on the metric of militancy in opposition to capital's unchecked ambitions, the greens often rate high. Environmentalists have regularly put up some of the fiercest battles against the prerogatives of property and money of any movement in America. This has been especially true in the Bay Area, where our militant particularists have refused to give in to housing developers, freeway builders, and electronics makers.

A similar criticism is made from the racial left these days. Upper-class white environmentalists are not radical enough because they don't grasp the oppression of the class system and the racial order.[31] Again, this is true to a considerable extent. But even without that understanding, greens have commonly been the fiercest opponents of the production system that so glibly poisons the earth (and people of color in its wake), of the highway network that spews pollution over urban neighborhoods, and suburban sprawl that starves inner cities of their vitality.

A fairer comparison is not to the ideals of the far left, however meritorious, but with other major opposition movements to the untrammeled power of business and industry in America. Consider the role of labor unions, for example. It is no less true that unions are a place where radicalism, reformism, and complicity with capital have always mixed uneasily. The labor movement has brought clear benefits to working people, despite its failure to live up to socialist hopes, but it has all too often been weak, craven, and corrupt as well.[32] That failure does not make unions any less necessary to any project of ameliorating American capitalism in all its ferocity.

It is not easy to pin down the ideological coloration of the Bay Area's progressive environmentalism. It has drawn on many intellectual currents critical of the prevailing order. For some, this includes classic socialism, for others anarchism, for others antimodernism, and for yet others religious faith. One encounters such ideological cross-dressing all the time. Gary Snyder, an inspirational writer of deep green persuasion, combines Zen, bioregionalism, and Wobbly sentiments in a seamless whole. Kenneth Rexroth, whose bohemian soirées launched the Beat movement in San Francisco, loved backpacking in the Sierra as much as he loved anarchism and Chinese poetry. But, in the end, green is one of the primary colors in the flag of the Left Coast, where politics and culture have long been out of step with the rest of America.[33]

Standard histories of American environmentalism are remarkably place-less—a fatal inattention to geography. Even though they inevitably make reference to the special role of the West as the frontier of American con-quest and of nature in the national imaginary, they forget that San Francisco dominated the West for a century.[34] It should not surprise us, then, that the Bay Area had a significant role in the national nature play.

In fact, the San Francisco Bay Area deserves special credit in the history of conservation and environmentalism. It was the birthplace of wildlands preservation with Muir and the Sierra Club in the 1890s. It gestated the National Park Service and the recreational parks under Stephen Mather and Horace Albright in the 1910s. It became the first place to organize mass movements for environmental protection under the Sierra Club and Save San Francisco Bay Association in the 1960s. And it led the way in the fight for clean air, clean water, and a nontoxic environment behind Citizens for a Better Environment and Silicon Valley Toxics Coalition in the 1980s.

One example should indicate the problem of inattention to place. In Bob Gottlieb's fine history of modern environmentalism, he credits Gary Snyder with coining the phrase "earth household" without placing Snyder in Northern California; he puts Denis Hayes, the organizer of Earth Day, at Harvard Law School while missing the fact that he was Stanford student body president; and he cites Peoples Park as "especially suggestive of how environmental themes emerged in [the 1960s]" but never links this to the qualities of Berkeley.[35]

The point is not to generate a list of "firsts" by Bay Area greens in the national accounts, however. Paying attention to place is not the same as civic hagiography. It is an essential part of the larger project of environmental history. Ideas and practices may reverberate nationally, but that doesn't mean they come out of nowhere. Whatever their larger compass, they still must be based in everyday life and landscape, in personal and collective geographies. The Bay Area's larger achievements are grounded in a place uniquely saturated with a green public culture. And that has redounded to the benefit of a larger national, and international, movement.

1

OUT OF THE WOODS *Stirrings of Conservation*

The opening chapter in the greening of the Bay Area takes place at the end of a half century of unadulterated exploitation of nature. The mining era's toll on California is one of the greatest tales of despoilation in the annals of modern environmental history, one in which entire forests were leveled, whole mountains picked up and dumped in the valleys, rivers moved from their courses, and wildlife annihilated. Along with this went the elimination of peoples and cultures from history. And it all took place at breathtaking speed and in one of the world's most distinctive landscapes. The extent of the destruction was already clear by the 1860s, and public sentiment against it would grow in subsequent decades.[1]

That plunder would have meant little, however, if California's people had not changed in the interim. San Francisco, the Queen City of the West, ballooned to a quarter million people during the mining era, a third of a million by the 1890s. It had become settled, wealthy, and bourgeois and had grown a thick layer of civic culture marked by universities, scientific academies, and museums. The city's leaders had great ambitions for both imperial expansion and civic improvement by the end of the nineteenth century, leavened by the rise of the United States to the first rank of industrial powers. And the region's upper classes, along with those in eastern cities, were nurturing a newfound love for travel, outdoor recreation, and natural scenery.

This is the era of the founding of the Sierra Club, Yosemite National

Park, and the first Redwood Park, followed by the battle for Hetch Hetchy Valley and the birth of the National Park Service. These events are touchstones of American environmental history. They establish San Francisco as one of the hearths of conservation and nature appreciation in the United States, along with Boston, home of the Appalachian Mountain Club, the offices of Frederick Law Olmsted, Senior and Junior, and the first metropolitan park system, and New York, birthplace of Central Park, the National Audubon Society, and the Boone and Crockett Club.[2]

THE VOICE OF THE WEST

The Sierra Club was the pioneer preservationist organization in the United States, founded in 1892 to agitate for Yosemite and Sequoia National Parks and generally advocate for the Sierra Nevada of California. It was established by John Muir, William Keith, Warren Olney, and a group of Berkeley and Stanford professors, including Joseph LeConte and William Dudley, and soon enrolled 300 prominent Bay Area businessmen, scientists, and professionals. The club's inspirational leader was Muir, who came to California after the Civil War—a refugee from Wisconsin farm life and midwestern industry. Spurred on by Robert Underwood Johnson, editor of *The Century*, Muir became the most influential voice for the natural wonders of the American West, writing in national magazines such as *Scribner's* and *Harper's*.[3]

Yosemite Valley was the country's first great scenic park—those defining places of American nature worship and national identity—granted by the federal government to the state of California in 1864. Yosemite served as an inspiration for the creation of Yellowstone in 1872 (officially, the first national park) and the Adirondack Forest Preserve of the state of New York in 1883 (the model for the national forest reserves a decade later). When Muir and his friends first gathered in 1889, it was to promote a larger park at Yosemite and better park administration, as well as federal protection for the finest groves of giant sequoias farther south. Yosemite and Sequoia became the second and third national parks the following year. Muir's beloved Kings Canyon would not win national park status until 1940, rounding out the Sierra Club's first half century.[4]

Muir and the Sierra Club were hardly a voice in the wilderness for the Sierra Nevada. Fulsome paeans to Yosemite, the big trees, and the High

Sierra had been coming out of San Francisco since before the Civil War. These were voiced by a generation of ecstatic Californians such as James Mason Hutchings (editor of *California*), Thomas Starr King (San Francisco's leading minister), Joaquin Miller (the state's poet laureate), and Clarence King (author of *Mountaineering in the Sierra Nevada*). They were backed, in turn, by the pioneering landscape photography of Carleton Watkins and Eadweard Muybridge. Lacking a mythology comparable to that of the *Mayflower*, Californians seized on the state's natural splendors to anchor themselves ideologically. More practically, town leaders in the San Joaquin Valley agitated for protection of the Sierra for tourism and water supply, and the powerful Southern Pacific Railroad backed their effort in order to increase ridership.[5]

By the 1890s, sentiment to halt the unrelenting pillage of the American forests was running high in industrial cities dependent on wood and water, and the U.S. government made a dramatic about-face to retain vast tracts of public lands as forest reserves. Meanwhile, scientific management of timber harvesting had become all the rage, under the influence of German and British foresters. At first, the Sierra Club and the scientific managers saw themselves as allies in conservation. But they soon parted ways because of Muir's belief that national parks should be protected from all commercial plunder, whether by logging, grazing, or mining. This was not just a matter of spiritual connection to nature but of the rediscovery of the need for a genuine American "commons"—land for all, beyond the reach of the all-powerful market. By the time Chief Forester Gifford Pinchot excluded Muir from the historic Governors' Conference on Conservation in 1908, "preservationism" was a heresy to be suppressed.[6]

The schism came to a head over the plan to turn over Hetch Hetchy Valley in Yosemite to the city of San Francisco for a water-supply reservoir. Muir's greatest campaign was to save Hetch Hetchy, whose flooding he regarded as a desecration of holy ground in pursuit of the almighty dollar. He was ably assisted by William Colby, organizational pillar of the Sierra Club for its first fifty years. But the club and local opinion were split over the project, and Muir lost.[7] The Hetch Hetchy battle secured John Muir's national legacy but killed him in the end. After Muir's death, the Sierra Club retreated from political confrontation and instead led hikes for the Bay Area's well-to-do. Inspiration for the further protection of wildlands would pass to the Wilderness Society during the 1930s and '40s.

Muir's voice reached a national audience like none other had. He was marked by oratorical brilliance and unrelenting advocacy. What made him especially effective was the way he transcended time and place with his calls for radical redemption of American society through appreciation of nature. Muir's spiritual attachment to the woods and the mountains harkened back to the great New England transcendentalists Henry Thoreau and Ralph Waldo Emerson; his aesthetic delirium, the English Romantics John Ruskin and William Wordsworth; and his absolute conviction and moral fervor, the evangelicalism of his father, a fierce Scots Presbyterian. Pinchot is remembered by historians; Muir's name still resonates in popular memory.

Muir was not a throwback, however; he was a thoroughly modern man, accomplished in the mechanical arts and natural history. He was the first to see that Yosemite had been carved out by glaciers and that Yellowstone was a geologic laboratory at work, and he was a fine amateur botanist. Although he contributed to the myth of American wilderness—a convenient erasure of millennia of human habitation—he was among the first to help the conquering host of Euro-Americans to appreciate the land they had so ruthlessly seized. He wanted people to enjoy places such as Yosemite and called the parks "the people's playgrounds." Above all, he had a journalist's touch for popular appeals at a time when mass-circulation magazines were beginning to have a major impact and writers such as John Burroughs were educating Americans about the joys of going back to the woods.[8]

Muir was more than a mountain man. He lived and worked for more than forty years at his father-in-law's ranch near Martinez, along the Sacramento River (the ranch is now a national historical monument). In the city he found allies, from artists to railway moguls, and men and women who would carry forward his ideals. Bay Area environmentalism took a radical turn with Muir, exceeding that of the higher circles of New York and New England conservation. The idea of preservation took root in the local soil and led a thousand flowers to bloom in the barren landscape of commercial exploitation of nature.

After World War II, the Sierra Club again took on the mantle of Muir's uncompromising preservationism. Under a new generation of directors, such as Ansel Adams, Richard Leonard, and Martin Litton, and especially through the dynamic leadership of David Brower, the club led the fight against violations of the scenic grandeur of California and the American West by dam builders, lumbermen, and ski promoters. During struggles ranging from the fight to save Dinosaur National Monument in Colorado in the early 1950s to the battle to save the Grand Canyon in the late 1960s, the loss of Hetch Hetchy was never far from their minds. Brower proved a brilliant publicist in the tradition of Muir, and his Exhibit Format books and full-page newspaper ads captured the national imagination for the cause of rivers and wildlands. The club's efforts helped expand the nation's pristine reserves by millions of acres through the Wilderness Act of 1964.[9]

As a result, the Sierra Club grew into a mass political movement, shedding its traditional clubby atmosphere. Its membership shot up from a few hundred to thousands by the mid-'50s, then catapulted from 16,000 in 1960 to 27,000 in 1965 and 113,000 in 1970. In the process, it outgrew the Bay Area, going from four local chapters to two dozen across the country. It became the flagship of modern American environmentalism, with a membership of nearly one million nationally by the end of the twentieth century, while retaining a vital tradition of grassroots activism and control. The growth of the Sierra Club propelled the "geography of hope" behind the tremendous achievements in U.S. land conservation in the postwar era. The club's anchor remains the Bay Area, where Muir's radical legacy is strongest.[10]

Muir's conflicting impulses also live on in the Bay Area. On the one hand, local environmentalists are steeped in premodern and antimodern ideas, such as Romanticism, Transcendentalism, bohemianism, communalism, and organicism. These run against the grain of the American commercial mainstream by arguing for the virtues of a simpler life, return to the land, and rejection of crass materialism. On the other hand, they have just as eagerly adopted some of the most prominent Modernist tendencies in American thought, such as ecological science, New Deal programs, and city planning.[11] We shall see how this ideological brew manifests itself over the years.

THE MAGIC MOUNTAIN

The Bay Area's adoration of the outdoors, mountains, and parklands was not confined to the faraway Sierra Nevada. It flourished closer to home, on Mount Tamalpais in Marin County. San Franciscans developed such a love affair with Mount Tam (as it is known) that it was regularly compared with Japan's Mount Fuji for its sublime vistas and spiritual effects—despite being a lowly 3,500 feet high. Each year hundreds of walking expeditions set off for the glorious views from the summit. Excursions became more common with the opening up of ferry service, the North Pacific Railroad, and the Blithedale Hotel in the 1870s. Things stepped up again with the rail spur and founding of the new town of Mill Valley in 1890, followed six years later by a narrow-gauge railway up the mountain to a hotel and restaurant at the summit.[12]

This local enthusiasm was in keeping with the worldwide mania for walking and mountaineering coming out of Britain, as well as with the rage for camping and taking to the woods on the East Coast led by the Appalachian Mountain Club and the Boone and Crockett Club. Hiking clubs sprang up all around the Bay Area, including the Berkeley Hiking Club, Contra Costa Hiking Club, Tourist Club, and California Alpine Club. The Sierra Club was just one of many such groups in the beginning. Fishing and hunting were other burgeoning recreational activities typical of American urbanites of the late nineteenth century.[13]

The adulation of Mount Tam was such that soon talk was rife about the possibility of creating a Mount Tamalpais National Park—which would have been one of the first parks in the country. It is hard to imagine such an idea being taken seriously today, but a Mount Tamalpais National Park Association was created in 1903. A mass meeting was held at Marin's Lagunitas Country Club to promote the idea, featuring U.S. Chief Forester Gifford Pinchot (the father of forest conservation), the chief of the U.S. Biological Survey, James Phelan (former mayor of San Francisco), and David Starr Jordan (president of Stanford University). The chief promoters of the park were Sidney Cushing and William Kent, builders of the narrow-gauge railway, who would have been well aware of the flourishing trade in railroad tourism across the West from Yellowstone to Yosemite, especially the success of the Del Monte Hotel on the Monterey Peninsula.[14]

William Kent was more than an investor and land developer, however. He was, first of all, the largest landowner in Marin County. Kent's father, Albert, had been a Chicago meatpacker who moved west in 1872 after a camping trip on the North Coast; he bought an old Mexican rancho and founded Kentfield. Albert was a great outdoorsman, and Will was a hunter and marksman, in keeping with the time. But Will's heart was still in Chicago, where he had grown up. He became a wealthy businessman there in his own right, engrossed in that city's rich civic milieu (he was friends with Jane Addams, John Dewey, and Peter Finley Dunne, among others). Kent became a leading reform Republican in Chicago, serving as alderman and president of the Municipal Voters' League. He donated land to Florence Kelley of Hull House to create the first children's playground in Chicago and subsequently became a board member of the National Playgrounds and Recreation League. After he moved west to help his aging mother, Kent became embroiled in earthquake reconstruction and antigraft politics in San Francisco, and he decided to settle again in Marin.[15]

Kent soon converted to the Bay Area conservation ethic, joining the Sierra Club. Kent is best known for saving Muir Woods from the lumbermen and a local water company. After the national park idea had fizzled out, the last prime stands of redwoods on Mount Tamalpais along Redwood Canyon on the peak's southwest flank were in danger of being sold off for logging and then inundated for a reservoir. Kent bought the grove instead and wished to make a public donation of it. On the suggestion of Frederick Law Olmsted Jr., Kent decided to turn it over to the federal government. He had his friend Gifford Pinchot urge President Theodore Roosevelt to declare it a national monument in honor of John Muir, which was done in 1908. It wasn't quite the full-fledged national park Kent had once envisioned, but Muir Woods is still the most-visited grove of redwoods in California—and one of the most well-tramped group of trees in the world.

Kent followed up the Muir Woods coup by helping establish the Marin Municipal Water District in 1912, to protect the northern flanks of Mount Tam and to supply water to his properties. He was in on the founding of the Tamalpais Conservation Club, which later secured a state park at the summit (with another gift of Kent land). Elected to Congress, Kent achieved prominence as an Independent Progressive from 1910 to 1916. He introduced the first bill for a Redwood National Park in 1913 and carried the National Park Act in 1916. He was also a founder of the Save the Redwoods League. He was a most flexible capitalist, comfortable operating both public and private enterprise and working at any and all scales of government.

Kent's environmental legacy is marred by his sponsorship of the Raker Act in 1913, whereby Congress turned over Hetch Hetchy to the water men. This was symptomatic of the position among San Francisco Progressives, led by James Phelan, whose goal was the recovery of their city after the 1906 earthquake and who hated the Southern Pacific, which supported Muir's campaign for Yosemite. Muir denounced Kent's treachery and broke off all further relations. Nonetheless, William Kent was the father of conservation in Marin County, where the Kent family remains a force to this day, and in championing a park for Mount Tam, he laid one of the cornerstones of the Bay Area greenbelt.[16]

Another inspired Marin conservationist was Laura Lyon White, wife of Lovell White, a prominent San Francisco banker. Lovell was president of the Tamalpais Land and Water Company that developed Mill Valley, so the Whites moved to a house they had built there in 1890. Laura Lyon White was the embodiment of women's Progressivism: active in the California suffrage movement of 1896, founder of the California Club in 1897, advocate for women's representation on public bodies, and campaigner for better working conditions for women (as well as aid to schools and juvenile justice courts). At the same time, she was chair of the Outdoor Art League, founder of the Marin Outdoor Art Club, on the San Francisco Playground Commission, and proponent of civic beautification and city parks. She refused the presidency of the California Federation of Women's Clubs, to her great credit, because of its exclusion of colored women's clubs.

On the larger canvas of conservation, White served as president of the Sempervirens Club, which saved the first redwoods in California, then led the battle to save the Calaveras Big Trees, carrying a petition with 1.5 million signatures to the White House in 1905 (the trees were included in Yosemite National Park in 1909). She and her husband turned over land in Redwood Canyon for part of Muir Woods National Monument. In White, one sees a confluence of interests that were characteristic of women conservationists of the first half of the twentieth century, especially the junction of gardening, civic works, and nature preservation. Yet the role of such women is usually underappreciated, and White's name is almost never mentioned in standard environmental histories. If John Muir and William Kent were founding fathers of Bay Area conservation, Laura Lyon White was its founding mother.[17]

CUT AND RUN

Bay Area preservationists may have been inspired by the view from mountain peaks, but they soon turned their eyes toward the state's grandest trees: the coast redwoods and giant sequoias of the Sierra. Redwood groves came to be seen as cathedrals of a transcendent god, living fossils proving the depth of geologic time, and ecological wonders worthy of scientific study. They were widely portrayed as America's answer to Gothic cathedrals and the great monuments of Europe in age, size, and power to smite the observer with the weight of the past and the force of religious conversion.[18]

People think of the Redwood Empire as something far away in Humboldt and Del Norte counties, 200 miles up the northern California coast. But the most readily accessible portion of the redwood country is in the hills surrounding San Francisco Bay. Here the trees grow nearly as large as they do along the Eel and Mad rivers up north, and the extent of the forests is substantial. The main stands are on the western slopes of the Coast Ranges, where rainfall averages 35–50 inches per year and can go as high as 100 inches. The Pacific coast timber and lumber industry started here. Yet even longtime residents of the Bay Area find it hard to believe that Redwood City began as a lumber port, not as a land speculation with a fancy name.

Cutting began around Mount Tamalpais during the Mexican era, when John Reed's sawmill at Mill Valley was constructed in 1836. The Gold Rush of 1849 brought a brief lumber boom to coastal Bolinas, and Samuel Taylor cut timber and milled paper on Lagunitas Creek on the mountain's northern flank after that. But Marin was soon cut over, and the action moved up to Sonoma County. The East Bay hills behind Oakland had a patch of redwoods as imposing as any in height and girth, which were used by the Spanish to guide their ships through the Golden Gate, but these were wiped out by the mid-1850s. Naturalist William Gibbons of the California Academy of Sciences documented the enormous stumps and brought John Muir and Alfred Russell Wallace to view them in the 1890s, mourning the lost opportunity to create a park on the site.[19]

The forest lay thickest on the slopes of the Santa Cruz Mountains south of San Francisco, and this became the principal redwood-logging region in California from the 1850s to the 1880s. A significant industry remained in San Mateo and Santa Cruz counties into the twentieth century. The eastern slopes, the most accessible, were the first to be cut. Timber coming off the ridge was milled in Woodside and Portola Valley, then sent down to wharves at Redwood City and Ravenswood to be floated up to San Francisco, which was built out of peninsula redwood. The first water-powered mills on the Pacific coast were installed there in 1850, and soon there were some fifty little mills in the area, plus a tannery operation using bark from tan oaks cut amid the redwoods. By the mid-1870s the stands were exhausted.

Then the center of logging jumped over the crest into the redwood and Douglas fir forests of the western slope of Santa Cruz and San Mateo counties. Scores of mills popped up along the creeks from Half Moon Bay south to Santa Cruz and deeper in the mountains at La Honda, Felton, and Boulder Creek. These included the highly successful operations of Purdy Pharis, "the shingle king," and William Waddell, who built the largest timber wharf on the coast, at Point Año Nuevo north of Santa Cruz. Neither ended well, with Pharis shooting himself and Waddell done in by one of the few remaining grizzly bears in the Bay Area. Charles Hanson, the one substantial tycoon to come out of this milieu, moved his operations up to Tacoma, Washington, in the 1880s—helping to open up the Northwest lumber industry—while continuing to live in Redwood City.[20]

Survival of the redwoods would not have become a burning issue had the forests been cut less ruthlessly. This was a fully modern, industrial assault in the manner of the timber industry around the Great Lakes or the Alleghenies, whose devastation triggered a national reaction after the Civil War. Already in the 1860s and 1870s, many Californians feared the exhaustion of the timber supply. Josiah Whitney, Asa Gray, and E. W. Hilgard helped found the American Forestry Association in 1875, and California established a state Board of Forestry in 1885—the first of its kind in the country—to regulate cutting.[21]

The federal Forest Reserve Act of 1891, and subsequent creation of the national forest system, set aside large areas of the public domain, but by that time there were no redwoods left on public lands to reserve; all had been privatized by the 1870s. The first two redwood parks, Big Basin and Muir Woods, had to be bought back from big capitalists: a lumber company in which Timothy Hopkins was the principal investor and Francis Newlands's timber and water syndicate, both derived from Central Pacific Railroad fortunes. There were many small timber operators at work but no significant number of common folk or native people living off the forests whom parks would displace as they did at Yellowstone, Yosemite, and the Adirondacks.[22]

BAY AREA BIG AND TALL

The reaction against redwood logging began at the end of the nineteenth century, and it started over the remaining trees around the Bay Area. Public concern for the loss of the groves of giants emerged for the first time in the 1880s, led by Ralph Smith, a newspaper editor from Redwood City; but the campaign faded after Smith was shot by a man he had criticized in his paper. It revived again under the leadership of a group calling itself the Sempervirens Club, formed in 1900. Their campaign focused on Big Basin in northern Santa Cruz County, which was in immediate danger of being logged. They succeeded in having the core Big Basin groves purchased in 1902 as the California Redwood Park. With the reversion of Yosemite to the federal government, Big Basin became officially the first state park—"to be preserved in a state of nature."[23]

Sempervirens Fidelis

The guiding spirit behind the Sempervirens Club was Andrew Hill, a photographer and painter from San Jose, who rallied the South Bay behind the idea of saving Big Basin. Hill was inspired by the majesty of the biggest redwoods, but there were other reasons for saving redwoods besides the aesthetic and transcendental, including the scientific, commercial, and social reformist. As a result, Hill was able to mobilize a broad political coalition to save Big Basin.

To begin with, the Sempervirens Club had the eager support of Professor William Dudley of Stanford University, who conducted the first botanical studies of redwoods. Dudley had marked Big Basin as an ideal park site a decade earlier. He easily won over Stanford president David Starr Jordan to the cause, and the first meeting to rally support was held on the campus. Recreationists were quick to join in, as well; by the 1880s, people from the towns were pouring into the Santa Cruz Mountains to hike, camp, and fish. W. W. Richards, wealthy secretary of the 20,000-strong California Game and Fish Protective Association, brought along the upper-class sporting set. The religious were led into the fold by the Reverend Edwin Sidney Williams, Congregational minister of Saratoga, California, and Father Robert Kenna, Jesuit president of Santa Clara College, who saw the park as a beneficial influence on the working class. The San Jose *Mercury*, chief voice of the press in the Santa Clara Valley, enthusiastically backed the plan for Big Basin.

Women played a major role in saving Big Basin, and the Sempervirens Club was a rare instance for the time of a mixed association of men and women. Author Josephine Clifford McCrackin, a pioneer writer on nature and culture in the Southwest, lived in the Santa Cruz Mountains and became the club's main literary voice, backed up by Carrie Stevens Walter, a writer for *Sunset* magazine, and fellow San Jose Club woman Louise Jones (who had risen from working-class origins). Laura Lyon White of San Francisco and Mill Valley became president of the club in 1903 and began establishing chapters all over the state, recruiting big names such as Phoebe Hearst, Joseph Knowland, and James Phelan to an honorary board. One suspects that this feminine input is why the Sempervirens Club has ended up so little known as one of the founding institutions of U.S. forest preservation, merely a footnote to the histories of the Sierra Club and Save the Redwoods League, both robustly male.[24]

There were, in addition, local businessmen who could be rallied on the idea of a Big Basin park on the practical interests in profit. One selling point was that it would bring tourist dollars to Santa Cruz and San Jose. The Santa Clara and Santa Cruz Boards of Trade enthusiastically backed the plan for this reason. Hopes for tourism grew out of the success of a few private groves in the Santa Cruz Mountains that had been making money as minor attractions since the 1860s—though these were more in the spirit of P. T. Barnum circus shows than portals to the sublime.[25] After 1880 one could ride the narrow-gauge railway from San Jose to Cowell's Grove near Felton, and plans for vacation homes and a hotel began to spring up.

A more distant reason for concern was San Francisco's ever-growing thirst, and the Spring Valley Water Company responded by wishing to grab the watersheds on the western slope of the Santa Cruz Mountains. The first Sempervirens Club president was Charles Reed, San Francisco Supervisor and notorious flack for the water company. Finally, there was a (false) hope that maintaining the forest to the west would keep the rain falling on the farms of the Santa Clara Valley. For all the preceding reasons, the bill for a park was carried through the Legislature by the Southern Pacific machine and won over the opposition of a doubtful governor and resistance from the timber industry.[26]

Once a modest park of 3,500 acres was established, interest in Big Basin died down. Andrew Hill kept after the state, and bits and pieces were added, bringing the park to almost 10,000 acres by the 1920s. But this was far from the 30,000 acres stretching from the skyline to the sea that club founders had originally envisioned, and after Hill died in 1922, the Sempervirens Club languished. In a sign of the changing times, subsequent club president William Flint quit in 1927 to pursue land-development projects with James Irvine in southern California. The club's guiding light after that was Herbert Jones, a state senator from San Jose and son of Louise Jones, who was too busy with agriculture and water-supply issues in the Santa Clara Valley to do much for the redwoods.[27]

Riches and Redwoods

The redwood rescue effort moved up the coast to Humboldt County, where cutting was proceeding rapidly, especially after Highway 101 was finished in 1915, opening up the interior to the lumbermen. A new group, the Save the Redwoods League, arose to lead the way in redwood preservation. The league became the country's most prominent conservation group between the world wars, taking the mantle from the Sierra Club after the death of Muir in 1914. It also moved conservation to the right, along with the rest of American politics, in the aftermath of the Progressive era.[28]

Founded in San Francisco in 1918 by an alliance of Bay Area and New York luminaries, the Save the Redwoods League grew within two years to a national membership of 2,000 of the choicest of America's elite. While easterners such as Madison Grant (a founder of the Boone and Crockett Club) have gotten much of the attention in histories of the league, the guiding lights of the group were local: Newton Drury (executive secretary), Franklin Lane (first president), John Merriam (second president), Arthur Connick (third president). The league was even more upper crust than the Sierra Club, though membership in the two organizations overlapped. League members were mostly businessmen close to the centers of national power. Lane was city attorney of San Francisco and then Secretary of the Interior under Woodrow Wilson; John Merriam was a professor of paleontology at University of California, Berkeley, and then director of the Carnegie Institute in Washington, D.C.; Drury was secretary to the president of the University of California and also was an advertising executive; Connick was a banker from Eureka and San Francisco. They were joined by the likes of J. D. Grant, a timber, oil, and electricity capitalist; Stephen Mather, a borate millionaire and director of the National Park Service; and William Kent, a landed aristocrat and congressman; as well as such public figures as UC presidents Benjamin Ide Wheeler and Robert Gordon Sproul and prominent San Francisco bankers William Crocker and James Moffitt. These men were mostly Republican (with some Wilson Democrats) and frequently active in Progressive civic organizations.[29]

To its credit, the Save the Redwoods League succeeded in making the tall trees a national cause célèbre. Drury ran the public relations campaign in the West, Mather in the East. Since all two million acres of redwood forest were privatized, forest reserves were out of the question and repurchase was a necessity to keep groves out of harm's way. By the end of the 1920s, the league had secured such prime stands along the North Coast as the Richardson, Rockefeller, Jedediah Smith, Prairie Creek, and Del Norte Coast groves. The league's strategy was private philanthropy, in the manner set in motion by Andrew Carnegie and John D. Rockefeller (whose son was a major donor to the cause). They gave of their own wealth and that of their rich friends to establish a green branch of American philanthropy—setting a precedent that continues to this day. By 1960 they had purchased upward of 60,000 acres; today the figure is close to 180,000 acres, at a total cost of more than $100 million.[30]

The league struck a line that differed from the Sierra Club's, even though it shared many of the same transcendental ideas about nature's glories and some of the same disdain for the lower classes (Muir couldn't hold a candle to Madison Grant, however, who was one of racial supremacy's chief advocates of the day). The league's political strategy was more deeply wedded to private property and capitalist exploitation of timber and strongly averse to taxation, government, and too much democracy. The league's members took on the organization and funding of redwood redoubts as a private responsibility, not an exercise in responsible government.[31] This conservatism put strict limits on what they could achieve—something less than 5 percent of old-growth redwoods were secured. Moreover, the Great Depression and New Deal tax increases brought the philanthropic approach crashing down and made way for greater state intervention and popular participation (see chapter 3).

A Death in the Family

After World War II, redwood cutting resumed with a vengeance in the North Coast counties (Humboldt, Del Norte, and Mendocino). Powered by bulldozer and chain saw and driven by the demand for lumber to feed the postwar housing boom, the timber companies felled the remaining red-woods in rapid-fire fashion. In the 1950s, cutting proceeded at a rate treble that of previous years, peaking in 1958. Then came the era of clear-cutting in the following decade, blessed by the state Board of Forestry. Agitation began to build for reform of California's timber cutting, leading to the Forest Protection Act of 1973.[32]

By the early 1960s, Bay Area conservationists were calling for the last old-growth forests to be saved in a Redwood National Park. They hoped for the strong support of Interior Secretary Stuart Udall of the Kennedy administration, but Udall wavered. Disillusionment with the government agencies conservationists had long trusted was growing. The redwoods were the opening salvo in the timber wars that would sweep over the Sierra Nevada and the Pacific Northwest for the rest of the century.[33]

The Sierra Club broke with the Save the Redwoods League over the Redwood National Park. They sought the protection of full watersheds instead of just prime groves—which were vulnerable to being swept away in floods created by logging practices upstream. In particular, the club, led by David Brower, Martin Litton, and Edgar Wayburn, wanted the still-private Redwood Creek forest in its entirety, while the league, led by Newton Drury, wanted the groves it had donated to the state to be turned into a national park. Brower put out a dramatic book on the devastation of the redwoods and made his first use of national newspaper ads as a political tactic in support of the Redwood Creek park.[34]

In 1968 a rump Redwood National Park of less than 60,000 acres was established, almost half of which came from three state parks. The major addition was a narrow "worm" along Redwood Creek consisting of the world's tallest trees, surrounded by barely 10,000 acres of old-growth forest. But the Sierra Club would not give up, even as the cutting continued unabated right up to the worm's boundaries. Congressman Phillip Burton (D–San Francisco) finally pushed through a new bill in 1978 that doubled the park's size. It was the most expensive national park ever, with the least return. The Redwood Creek watershed had been almost entirely leveled, the world's tallest tree had been cut in an act of sweet revenge by Arcata Corporation, and the rest of the tall grove washed away in a flood. Victory was bittersweet.

Nor did that end the controversy. The redwood wars flamed up again a decade later, after Maxxam Corporation, a junk-bond–financed conglomerate, bought up Pacific Lumber Company and its Headwaters Forest. Maxxam planned to liquidate the last remaining stands of private, old-growth redwoods, which it saw as underexploited assets, in order to pay off its bonded debt. A new, more militant environmental group, Earth First!, joined the fray, with its tactics of tree spiking and tree sit-ins. Then Judi Bari entered the scene, fresh from New York, and tried to bring together the radical greens and displaced timber workers. She organized Redwood Summer in 1990, modeled on the civil rights movement's Mississippi Summer, and threw the timber industry into a fit. She and her friend Darryl Cherney were nearly assassinated by a bomb planted under their car seat while they were campaigning in Oakland. They were immediately arrested on trumped-up charges of having built the bomb themselves. Later, protester David Chain was killed when the tree he was living in was cut from under him. Apparently, saving redwoods is more fraught than the old conservationists ever imagined.[35]

The events of a century ago gave the Bay Area a good start on the road to a regional culture of conservation and preservation. Both conventional and radical ideas of nature protection had set roots among the local bourgeoisie, and the first parks had been set aside in the Sierra and at Muir Woods and Big Basin. Parts of the North Coast redwoods, Mount Tam, and Mount Diablo became state parks, though another fifty years passed before the Bay Area got its own national park. Despite defeat at Hetch Hetchy and in the redwoods, the era between the world wars would see expansion and regularization of state and regional parks, and leadership would pass to a new generation beyond John Muir, William Kent, Andrew Hill, and Laura White. Before moving to that story, however, we need to consider another kind of contribution to the greenbelt around the bay: resource-intensive activities such as farming, mining, and water supply.

2

FIELDS OF GOLD *Resources at Close Quarters*

City and country are necessarily interwoven. Urban demand calls
forth the products of the countryside. Cities are encircled by neck-
laces of farms, reservoirs, and quarries supplying food, water, and
building materials. San Francisco, Oakland, and San Jose's immediate hinterlands have lain *within* the Bay Area, which has been intensely mined for
natural resources and agricultural commodities. Having already looked at
the redwood timber industry, we shall now consider agriculture, water, and
mining.

As cities grow, they do not expand neatly into a previously settled countryside like a wave from a stone thrown in a pool. Instead, city and countryside develop in tandem. This is especially true of California, which
"became the scene of ruralized urbanization, a rush of simultaneous town
founding and orchard planting."[1] San Francisco, Oakland, and San Jose had
farms within their city limits well into the twentieth century, while the rural
areas were dotted with towns such as Vallejo, Petaluma, and Gilroy.

With the coming of suburban sprawl, the built-up areas advanced in
bursts that look more like solar flares than tidy suburban rings (see map
4). These flares project deep into previously rural areas, especially along
transportation corridors such as Highway 101 through the North Bay or
Highway 24 beyond the East Bay hills. This flaring is especially marked in
a place with the rolling topography of the Bay Area, where development

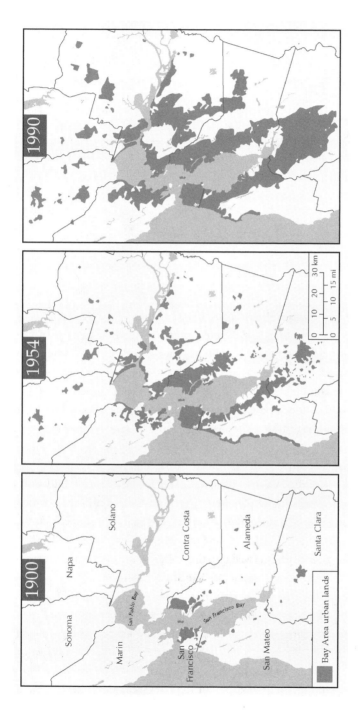

MAP 4. Bay Area Urban Area, 1910, 1954, 1990. Drawn by Darin Jensen. Courtesy of Bay Area Open Space Council, GreenInfo Network, Greenbelt Alliance.

slinks along coastlines, up valleys, and over passes, all the while lapping at the foothills of the Coast Ranges.

The city's embrace of the rural means, too, that elements of the country-side are incorporated into the urban landscape itself. The country is not entirely bulldozed away. Some things are mere remnants, such as old farm-houses or windmills, which can be picturesque or forlorn reminders of the past. But others are part of present-day land-use patterns, operating like mitochondria in the cells of more-complex life forms. These patterns include urban nodes such as Redwood City, street grids such as at Alvarado in Union City, and turnpikes such as old Monterey Road in San Jose. These are an enduring part of the city, living on as neighborhoods, commuting routes, property lines, and political identities.

AGRICULTURAL HEARTLAND

While the term "greenbelt" evokes parks and woodlands, more than half the open space around the Bay Area is still agricultural. It used to be far more. Of a total of 4.5 million acres in the Bay Area in 1980, farmland constituted 2.7 million (2 million in rangeland, 700,000 in cropland). As farming con-tinues to be pushed outward, it is easy to forget how vigorous and substantial agrarian production once was: miles upon miles of orchards and vineyards on terraces, hillsides busy with cattle and sheep, vegetables and berries on the flats, hay and forage in the seasonal wetlands, and chickens by the millions. The Bay Area's agrarian economy did not hit its peak until the 1920s.

Agriculture took off under the impulse of local demand during the 1849 Gold Rush. Farms grew up along the transport and settlement route from San Francisco to the Mother Lode mining region in the Sierra foothills to the east. Agriculture encircled the bay's margins, then moved up the fertile valleys of the Coast Ranges—Sonoma, Napa, San Ramon, Livermore, and Santa Clara. Later, agriculture moved into the San Joaquin Valley, the Central Coast, and Southern California. Most people think that because the San Joa-quin and Imperial valleys are the heart of agribusiness today, it was always so.

Nature influenced this geography: river transport, ample sunshine, good soils, winter rains, and, most importantly, the location of gold. Northern California had the money, consumers, workers, farmers, merchants, and riverboats. As Rodman Paul observes, "When a new American agriculture began, its birthplace was not in Southern California, where Hispanic settlement had started, but rather in the central part of the state." The high rank of Alameda and Contra Costa counties in the state's nineteenth-century farm economy would surprise people today, as would a contemporary observation that Solano County was "one of the most wealthy, populous and large productive agricultural counties in California." Better known for their agrarian prosperity are Santa Clara and Sonoma counties, which remained in the top echelon of California farmlands until the 1950s.[2]

Grains—wheat and oats—were the first crops of choice for the prime valley bottoms. Yields were tremendous and quality excellent. Planting spread like wildfire. It began in the Santa Clara Valley, then swept up the Petaluma River, over the Montezuma Hills, and thence up the west side of the Sacramento Valley (the biggest production area by 1870). The wheat boom overshadowed all else for thirty years, with a backdrop of cattle and sheep on the hills and cows along the cooler coast north of the Golden Gate.

Truck gardening and fruit farming quickly took hold under the stimulus of high food prices. Apples concentrated in western Sonoma County, plums and olives in the Napa and Sonoma valleys, pears and peaches along the riverbanks near Sacramento, plums and cherries around San Jose, apricots and vegetables in southern Alameda County. When the fruit boom hit in the 1880s, orchards and vineyards went in everywhere, displacing grain and pushing grazing onto the hills. The Santa Clara Valley became the prune capital of America, while Sonoma was top-ranked in vineyards. Contra Costa County grew walnuts and apricots. Solano County nurtured soft fruits. The South Bay became the leading seed and nursery area of the state.

This was an economically rational landscape in the sense that small, intensive, and high-value farms lay nearest to San Francisco. Some farms and dairies were no farther away than Cow Hollow and Visitacion Valley or Oakland's Fruit Vale and Temescal district. Meanwhile, the bonanza wheat farms were pushed out to cheaper lands at the far ends of the Central Valley, far from the bay-delta core.[3] Yet more important than local markets, in the long run, was the capital to promote modern farming. Mining and mercantile wealth piled up in San Francisco and smaller cities, its owners eager for investment outlets. Land was transformed into real estate to be bought, sold, and subdivided. A wage labor force was assembled from immigrants and tramps in permanent migration. These became the hallmarks of California agribusiness.[4]

California's unending agricultural revolution had begun, and its secret was not the pull of urban demand but the capitalist drive to develop the forces of agrarian production to their fullest. Thirst for land was so intense that capital made its own supply. Investors diked and drained a half million acres in the Sacramento–San Joaquin Delta (97 percent of the original area of freshwater marshes). The planting and productivity grew so rapidly that in a generation output exceeded what local markets could absorb. Wheat, dried fruit, and canned goods followed gold and silver into global commerce.[5]

A remarkable instance of this propulsion of farming beyond every limit is the extraordinary cluster of chickens that sprang up around Petaluma— known from 1900 to 1950 as the Egg Basket of the World. With 3,000 farms, 2 million chickens, and 100 million dozen eggs shipped in 1930, Petaluma represented the first instance of mass breeding and feeding in the world, a revolutionary step in livestock raising. Lyman Bryce invented the modern incubator there in the 1870s and shipped the first mail-order chicks; by the 1920s, one giant incubator could hold 2 million eggs at a time. Feed consisted of local farm by-products from dairies, orchards, and fields, along with sardine meal shipped from Monterey rendering plants. When more intensive, confined poultry factories came along after World War II and the San Joaquin Valley and eastern producers stole Petaluma's thunder, hundreds of chicken coops were left to fall into picturesque states of dilapidation.[6]

The farm landscape lives on to this day around the outer fringes of the Bay Area, sometimes in plain sight and other times hidden. Wine is big business in the North Bay. Milk cows still graze at Point Reyes, behind Stanford University, and in the East Bay regional parks. Cattle ranches continue to operate on the rolling hills east of Mount Diablo and south of Mount Hamilton. A few flower growers persist around Pescadero, and Half Moon Bay sponsors the West's oldest giant-pumpkin weigh-in every October. A drive through the back roads of Orinda and Danville reveals patches of remnant plums or walnuts. More surprising, state workers trying to suppress a Mediterranean fruit-fly infestation in the Santa Clara Valley were stunned by the number of backyard fruit trees, many left over from prior orchards.

AGRIBUSINESS GOES TO TOWN

With this immense dynamism in the countryside, the urban economy came to be powered by a vast agro-industrial complex from the 1880s to the 1940s. Agriculture rebuilt the Bay Area into a system of farms and towns feeding into San Francisco and Oakland. At the same time, elements of agribusiness were located deep within the larger cities, including the canneries of San Jose and the grain silos astride the port of Oakland. And the infrastructure of the agrarian era inscribed patterns on the land that were deeply etched in the urban landscape even as new industries and suburbia swept over the region after World War II. Three dimensions of this agribusiness geography can be highlighted: transport, processing, and marketing.

Pathways

The road and rail network around the Bay Area was built chiefly to accommodate agrarian trade. Water transport came first, as boats plied the bay and rivers from wharves at Alviso, Hayward, and dozens of other sites and seagoing vessels set off from San Francisco for New York, Liverpool, and Canton. Many of the main roads throughout the Bay Area—almost all still functioning—were laid down to link farm towns to the wharves and to urban centers—for example, Alvarado–Niles Road in Union City, Winton Avenue in Hayward, and Mountain View–Milpitas Road. The first railway lines, from San Francisco and Oakland south to San Jose, were advancing into farmlands, and a multitude of spurs extended the system to individual canneries and packing sheds.

The most extraordinary site created by transport was Port Costa, along the steep banks of the Carquinez Straits, where the railroad, riverboats, and seagoing ships met to transfer grain. The wharves and warehouses ran for 2 miles along the shore. Established by Isaac Friedlander in the 1860s, the port flourished for a quarter century before declining with the wheat trade in the 1890s.[7] Horses and mules were a ubiquitous means of urban locomotion up to 1920, and signs of their presence were everywhere: stables, troughs, hitching posts, carriage and wagon works, harness makers, feed dealers—and manure.

Processers

As important as transport were the techniques of processing and preservation that made long-distance shipping possible: milling, canning, drying, fermenting, and refrigeration. Flour and sugar mills were first, and they encircled the bay at such places as Vallejo, Alameda, Alvarado (Union City),

San Francisco, and Crockett. By the 1870s, canneries became a marked feature of the urban landscape as fruit and vegetable canning grew into the region's leading industry. Clusters of canneries arose in San Francisco, Oakland, and San Jose and along the Contra Costa shore. These packed everything from apricots and stewed prunes to tomatoes and asparagus. The pioneer canneries were situated right next to the orchards and fields, as in the case of J. Lusk of North Oakland or Tom Foon in Alviso near Chinese and Japanese truck farms. Drying yards, packing sheds, and warehouses were ubiquitous in the rural areas and small towns.[8]

Wine-making was established in the Bay Area by European immigrant pioneers including Agoston Haraszthy, Gustave Niebaum, and Charles LeFrance, who built magnificent wineries and caves in Napa, Sonoma, and Almaden in the 1860s and '70s. But the wine was mostly ordinary. After the cataclysm of 1906, when 15 million gallons perished in San Francisco, the merchants and bankers created California Wine Company, which built an enormous winery constructed at Point Molate (Richmond): Winehaven. This crenellated fortress was for a brief moment of glory the largest winery in the world.[9]

A common element of the agro-industrial landscape around the bay was the independent dairy, such as Berkeley Farms or Peninsula Creamery. Cheese and butter were churned out from small plants from Marin to Gilroy. Slaughterhouses, meatpacking plants, and tanneries were concentrated at Islais Creek, South San Francisco, Emeryville, and East Oakland. Millions of head of cattle were driven up from the Central Coast and San Joaquin Valley to meet their makers in the cities.

Promoters

Food sales were vital to the success of the whole agrarian economy, and the more that could be sold, the better. Selling and promoting foodstuffs began with merchants and packers, such as San Francisco's Rosenberg Brothers (dried fruit) and Grundlach-Bundschau (wine), who organized the export trade. They were joined at the turn of the nineteenth century by growers' cooperatives, which became the dominant marketers by the First World War. The largest in the Bay Area was the prune and apricot coop Sunsweet, which ran dozens of packinghouses around the region, especially in the Santa Clara Valley. Apple warehouses were a common feature in Sonoma County, especially in Sebastopol, home of the Apple Growers Union. Then, there were the great wholesale markets, located near the ports, serving retailers and restaurants in San Francisco and Oakland; remnants of these still function.

The agro-economy reached deep into the urban fabric with that fixture of food sales, the grocery store. Groceries were the biggest segment of retail sales until well into the twentieth century, and grocery stores were on corners throughout the cities. Oakland's Safeway became the second-largest grocery chain in the country. Lucky, Cala, Purity, and Mutual were other early chains in the Bay Area. Supermarkets, born in Los Angeles between the two world wars, came north after 1945 and took over in the auto-age suburbs. Most of the stores of that epoch still stand.[10]

The agrarian landscape of the Bay Area was once thick with farms, railroads, packinghouses, and chicken hutches, reflecting the prosperity and accumulated capital of the agro-industrial economy. Much of this landscape is now gone, buried beneath miles of asphalt and housing tracts. We should not idealize the agrarian order of that time as something apart from the city, nearer to nature, or socially idyllic. It was none of those. It was joined to the city at the hip, a radically altered ecology of monoculture, a world answering to the bottom line. It even had dramatic adverse environmental consequences, such as sinking the land as much as 20 feet in the Santa Clara Valley through overdraft of groundwater.[11]

An inscription carved in bold glyphs on the side of Hilgard Hall on the Berkeley campus of the University of California reads, "To rescue for human society the native values of rural life." This marks the College of Agriculture, willing servant to California agribusiness since 1875. From that college has issued a steady steam of aid, advice, and research in support of bioengineering the countryside. It is an iconic site for the marriage of city and country. But the noble words ring false in a place where rural values had little to distinguish them from those of the capitalist city.

AGRARIAN LEGACIES

Agriculture has been driven from the central Bay Area in dramatic fashion over the last fifty years. The bulldozer armies of the night have turned the fields and orchards into city streets, shopping plazas, industrial parks, and houses as far as the eye can see. The fingers of the city have reached out and grabbed the ranches and farms by the throat, slowly choking them to death. The land hunger of the city increased dramatically with the postwar style of space-extensive suburbanization. From Cupertino to Fairfield, familiar valleys and hillsides disappeared with startling speed. Roughly 750,000 acres of Bay Area farmland disappeared between 1945 and 1980, one-fourth the total; about 200,000 acres went directly into pavement and structures. The average consumption of land during this period was more than 6,000 acres per year—most of that in the Santa Clara Valley.[12]

The rapid paving of farmlands after the Second World War led to a reconsideration of state and local land-use policy. The first volley in farmlands protection was the Greenbelt Act of 1955, aimed at land conversion in Santa Clara; it proved ineffective. It was superceded by the California Land Conservation Act of 1965, known popularly as the Williamson Act. This act allows farmers and ranchers to contract with county governments for ten years of property-tax relief, on the promise not to convert to urban uses. While the Williamson Act was popularly linked to saving open space from urban sprawl, its deeper motivation was tax relief for large landowners— part of a long run-up to Proposition 13. It proved wildly popular with farmers, who signed up half the prime farmland in California by the end of the 1970s, including some 2,200 square miles in the Bay Area. Thereafter, Williamson Act contracting leveled off, no doubt because of the general property-tax reductions granted by Proposition 13 in 1978.[13]

Most of the acreage covered by the Williamson Act is in the Central Valley or the distant fringes of the urban conurbations; only 5 percent is at the near-urban fringe. Furthermore, owners wishing to develop their land have been too easily able to withdraw from the promise not to convert to urban uses— after gaining substantial tax breaks—because of cozy relations with county officials. Such withdrawals have been highest in rapidly growing urban counties such as Contra Costa. Environmentalists won a strong ruling from the State Supreme Court on contract enforcement, but it made little difference.

In the 1980s People for Open Space revived the issue of disappearing farmland, issuing a report, *Endangered Harvest*, and launching a campaign to create an Agricultural Lands Conservation Commission. Farmlands protection flowed easily from the group's part in saving the Napa Valley in the 1960s (see chapter 8). More broadly, it incorporated working agricultural reserves as an integral part of the greenbelt vision—not just woodlands, scenic spots, and wildlife habitat. This vision reverberated with changing environmentalist sentiments toward maintaining working rural landscapes as well as wilderness. But it was ultimately unsuccessful.[14] As a result, the greens turned to other tactics for protecting farmland, such as stronger general plans, urban limit lines, and purchase of easements (see chapters 6 and 7).

The replacement of farmlands by the urban hardscape should not be thought of as a dead loss. The geography of agribusiness has adapted as the city expanded. Cherries and apricots moved to the east side of the San Joaquin Valley, plums to Fresno County, and prunes to north of Sacramento. Berries were displaced from the northern Santa Clara Valley to the Pajaro Valley, now the national center of production. Cut-flower growers moved from the southeastern bay rim to the San Mateo coast to Mexico, while seed growers moved down to Santa Maria and now Costa Rica.

Not all the farms were displaced by urbanization, either. Agribusiness had been remaking its own landscape for a century: wheat replaced cattle in valley lands by the 1860s; fruit and vegetables elbowed wheat out of the way by the 1880s; wharves fell into disuse with the coming of the railways; and railway spurs grew weeds with the arrival of trucks. Many local growers simply lost out to competitors elsewhere, such as potato operations in Idaho and Kern County, asparagus growers in the Imperial Valley, and apple orchards in Washington State. So it's not the loss of a golden age of agriculture that hurts the most; rather, it is the useless sacrifice of fine soils, rural wildlife, and beautiful scenes for notoriously space-wasting forms of living, to accommodate the American thirst for big houses, big cars, and big parking lots.[15]

Urban food processors continued to function even as nearby farms were swallowed by the city. The last slaughterhouse in South San Francisco did not close until 1957. Oakland's Gerber's and Granny Goose factories survived into the 1990s by contracting for fruits and vegetables from the Central Valley. Spreckels and Holly Sugar moved out long ago, but C&H Sugar at Crockett marches on to this day, processing cane sugar from Hawaii. Coffee-roasting operations in San Francisco, such as MJB and Folgers, found that rising land values made it more profitable to sell and move (Hills Brothers, for example, built a new factory in Salinas). Safeway relocated its gigantic regional warehouse in Richmond to Tracy in the 1980s, to beat the traffic and the unions.

As farmland disappeared, however, most of the infrastructure of agrarian industry and commerce—packing sheds, dairies, slaughterhouses, and warehouses—fell into disuse and decay. All that's left of Port Costa is pilings that jut into the Carquinez Straits at low tide. Some magnificent throwbacks still stand, through reuse and sheer inertia. Winehaven is still there, along with its worker cottages, because the Navy used it for decades. The Cow Palace in Visitacion Valley, originally conceived for livestock shows and agricultural fairs, was already obsolete by the time it was finished by the WPA in 1941; but it has survived by hosting grand national rodeos, sports and boat shows, and national political conventions. The giant apple warehouses of Petaluma sat vacant for years because the cost of demolition was prohibitive and land in weary industrial districts little in demand.

The return of retailing, offices, and high tech to parts of the inner city in the late twentieth century triggered redevelopment of many former agro-industrial sites. Dubuque's packinghouse in South San Francisco fell to airport-related warehouses. Del Monte's Fruitvale cannery was knocked down for an outlet mall, and another in Emeryville disappeared beneath the empire of Pixar. Winehaven has been slated for an Indian casino by the city of Richmond. Compliant local governments are happy to jump in to clear away the past to compete for today's tax revenues.

Fortunately, some fine remnants have risen from the dead by transfiguration into malls, offices, and lofts. The first warehouse anywhere to be converted into chic consumer space was San Francisco's Ghirardelli Square in 1964, carved out of a former chocolate factory. The nearby Breuher-Schweitzer Malt Company factory sat derelict for twenty-five years until housing prices made it feasible to convert to pricey condominiums. The Moorish architecture of Berkeley's Heinz cannery was saved through the intervention of preservationists and now serves as mixed retail space. The Italianate Hills Brothers roasting plant on San Francisco's waterfront is now an office complex. The Pruneyard in Campbell, an oxymoronic shopping center, was formerly Sunsweet's biggest packinghouse. Lofts now inhabit Safeway's old headquarters in Oakland. The agrarian past peeks through the urban present in unexpected ways.

URBAN WATERSHEDS

The extensive watershed lands of the San Francisco Water Department, the Marin Municipal Water District, and the East Bay Municipal Utility District (EBMUD) are anchors of the Bay Area greenbelt. EBMUD still holds 26,745

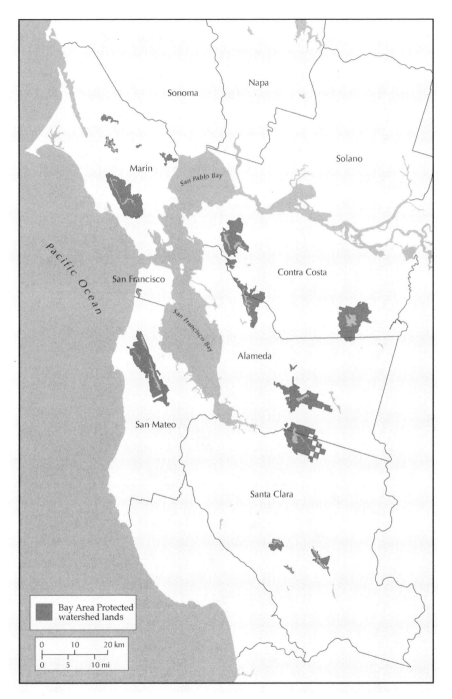

MAP 5. Bay Area Protected Watershed Lands, 2005. Drawn by Darin Jensen. Courtesy of Bay Area Open Space Council, GreenInfo Network, Greenbelt Alliance.

acres in the East Bay hills. Marin Municipal Water District owns 21,792 acres on the north flank of Mount Tamalpais. San Francisco Water Department watersheds are enormous: 22,000 acres on the Peninsula (the biggest landholding in San Mateo County) and 35,000 in Alameda County, north of Mount Hamilton (4,209 feet). Contra Costa Water District maintains 19,000 acres on the east side of Mount Diablo at Los Vaqueros reservoir. Santa Clara Valley Water District and Sonoma County Water Agency's holdings are limited to 1,000 acres each, surrounding large reservoirs.

These watersheds began as urban feeder lines reaching into the countryside, acquired for straightforward reasons of water supply and private profit.[16] They were, nonetheless, the largest pieces of land protected from development in the early twentieth century, at a time when parks were still few and far between. Watersheds were an extension of the landscape of extraction, but with the crucial difference that they were kept relatively pristine to protect drinking-water quality. Turned to recreational or habitat uses later on, they have become key links in the necklace of open lands overlooking the bay, amounting to more than 125,000 acres.

Unquenchable Thirsts

Up to 1930, bay cities depended entirely on local water sources to serve a population passing the million mark by 1920. After the rivulets of San Francisco were tapped in the 1850s, the Spring Valley Water Works (later the Spring Valley Water Company) began acquiring water rights in San Mateo County. It built a succession of dams along the creeks of the San Andreas rift valley from the 1860s to 1880s (including the world's first concrete dam), with flumes and tunnels leading back to the city. San Andreas and Crystal Springs reservoirs were the final result. Spring Valley then jumped over to the creeks and aquifers of southern Alameda County. Marin had several small private companies on the slopes of Mount Tam, the oldest being Dixon Water Company supplying San Rafael and the largest the Marin Water and Power Company, on the north slope, and the North Coast Water Company, on the south slope.[17]

Oakland turned to Anthony Chabot's Contra Costa Water Company, formed in 1866. The first storage reservoir was tiny Lake Temescal (built in 1868), followed in short order by what is today called Lake Chabot. Three more reservoirs were added in the early twentieth century: San Pablo, Upper San Leandro, and Lafayette. Meanwhile, the company morphed into

Peoples Water and then East Bay Water. Along the way, it garnered approximately 40,000 acres of watershed in the hills. San Jose depended on well water distributed by the San Jose Water Works and Federal (now California) Water Service.[18]

After 1900, San Francisco and Oakland were the last two large American cities still relying on private water companies. Fortunes could be made on water, and competition was fierce to win municipal contracts; in Oakland the story reads like the Owens Valley water wars, with holes drilled in flumes, wells spiked, and a sewer connected to a city water main. Yet the private water companies struggled to keep up with burgeoning urban demand. In order to secure water rights, they had to acquire watersheds well ahead of need. Speculation in such land brought ruin on two of the region's greatest capitalists—Billy Ralston of San Francisco and Francis Marion Smith of Oakland—but left vast acreages in the hands of the water companies.

Local streams and aquifers were ultimately not up to the task of supplying the burgeoning population, because of small watersheds, variable rainfall, and indifferent geology. Surface water was uncertain and often of poor quality. Excessive groundwater pumping caused land subsidence and saltwater intrusion. Hence, by the 1890s municipal ownership was already being bruited and distant sources eyed. The Progressive takeover of state politics under Governor Hiram Johnson in 1911 led to two enabling acts for urban water-district formation. But it took many more years to overwhelm local supplies, public lethargy, and water-company opposition. This is less a tale of certain triumph of public good over private gain, more a case of slouching toward River Jordan.

Water Mains

Each part of the Bay Area found its own fix. Marin got the first state charter for a municipal water district, in 1912, and, using a bond issue, began buying up private suppliers (some two dozen). San Francisco grabbed the Tuolumne River in Yosemite and bought out Spring Valley Water Company in 1930. The fight over congressional approval of the wanton seizure of Hetch Hetchy to slake the city's thirst is justly remembered as one of the darkest moments in American environmental history. Oakland, Berkeley, and Richmond formed the East Bay Municipal Utility District in 1921, took over the East Bay Water Company, and reached out to the Mokelumne River.

San Jose came under the umbrella of the Santa Clara Water Conservation District in 1929 after a protracted campaign and several defeats of plans to replenish groundwater that had been badly overdrawn by farmers. The district built reservoirs, such as Anderson and Lexington, in the surrounding foothills and began percolating water back into the streambeds, but unlike the other water districts, it never took over the private retailers and small municipal systems that dotted the South Bay. Sonoma County Water Agency, created in 1949, utilizes the Russian River and a large system of wells in the deep gravels of the river valley; it sells to Santa Rosa and other cities. Contra Costa Water District has served the northern tier of the county since 1936.[19]

EBMUD was the first to import water from the Sierra (100 miles), in 1929, while San Francisco's much-delayed Hetch Hetchy project opened in 1934. The Santa Clara Valley and northern Contra Costa County began receiving water from the Central Valley in the early 1950s through the Contra Costa Canal and the South Bay Aqueduct. These were supplemented by the San Felipe project in the 1980s, after a bitter fight and a close vote. Sonoma County's supply was bolstered by imports from the Eel River in the 1950s and damming the upper Russian River in the 1980s (see chapter 8).

These water systems are etched clearly on the landscape, in distant dams, aqueducts, and two dozen local storage reservoirs. Many of the latter form substantial lakes in the hills, such as San Leandro, Anderson, and Crystal Springs reservoirs. The water districts' storage tanks, pumping stations, and hydrants pop up all through the urban field, and thousands of miles of delivery pipes and sewers trace a subterranean geography of water supply that rarely comes to light.

Unintended Consequences

The legacy of local watersheds is the opposite of that of Hetch Hetchy. By and large, they have been an unmitigated benefit to the region as open space. Marin Municipal Water District immediately threw open its lands to public recreation, continuing the tradition that began with day hikers on Mount Tamalpais, and still has the most-active trail and public-use program of all the water agencies (Mount Tam was the birthplace of mountain biking in the late 1970s). EBMUD allows grazing and hiking on its lands and leases recreation areas on some of its reservoirs to the East Bay Regional Parks District. Santa Clara Valley Water District, Contra Costa Water District, and Sonoma

County Water Agency reservoirs are popular for weekend recreation. The largest recreational lake in the Bay Area is Lake Berryessa, in Napa County, a federal facility that provides irrigation water for Yolo County to the east.

San Francisco, by contrast, is the curmudgeon of watershed proprietors. The city water department (Public Utilities Commission) has refused all public access to its domain in order to safeguard water quality, leaving its watersheds among the wildest of Bay Area open spaces. That meant, ironically, that these lands later came to be recognized as important wildlife habitat. Pressure to open them up to the public is intense, and the city has reluctantly allowed the Bay Area Ridge Trail to go through its land with limited hiking access.[20]

This does not mean that management of local watersheds is without blemish. There was a fierce battle over EBMUD's plan to put a reservoir into Buckhorn Creek valley in the East Bay hills in the 1980s. The proposal was killed by the environmental majority on the district board, led by Andy Cohen (now with the San Francisco Estuary Institute). Another ill-fated proposal for watershed lands came in the mid-1990s, when the administration of Mayor Willie Brown of San Francisco tried to expand quarrying activity on the city's property near Sunol. That ran into such trenchant public opposition both in the city of San Francisco and in southern Alameda County that the project ran aground.

Contra Costa Water District succeeded where the others failed, winning approval for the Los Vaqueros reservoir on the eastern flank of Mount Diablo in the mid-1980s. Los Vaqueros was part of the State Water Project expansion that was killed in 1982 by the Peripheral Canal referendum (see chapter 5). The local water district brought it back from the dead so that more water would be pumped out of the delta in winter to supply summer peak demand on the fast-growing periphery of Contra Costa County. It got the go-ahead from county voters over ardent environmental opposition led by Seth Adams. But a deal was cut that left thousands of acres of protected watershed lands, snatching an element of victory from the jaws of defeat.

Rivers of No Return

In a delicious turnabout, watershed protection has been reversed, from a focus on upstream lands to rescuing downstream creeks. Dozens of creeks draining into the bay have been languishing forgotten in culverts beneath city pavements for decades. The Bay Area is especially rich in small streams flowing out of the hills, which once provided spawning grounds for an extraordinary wealth of anadromous fishes, including chinook salmon and steelhead—the seagoing version of the world-renowned rainbow trout, first identified and named in an East Bay creek. Less than half still support trout, and the Central Coast steelhead is a threatened species.

The East Bay has at least two dozen such watercourses, the largest of which is Alameda Creek running down Niles Canyon. The South Bay has several, notably Guadalupe River and Coyote Creek. San Francisco, surprisingly, has several spring-fed creeks such as Mission, Lobos, and Islais. Marin has a few, including Cascade, Miller, and Novato creeks (as well as coastal streams such as Olema and Walker creeks). In the North Bay, the Napa and Petaluma rivers dominate the scene. San Mateo County has substantial bayside creeks, such as San Francisquito, San Mateo, and Redwood (as well as a number of important streams flowing to the Pacific, such as Pilarcitos, Pescadero, and San Gregorio).

The few wild gashes still remaining in the city's cement resonate strongly with ordinary folk. Open streams run through some of the most heavily used parklands: Oakland's Dimond (Sausal Creek) Park, Berkeley's Cordonices and Live Oak parks, and San Jose's Coyote Creek parks, for example. Witness, too, the tenacious efforts of common citizens to protect local creeks. A precedent-setting case was African-American mothers in North Richmond who fought to save Wildcat Creek as a play area for their children. With technical help from Ann Riley and Phillip Williams, they stopped a channelization project of the Army Corps of Engineers (in revenge, the Corps bulldozed the riparian vegetation before departing). Not only did the Richmond struggle establish the legitimacy of streambed and habitat restoration as a tool of flood control, it stimulated the organization of Friends of the Creek groups, which have sprung up on just about every stream in the Bay Area since then.[21]

Another precedent was the "daylighting" of a portion of Strawberry Creek in Berkeley in the 1980s. Doug Wolfe, a city landscape architect, and Richard Register, a local activist, put forth the radical idea of digging up a stretch of culverted creek as part of a new park on land the city had just acquired from the Santa Fe Railroad. The idea caught on with Carol Schemmerling, a Berkeley Parks and Recreation commissioner, as well as David Brower (a Berkeley resident) and others, and Strawberry Creek Park was born.[22]

The Urban Creeks Council was begun by Carol Schemmerling and Ann Riley in 1982, inspired by the Wildcat Creek restoration and Strawberry Creek Park. Schemmerling and Register had the idea of educating people about the waters beneath their feet by stenciling "no dumping, drains to bay" on curbstones all over town—a practice now copied all over the world. The council leads educational walks along remnant gullies and glades and undertakes daylighting and restoration projects. Basic restoration minimally includes stopping leaking septic tanks and sewers, cleaning refuse and barriers from streambeds, and returning straightened watercourses to sinuous curves.

Berkeley adopted a pioneering ordinance in 1989 to limit further building within 30 (now 25) feet of creekbeds, which has riled many homeowners because no one is sure where all culverted creeks are or who bears liability for them.

Register developed a bold plan to tear up Center Street in Berkeley and make Strawberry Creek a centerpiece of downtown, while using downtown development-rights transfers to buy up the creek right-of-way all the way to the bay. But the city balked.[23] Napa has been more adventuresome, allowing a major reengineering of the Napa River's sinuous bed and rezoning of the floodplain. San Jose has recovered a stretch of the Guadalupe River near its downtown for a park. Now city planners are regularly considering projects to bring urban creeks back into the sunshine.

Stream protection had become a major preoccupation of land stewards throughout the Bay Area by 2000. Setback limits on development near watercourses had become part of most county general plans. The state of California began sponsoring watershed initiatives to balance water use and stream flow among competing interests. Grants for restoration projects have flowed from foundations and state and federal programs. The biggest such effort is the San Pablo Bay Watershed Restoration Program, launched in 2000 by the Bay Institute, Corps of Engineers, and California Coastal Conservancy. As a result of all this activity, chinook salmon and steelhead are making a comeback in many places, such as Lagunitas Creek and the Napa River.

THE MINE IN THE GARDEN

Mining was the main business of early American California. The Mother Lode region was in the Sierra foothills and the Comstock Lode lay east of Tahoe, but the main beneficiary of the billions in gold and silver was the city of San Francisco. This has led to the impression, solidified by Gray Brechin's *Imperial San Francisco*, that the city's resource supply lines have always stretched over long distances.[24] Yet the bay region, close at hand, is very much a resource hinterland for the city, and every inch of the surrounding hills and vales has been combed for minerals and materials. The bay region has been mined for all it's worth.

All That Glitters

Mining inscribed another geography of rural production, along with farming, timbering, and water supply. Across the middle ground of the region are scattered sites of extraction and accompanying towns, most of which are now defunct. They have etched their legacy into the urban landscape. That legacy is not always pretty. Plunder of the countryside has left ugly scars and relentless pollution. Yet, in many cases, brownfields have been turned back to green as unexpected urban parks and byways.

The most lucrative mineral sites were the quicksilver (mercury) mines. Mercury was the catalyst for precious-metal mining because it adheres to gold and silver particles, separating them from gravel and dirt. Gold and silver may have been far afield, but the principal deposits of mercury were right at hand. The biggest mines were at New Almaden, south of San Jose, and New Idria, in San Benito County. Production from the New Almaden district totaled $54 million (in current dollars) from 1850 to 1919, more than any comparably sized gold district in the state, and the New Almaden mine gushed forth more wealth than any other mine in California. New Almaden is now a county park.

In the North Bay, clustered around the junction of Napa, Sonoma, Lake, and Colusa counties, were more than 100 smaller mines, most notably Oat Hill and Great Western. Here the quicksilver was nearer the surface and could be extracted by smaller operators. When the shift to cyanide processing in the 1890s cut deeply into demand, the industry survived by exporting cinnabar to China for dyes. Total state mercury production by 1919 added up to $140 million (in 1940 dollars).[25] Most remaining mercury mines were shuttered by environmental laws in the 1970s. The last Bay Area mine was the Knoxville in northern Napa County, which Homestake Company winnowed for gold, silver, and mercury until the early 2000s.

The toxic residue of mining lives on around the bay; mercury gets into the waterways, where it is converted to lethal methyl mercury, which moves through the food chain. There's no way to get rid of it except by gradual burial beneath sediments, beyond the biotic zone. The biggest source of mercury is New Almaden, which still contaminates Alamitos Creek, the Guadalupe River, and the South Bay. Mercury leaches down Walker Creek into Tomales Bay and off the east side of Mount Diablo. It comes down Cache Creek from North Bay mines into the Sacramento River and thence into the bay.[26]

Other metals were taken from the Bay Area in small quantities. Ochre and mineral dyes from Napa and Sonoma were valued before the advent of modern coal-tar extracts. Later, the main minerals exploited were chromites and manganese (used in metallurgy), found along the border of southeast Alameda County and northeast Santa Clara County (most notably at Tesla). Magnetite was widespread, coming out of Napa, Sonoma, Alameda, and especially Santa Clara counties. Huge quantities of sulfur (and a little copper and gold) were taken from pyrites at the Alma and Leona mines in the Oakland hills and used to make sulfuric acid by Stauffer Chemical Company. Gypsum has long been taken from southern Napa County, for many years by Kaiser Gypsum and now by Syar Industries. Bleaches and paints were manufactured using such humble minerals as chloride salts and lime.[27]

Before hydroelectricity was brought down from the Sierra and oil and natural gas were brought up from Southern California, California's major energy sources were moving water, timber, whale oil, and coal (used for both burning and coal-gas extraction). Whale oil was processed in San Francisco and Point Richmond, and the Bay Area was for a time the world's largest whaling center. Almost all the coal came from the north slope of Mount Diablo, taken by the Black Diamond and Pittsburg coal companies. Total output up to closure in 1902 was four million tons, worth $20 million—not much by eastern standards, but crucial to fueling the state's nineteenth-century economy. Five little mining towns sprouted nearby, including Nortonville and Somersville. Silica for glass and metal casting was also taken out of Mount Diablo in large quantities.[28] Today, Black Diamond Mines Regional Preserve is part of the East Bay parks system.

Concrete Enterprise

Building materials are the very stuff of city-making. Being heavy, they are usually drawn from close by. Quarries for crushed stone used in roadways and for fill are ubiquitous. San Francisco once had several quarries, along a line from Mount Davidson to Twin Peaks. The steep cliffs of Telegraph Hill are largely a creation of the Gray Brothers, the city's main quarrymen, who were finally halted by public protest in the 1920s. The south end of Potrero Hill was chopped off for industrial fill around Islais Creek. Candlestick Hill was cut down to half its size and dumped into the bay to fill Hunter's Point and Candlestick (Brisbane) cove. Northern San Mateo still takes gravel and rock from the Brisbane quarry on the north side of Mount San Bruno.

In the East Bay, quarries in the hills of Richmond were used to fill the marshes for industry and shipyards. There were quarries in the El Cerrito hills, the scars of which are still clearly visible, and Berkeley's charming old walls were built from local stone in La Loma quarry. Oakland had quarries on Moraga Road and in Dimond Canyon; the longest running was Oakland Paving Company at Fifty-first and Broadway. East Oakland sports a gigantic scar from the Leona quarry, used for road gravel for most of the last century. Farther south, there are huge quarry scars along the East Bay hills at Hayward, Fremont, and Milpitas. In the South Bay, quarrying continues at Coyote Pass south of San Jose and at the Hansen quarry behind Cupertino, visible for miles. Over the East Bay hills, the DeSilva quarry still operates at Apperson Ridge *within* Sunol Regional Park. Kaiser's Pacific Coast Aggregates east of Pleasanton was the largest quarry in the Bay Area when it was still active. In the North Bay, McNear Quarry east of San Rafael is the longest in continuous operation in the region, and there are large quarries south of Napa, Petaluma, and Vallejo.[29]

Many of these quarries have been quietly folded into the urban fabric. Houses and gardens cling to Telegraph Hill's cliffs, which are ascended by charming stairways. Corona Heights Park nestles into a Twin Peaks quarry, while the nearby Simon-Fouts quarry is now a housing complex.[30] The Berkeley quarry became La Loma Park, Oakland Paving Company a shopping center, and the Leona quarry an enormous cluster housing complex.[31] Kaiser Aggregates pits are swimming holes at Shadow Cliffs Regional Recreation Area, and Sonoma's Salt Point State Park harbors an old quarry. Donating worn-out quarries as public parks often puts a nice tax-coda on years of wealth extraction.

Gravel operations in rivers have left dramatic aftereffects. California Building Materials and Niles Sand, Gravel, and Rock Company dredged

Alameda Creek for decades, leaving huge pits (now in a regional park). Gravel extraction greatly altered Coyote Creek and Los Gatos Creek in the South Bay. The Russian River has been mined unmercifully, despite its importance in restocking the aquifer for drinking water. The hydrologic profile has been so sharply altered that it has undermined highway bridges. Yet the extraction continues.[32]

Paving blocks and building stone were sought nearby, as well. Sonoma County provided the basaltic cobblestones for San Francisco's streets from a quarry in today's Annadel State Park. A quarry on Angel Island was used for stone to build the original Bank of California; limestone came from Pacifica Quarry; and Marin's McNear's Quarry yielded gray rocks. The Greystone quarry near New Almaden was used to build much of downtown San Jose and yielded the blocks for Stanford University's immense quadrangle. Harder substances such as granite, marble, and slate were mostly taken from the Sierra and ferried downriver.[33]

Humble brick was fired from local clays. By 1900 there were brick makers operating at San Jose, Pleasanton, Point Richmond, Port Costa, San Rafael, Glen Ellen, South San Francisco, and Twin Peaks–Corona Heights. Marin's McNeary quarry (now only gravel) on San Pablo Point still functions, though it's hemmed in by encroaching residences. Brick production passed $1 million in annual sales by 1900 and peaked at more than $5 million in 1925.[34] There was a substantial demand for terra-cotta, decorative tile, roof tile, firebrick, and clay pipe, which came from specialized potteries around the bay. One of the earliest stood on Mission Creek, another in the Inner Mission in San Francisco; these closed by the 1920s. For a century, pottery works could be found at Alameda Point, East Oakland, San Jose, Niles, Port Costa, Richmond, El Cerrito, and South San Francisco. None of the brickworks or potteries is still operating; changing fashion, technology, and seismic safety standards have ended most uses of brick and terra-cotta.

The greatest building material of the twentieth century was concrete. Reinforced-concrete construction overtook bricks in value after the great earthquake of 1906—which proved the durability of steel frame and concrete structures. The first Portland cement plant was built at Napa Junction (American Canyon), the second at Clayton on the north slope of Mount Diablo. But the best limestone deposits were found in the Santa Cruz Mountains, so the largest works were at Davenport on the coast and at Permanente Creek north of Los Gatos. Santa Cruz Portland Cement Company was California's premier producer in the early twentieth century. After 1920, oyster shells were used for making cement at Pacific Portland Cement's mill at Redwood City, which was producing more than one million barrels per year in 1950 (it was sold to Ideal Cement and finally closed in the 1970s). Henry Kaiser created Permanente Cement in 1940 to supply his Shasta Dam project, and it became the largest cement plant in the world, at five million barrels per year. Permanente (now owned by British interests) and Davenport (owned by RMC Pacific Materials) still operate today. Permanente's quarrying scar is the largest vertical feature of human occupation in Silicon Valley.[35]

The Bay Area has been well mined for its material wealth by the growing city, leaving many wounds in the land. Most of these have healed over and disappeared from view beneath the metropolis, and a few have been renovated as parks, part of the greensward. Others continue to mar the landscape and bleed poisons. Water supplies have been vigorously developed, leaving dams, reservoirs, and protected watersheds that have become valued recreational and wildlife resources over time. Agriculture has been a fundamental part of the region, and farmlands are still its largest open spaces. As a result, memories of plum blossoms and strawberry fields have not faded entirely from the hearts of the living. The wine country thrives better than ever, as we'll see (in chapter 8).

As the city grows, it paves over the countryside, leaving many urbanites with a sense of loss of easily accessible farms, streams, and woodlands. This loss is an important source of inspiration for land conservation and environmental protection. The intensive rural areas just beyond and alongside the city limits are a vital part of popular imagery of the countryside, and because they are often the most groomed and beautiful agrarian landscapes—like the old orchards of Santa Clara Valley or the vineyards of today's Napa Valley—they provide city dwellers with a vivid picture of Arcadian splendor. Ironically, the nearby countryside, more than any place else, is a creature of the city itself and already radically altered. As the environmental movement grew into a major force in the Bay Area by the middle of the twentieth century, such Arcadian images faced off against nightmare visions of urban sprawl. Sparks from this collision would burn their way into the politics of land in dramatic ways.

3

MOVING OUTDOORS *Parks for the People*

The reaction against the slaughter of the ancient trees and mountains had changed the public view of them and established the principle of protecting nature within the boundaries of parks. These protected places were still few and far between during the first two decades of the twentieth century: Big Basin, Muir Woods, Yosemite, and Sequoia. If more territory was to be held aside from exploitation and development, something more was needed than the yearnings of Muir and his generation for sublime nature. What was necessary was a shift in rationale toward the use of nearby countryside for civic recreation. A more capacious sense of parks for the people was needed.

This shift took place between the world wars. The result, by midcentury, was an explosion of state, county, and regional parks in California and across the country. Such parks were meant to serve city dwellers for outdoor recreation and restoration; that is, they were as much part of the urbanized countryside as reservoirs, brickworks, and berry fields. Parks did not yet represent the kind of rejection of urban expansion and intensive human land use that would come later.

A vital precedent for the spread of recreational parks in the countryside was the building of city parks in the nineteenth and early twentieth centuries. The most famous is, of course, Central Park in New York, designed by Frederick Law Olmsted and Calvert Vaux in the 1860s, but other big cities such as Chicago and Boston had impressive chains of urban parks and greenways by the end of the nineteenth century. These were complemented by neighborhood parks and playgrounds after 1900, as the virtues of recreation for the mass of city dwellers, especially children, came to be accepted wisdom.

San Francisco and Oakland joined the grand procession of city parks across the United States, creating green spaces that exist to this day and are among the best used of any in the region. In the twentieth century, the Bay Area played a more central role in the recreational parks movement, providing key leadership in the creation of the National Park Service, launching one of the largest state park systems, and coming up with a new government vehicle: the regional parks district.

SCULPTING NATURE

The jumping-off place for any history of city parks in the Bay Area is Golden Gate Park. It sprang from the nineteenth-century ideal of pastoral parks, which began with scenic cemeteries and then swept through all the big cities of the country in the post–Civil War era. The urban park movement's leading figure was landscape architect Frederick Law Olmsted, who had a hand in nearly every major park project, including those around the San Francisco Bay. Pastoral parks were widely supported by civic leaders in response to the rapid growth of commercial cities in the midst of a civilization that still imagined itself in agrarian terms. And they were monuments to civic pride in the same way as elaborate city halls.[1]

Design on the Dunes

San Francisco's Golden Gate Park was carved out of rough hills and sand dunes hard against the Pacific in the teeth of ceaseless winds and fogs. Imagined just after New York's Central Park, the project brought San Francisco into the mainstream of nineteenth-century urban planning and civic improvement. The inspiration was provided by Olmsted, who came to California to escape the trauma of the Civil War and drew up plans for a park east of Twin Peaks. Golden Gate Park itself was designed by William Hammond Hall and approved by the county supervisors in 1870. It was notable for being the largest of the pastoral parks—1,000 acres, compared to Central Park's 850—at a time when San Francisco ranked tenth in size among U.S. cities. The city was otherwise conventional in its adoption of the endless street grid from the bay to the Pacific, and it used the park as a means of settling competing claims over land titles so that the western half of the city could be opened up to development. Landowners happily anticipated the rise in property values that the park would bring to the distant margins of the city.[2]

Golden Gate Park encompassed the same vision as Central Park: pastoral release from the hard grid of the American city. The illusion of "bringing the country into the city" needed artful planning, Olmsted contended. This meant planting tall trees around the borders to shield the park from the city, arranging trees in masses around open meadows, adding lakes and quiet watercourses, hiding roads and pathways, and creating secluded areas that mimicked wildlands. Olmsted insisted on minimizing buildings, excluding monuments, and keeping out parade grounds, theaters, and other markers of urban recreation. These were to be anti-urban spaces, places of tranquility and quiet contemplation. Olmsted called his plan for Central Park "The Greensward." "The park," avers one historian, "was consciously designed to be the country within the city."[3]

Such idyllic tranquility, undulating on the swelling bosom of a pliant Mother Nature, entailed a massive reshaping of the land into hills, dales, paths, and lakes. At Golden Gate Park this transformation of the landscape depended on the skills of Hall and the hard labor of thousands of workers. With advice from a landowner who had witnessed dune control in France, Hall hit on an inexpensive way of stabilizing the blowing sand in the western half of the park. His efforts paved the way for the inspired landscaping of Chief Gardener John McClaren. Forest and dell emerged out of coastal scrub and dune grass, a perfect set piece of green Albion on the western shore. Golden Gate Park's plantings reflect the horticultural rage for exotic plants of the nineteenth century, and its botanical reserve, Strybing Arboretum, is a monument to worldwide collecting and the gardener's craft.

We are used to thinking of urban landscapes as radically altered from their natural state, but we're unused to the degree of manipulation in a place as green and pleasant as Golden Gate Park. Such greenswards—to borrow Olmsted's term—are as urbanized in their own way as brickworks or freeway interchanges, even if undertaken in the name of the idyllically rural. The park is no less a fantasy construct than Disneyland. Its design runs counter to present-day environmental sentiment, which favors less human impact and more respect for native vegetation. Nonetheless, over time, it has become naturalized in the eyes of San Franciscans, and few realize how heavy the human hand lies on the land.

Golden Gate Park reminds us that urban recreation—whether in a crowd or in spiritual isolation—is forever the point of open-space preservation. While the men who imagined Golden Gate Park were members or servants of the upper classes, the park did not long remain a pleasure ground for the rich in their carriages. The park was sensationally popular within a few years of opening, even before the western half was finished. By the 1880s, streetcar lines were bringing the masses to the park for walks, picnics, and horseplay, in the tradition of the amusement parks and beer gardens then in vogue around San Francisco. The people reclaimed their public space.[4]

Places to Play

The designers of large pastoral parks around the country mostly ignored the needs of the working class for accessible recreation. Jammed into the south of Market and Mission districts, San Francisco's working people and their children enjoyed little in the way of open space or playgrounds. By the end of the nineteenth century, the city finally began to offer play areas, thanks to Progressive reformers such as Alice Griffith, Laura Lyon White, and Rosalie Meyer Stern. For the reformers, physical activity was essential to the healthy development of the individual body, and children were wasting away in crowded city quarters. The first children's playground in the United States was built in Golden Gate Park in 1886, but it was used mostly by the middle class. The playground movement took the country by storm after the turn of the century, propagated by the Playground Association of America, formed in 1906.[5]

Along with playgrounds came the idea of neighborhood parks. No city took up the challenge more thoroughly than Chicago in the first decade of the twentieth century, and by the 1920s, the "small park" had become a staple of American city planning. Once again, the Olmsteds were in the thick of things, as the great man's sons, Frederick Law Jr. and John Charles, designed the first playgrounds in Boston and small parks in Chicago. The second Olmsted generation had no problem incorporating small units into a full array of parks, from playgrounds to parkways, pastoral parks, and "great outlying reservations." Small parks were also to become a way for developers to promote suburban tracts. San Francisco ended up with dozens of such parks. At midcentury, they would be taken up enthusiastically by landscape architects and city planners led by the Bay Area's Garrett Eckbo and Mel Scott; such parks became part of the gospel of modernist urban design and nearly ubiquitous—if not always well used—across postwar suburbia.[6]

By 1900, civic opinion had turned against the pastoral park idea. The old parks would be mightily altered by the addition of playgrounds, ball fields, and recreation halls. They were also claimed for such municipal functions as museums, parade grounds, and bandstands, as when the east end of Golden Gate Park was used for the Midwinter Fair of 1894 (San Francisco's answer to Chicago's Exposition of 1893). This reduced the rural effect desired by Olmsted Sr. and his generation, but the park's popularity only increased. Golden Gate Park remains a wonderful, if hard-working place, with the power to charm and delight millions of visitors annually.[7]

Parks of Merit

With all the glory showered on Golden Gate Park, little attention has been given to Oakland's city parks, which have played a significant part in inspiring Bay Area open-space and recreational advocates over the years. Frederick Law Olmsted Sr. laid out Oakland's Mountain View Cemetery in the same years that he did the preliminary designs for San Francisco's park, Yosemite Park, and the university campus in Berkeley. Mountain View was meant to be a pleasure ground in the style of the great pastoral cemeteries of Boston and Philadelphia a generation earlier. So Mountain View, begun in 1863, is the Bay Area's first city park—not Golden Gate Park (1870) or San Jose's Alum Rock Park (1872), as is usually claimed. At 226 acres, it is still one of Oakland's largest green spaces.

Similarly, few people know that Oakland's Lake Merritt—created by damming off the arm of San Antonio slough from San Francisco Bay in 1869—is contemporaneous with San Francisco's grand pleasure ground in the sand dunes—and equally unnaturally natural. Wholly forgotten is the remarkable fact that Lake Merritt was declared a waterfowl refuge by the state of California in 1870, thanks to agitation by a mayor who loved ducks.

It was the first and only wildlife refuge in the United States in the nineteenth century, even though bird and wildlife protection was less prominent in California than in the East at this time.[8]

Of course, the cemetery and the lake were just as much tied to civic boosterism and land speculation as Golden Gate Park. The developers of Piedmont in the 1870s benefited from streetcar traffic to the cemetery, and Mayor Samuel Merritt owned considerable property on the lakeshore. Unfortunately, Oakland's burghers were less committed to urban improvements than speculative profits, and Lake Merritt languished. By the turn of the nineteenth century, it was so degraded by sewage that it had lost its waterfowl and its status as a refuge.

Progressive Mayor Frank Mott revived Lake Merritt in the 1900s by building sewers, buying back the shoreline for Lakeside Park, and adding civic buildings—all in the popular City Beautiful style emanating from Chicago at the time. Two renowned city planners were brought in by Mott to propose new parks. Charles Mulford Robinson came in 1906, and Werner Hegemann followed in 1914. Both drew inspiration from ideas bruited earlier by Olmsted Sr. They looked to the hills behind the city as potential parkland and saw the wooded creeks and canyons as the natural openings to the higher ground, to be preserved for aesthetic and recreational purposes. But the tax-averse politics of Oakland could tolerate no more public works, and their ideas languished.[9]

Oakland, bursting at its seams, was too committed to making money to care about parks (though it did get a monumental city hall). The Realty Syndicate of Francis Marion Smith and Frank Havens built housing all along the lower hills. One development, Trestle Glen, was placed smack in the middle of Robinson's favorite park site up Glen Echo Creek. Nor did the city fathers give much thought to playgrounds and neighborhood parks. Oakland was becoming an industrial powerhouse and a working-class city with very little greenery or time to play. Nonetheless, the city gained two outstanding pieces of property in the upper hills. Joaquin Miller Park was deeded to Oakland by the poet's widow when she died in 1913 (425 acres today), and Redwood Park was purchased from the estate of Frank Havens in 1922 (its 1,800 acres were turned over later to the East Bay Regional Parks District).

Berkeley, like Oakland, consigned its hillsides to housing developers. The Hillside Club, formed in 1900 by Annie Maybeck, her architect husband Bernard Maybeck, and poet Charles Keeler, succeeded in turning back the grid, giving the Berkeley Hills their distinctive winding roads and pathways; but talk of a greenbelt in the hills came to naught. A fierce fight broke out over the Thousand Oaks district, which many people wanted preserved as a park for its rock outcrops and stands of old oaks. Instead, it was built out by developer Frank Spring. The university campus at Berkeley was another large piece of open space, with a significant botanical garden and a couple thousand acres of hillsides behind, but the administration couldn't resist the temptation to stick a football stadium into the mouth of Strawberry Creek canyon in the early 1920s. By the end of the decade, the entire East Bay had fewer than 1,000 acres of parklands.[10]

Oakland's final moment in the sun, after World War II, brought a revival of city parks under Parks and Recreation director William Penn Mott. Mott got the city to buy the 75-acre Isaias Hellmann estate in the hills and guided the building of Children's Fairyland on Lake Merritt, a little dream world of fairytale characters in cement that opened in 1950. Widely publicized, it came to the attention of Walt Disney, who visited, was charmed, and incorporated its sentiments into Fantasyland at his own park. After that, the lake and Children's Fairyland fell again into disrepair until the next wave of civic improvement late in the twentieth century. Meanwhile in Berkeley, the university turned its campus into an ugly jumble of Modernist buildings, destroying its great classical plan of 1900, and made upper Strawberry Creek canyon into a sculpture garden for the well-endowed laboratories of big-time science.[11]

WOODLAND PLAYGROUNDS

State parks around the Bay Area cover more than 170,000 acres in fifty-nine units (215,000 acres and seventy-three units including Santa Cruz County); this is the largest volume of open space under the management of a single entity in the region (see Appendix, Table 1). The state parks came about chiefly in two bursts: one in the 1920s and the other after the Second World War. We have already witnessed the birth of Big Basin, but only a handful of state parks had been designated by the First World War, and no administrative apparatus existed to manage them except the state forestry board—which had been caught selling off part of Big Basin for salvage timber. To understand the development of a state park system, however, it is necessary to go back to the origins of the National Park Service.

Serving a Nation

Just as the national park idea owes a great deal to Bay Area conservationists, so too does the creation of the National Park Service. There was no single bureau in charge of the parks before the First World War. The idea of a park bureau was floated by Secretary of the Interior Richard Ballinger in 1910 and legislation drawn up by East Coast Progressive J. Horace McFarland, president of the American Civic Association. When San Francisco's Franklin Lane became Secretary of the Interior in 1914, he appointed Adolph Miller, Berkeley economics professor, as assistant secretary for parks and Mark Daniels, a local landscape architect, as first general superintendent of the parks. Ballinger and Lane convened three conferences to mobilize support for a park service, at Yellowstone, Berkeley, and Yosemite. Agitation grew until Congress finally acted in 1916, on a bill introduced by William Kent and drafted by McFarland and Frederick Law Olmsted Jr.—whose role shows the affinity between the city park and national park movements.

The new National Park Service was almost entirely the work of graduates from the University of California. Stephen Mather, who had already been working with Lane to get the legislation, became its first director. He was assisted by Horace Albright, who succeeded his boss in the 1930s. Newton Drury of Save the Redwoods League took the helm in the 1940s, after which Conrad Wirth, appointed in 1952, observed that he was only the second non-Californian to have the job. William Penn Mott would pick up the Bay Area's mantle again in the 1980s.

Stephen Mather is the person most responsible for the character of the National Park Service and its tourist-friendly policies to attract a mass audience for the scenic wonders of America. Mather, who made his fortune selling Death Valley borax (he invented the trademark Twenty-Mule Team), put his mercantile talents to use in selling the parks to Congress and making them attractive to vacationers. Long before David Brower made nature books into propaganda, Mather and his publicist, Robert Sterling Yard, were putting out pamphlets, films, maps, and a magnificent photo portfolio to induce Americans to visit the national parks. In this, Mather was in tune with a budding tourism industry, which included railway companies, automobile associations, and local promoters. In particular, this self-avowed "auto crank" captured the spirit of the age in which families started going on vacation by car. Mather had the American promotional genius for turning playgrounds for the few into mass entertainment.[12]

State of the Parks

State parks became the rage in the 1920s, following on the success of the national park system. They were meant to complement the federal parks by saving scenic areas of secondary significance. Mather was instrumental here, too, calling the first National Conference on State Parks in 1921. The creation of a California state parks system was spearheaded by the Save the Redwoods League, which had found that buying large tracts of redwoods was a more expensive proposition than they had supposed.

To orchestrate the campaign, in 1923 the league set up a Committee on State Parks, headed by Duncan McDuffie, Berkeley real estate titan, Sierra Club director, and enthusiast for outdoor recreation, along with William Colby of the Sierra Club and J. S. Sperry, flour mogul. They were backed by the Sierra Club, Sempervirens Club, National Park Association, California Federation of Women's Clubs, and the California State Automobile Association, among others. But opposition, led by the timber industry, California Taxpayers' Association, and former governor George Pardee, induced a veto from Governor Fred Richardson in 1925. With a new governor elected in 1926, park advocates tried again. Reorganized as the California State Parks Council, they orchestrated charter legislation, a bond issue for land purchases, and a parks survey in 1927.[13]

McDuffie urged the council to join forces with Frederick Law Olmsted Jr., who carried out the California Parks Survey in 1927–28. Olmsted's recommendations were followed closely for years and served as an inspiration for states across the nation, as well. The bond issue of $6 million was to be matched by private donations, the kind of public-private partnership preferred by the council's upper-class donors. The newly formed State Parks Commission was chaired by William Colby and included such Bay Area luminaries as Madie Brown (mother of a future governor) and Joseph Knowland (owner of the Oakland *Tribune*). Newton Drury managed the acquisitions. With the collapse of the land market in the Great Depression, many landowners were desperate to sell, so purchases went forward rapidly from 1929 to 1934.[14]

TABLE A. State Park Expansion in the Bay Area, 1940–2000 (in acres, by county)

	1940s	1950s	1960s	1970s	1980s	1990s	Total
Alameda	3,478	963	443	3,897	187	8,968	17,936
Contra Costa	127	329	8,401	3,283	7,568	1,504	21,212
Marin	2,430	1,469	4,378	4,465	259	236	13,237
Napa	40	352	1,963	815	558	2,143	5,871
San Francisco	9	96	109				214
San Mateo	1,665	2,322	1,233	2,914	4,556	3,102	15,792
Santa Clara		12,161	959	102	27,366	16,684	57,272
Solano		124	241		84		449
Sonoma	47	56	10,026	13,216	484	5,522	29,351
Bay Area	7,796	17,872	27,753	28,692	41,062	31,159	161,334

NOTE: Figures from the California Parks and Recreation Department, compiled by Peter Cohen.

From a paltry five parks, the California state park system flourished, catching up with New York, Illinois, and Pennsylvania. The Bay Area gained several new parks along the way. Mount Tamalpais was first, in 1929, followed by expansion of Mount Diablo, San Gregorio and Pescadero beaches in San Mateo County, and Sonoma Coast and Armstrong Grove in Sonoma County. The state parks got a big boost from the New Deal's Civilian Conservation Corps and public-works programs, which built picnic areas, campgrounds, trails, and roads by the hundreds.[15]

Popular enthusiasm for outdoor recreation, delayed by the Depression and war, hit like a tidal wave in the postwar era. Public demand for car camping, fishing trips, and playgrounds in the woods propelled state park use to unprecedented heights. As California's population grew by half in the 1950s, park visits went up by 350 percent. The State Division of Parks and Beaches, backed by the Omnibus Parks Acquisition Act of 1945, launched another round of buying land. This provided for the purchase of Samuel P. Taylor, Stinson Beach, and Tomales Bay parks in Marin; Robert Louis Stevenson Park in Sonoma County (Mount St. Helena); Portola Park in San Mateo County; and Capitola Beach in Santa Cruz County. Olmsted was hired to do a new survey for park expansion, which he completed in 1950. New funding in 1955 prompted a third wave of growth under Newton Drury, now state parks director. This round included Henry Coe State Park in Santa Cruz County.[16]

The state park system could grow because funds were made available by New Deal–era policies. The Legislature allocated 30 percent of tideland oil royalties to the parks in 1938, raised to 70 percent in 1943. Nonetheless, the principal source of state money for land acquisition was still the general fund up to the mid-1960s (unheard of today): $15 million in 1945, $44 million in 1955. After that, major augmentations came from targeted state bond issues, including a $20 million bond act in 1964 and a $250 million bond act in 1972. These funds were further enlarged by monies from the Federal Land and Water Conservation Fund, derived from federal offshore oil revenues, set up in 1964 by the Johnson administration. California was the first state to take advantage of that fund. By 1980, after a fourth round of purchases, including such parks as the Forest of Nisene Marks, California had accumulated a million acres of parklands in 250 parks, including 20 percent of the coastline. "By most accounts," declared one historian, "it is the best state park system in the nation."[17]

Parklands in the Bay Area expanded by leaps and bounds, reaching 50,000 acres by 1970, 100,000 in the mid-1980s, and 160,000 by 2000, becoming a fixture of the regional greenbelt. They contain some of the most visually dominant and symbolically charged terrain, such as Mount Tamalpais, Angel Island, and Mount Diablo. They boast the largest single open space in the region: Henry Coe State Park in southeastern Santa Clara County (80,000 acres). They provide the largest parks in Sonoma and Napa counties: Annadel, Sugarloaf Ridge, and Robert Louis Stevenson state parks. They include most of the finest redwoods: Big Basin and Forest of Nisene Marks state parks in Santa Cruz County; Samuel Taylor State Park in Marin; Portola and Butano state parks in San Mateo. And they cover most of the best beaches along the coast.

Without the power and purse of the State of California, nothing like the present greensward could have come into being. Nonetheless, state parks are no longer the cutting edge of greenbelt planning. By the 1980s, tight state budgets and changing priorities in land management had led local conservationists to look elsewhere for leverage in open-space protection (see chapter 7). The state parks benefited from the shift to land-trust purchase and pass-through, and thousands of acres have been snapped up ahead of development and attached to existing parks. In a few cases, this

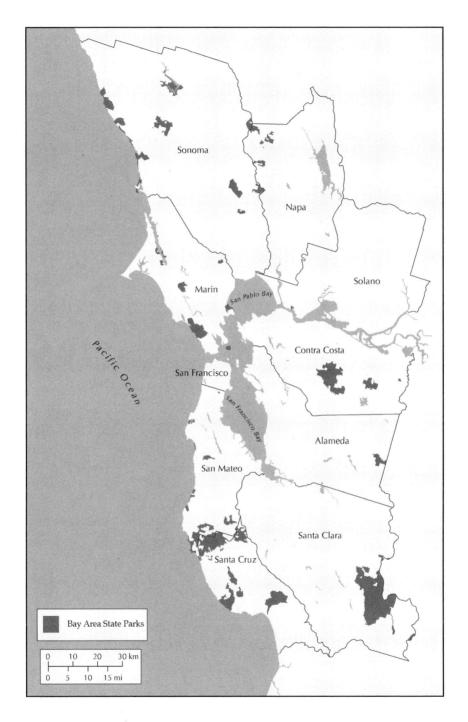

MAP 6. State Parks in the Bay Area, 2005. Drawn by Darin Jensen. Courtesy of Bay Area Open Space Council, GreenInfo Network, Greenbelt Alliance.

led to entirely new parks, as with Cowell Ranch State Park in Contra Costa County. Happily, a major renewal of parks bonds came in 2000 and 2002, leading to such major acquisitions of recent years as Montara Beach in San Mateo County and Sugarloaf Ridge in Sonoma County.

No doubt, the most contested of new parks has been the Eastshore State Park, a thin green line along the bayfront from Emeryville to Richmond, dedicated in 2002. Its origins lie in the attempt by a Chicago financier to turn a quick profit in the 1980s on former Southern Pacific and Santa Fe railroad rights-of-way, and it demonstrates once again the legacy of turning the bay over completely to industry in the nineteenth century. Southern Pacific–Santa Fe Land Company (later Catellus) proposed a string of hotels and shopping centers on the Berkeley and Albany waterfronts, which prompted a great row between environmentalists and African Americans in West Berkeley hoping to secure jobs from the developers. In the end, the cities rejected the overblown proposals, and the state stepped in to purchase some 1,800 acres. The East Bay Regional Parks District is managing the project of turning the ravaged shore into a people- and wildlife-friendly place, with the backing of Citizens for East Shore Parks, led by Norman La Force and Ed Cheasty. It has been no easy matter dealing with contending interests, including native-plant advocates, softball leagues, and windsurfers. Nor have development pressures abated, and park supporters have been desperately searching for the wherewithal to add properties at Albany's horse-racing track and Richmond's Point Molate slated for enormous hotel and casino complexes.[18]

THE EAST IS GREEN

County parks are a significant, if often overlooked, piece of the Bay Area greenbelt. They protect, in all, over 165,000 acres of land—almost as much as the state parks—and provide recreational spaces for millions of people (see Appendix, Tables 2 and 3). County parks draw less attention than city parks and playgrounds or state parks for the simple reason that counties are the odd man out when it comes to California government. Cities have strong identities, the state has all the money and power, special districts run schools and infrastructure, and counties get what's left over.[19] The exception to this obscurity is the innovative East Bay Regional Parks District created by two counties.

County parks began to sprout around the Bay Area after World War I. Sonoma County bought Armstrong Grove in 1917. Santa Clara County purchased Steven's Creek Park in 1924 and Mount Madonna Park in 1927. Santa Cruz County acquired the Welch Grove for a redwood park in 1930. San Mateo County established Memorial Park, a small redwood preserve on upper Pescadero Creek, in 1923. It took the more radical step in 1932 of establishing a Land Acquisition Fund to acquire several miles of public beaches at Pescadero, San Gregorio, and Año Nuevo. It also acquired Coyote Point, which juts into San Francisco Bay, in 1940. Marin County created Drake's, Stinson, and Shell beach parks from 1938 to 1943. Sonoma County purchased Doran Park and Sonoma Coast Park at the same time.

All the county parks along the coast were turned over to the state parks system in the 1940s, however, leaving the counties to rebuild. Very little would be done until 1970, when the state began sharing its parks-bond revenues with the counties. San Mateo County Parks Division (now Parks and Recreation) had been the first to recover, buying Huddart Redwoods in 1948, North County (Junipero Serra) Park a decade later, and Moss Beach Marine Reserve (now James Fitzgerald) in 1969. These are small units. Major expansion followed in the 1970s with Wunderlich, San Pedro Valley, Pescadero Creek, Sam McDonald, and Heritage Grove in the redwoods, followed by Edgewood Nature Reserve and Mount San Bruno in the 1980s, for a total of 12,000 acres. San Mateo continued to manage nine state beaches up to the 1970s, when it relinquished them back to the State Department of Parks and Recreation.[20]

Santa Clara County's parks department, created in 1956 (now Parks and Recreation) had long envisioned a necklace of parks along the hills surrounding the Santa Clara Valley but was starved for the funds to make it a reality. With a reawakening of green consciousness in the early 1970s, voters approved a charter fund (tax set-aside) for park acquisition (since renewed seven times). With these and state funds at hand, the county system expanded quickly, adding such major units as Ed Levin Park on Mission Peak, Joseph Grant on Mount Hamilton, Almaden Quicksilver Mines, and Sanborn-Skyline Park in the Santa Cruz Mountains. Today, Santa Clara County has an impressive twenty-eight parks covering 45,000 acres.

By contrast, Sonoma did not get a park agency until 1967 (called, oddly,

Sonoma County Regional Parks Department) and still has a relatively modest volume of holdings, around 5,000 acres (the largest unit is Hood Mountain). It chiefly caters to recreation through a scattering of forty small units that are more like a collection of city parks (for example, Helen Putnam Park near Petaluma). Napa County has so far refused all entreaties for county parks. Solano County has only three small units and a Parks and Recreation Commission. Most recent open space in the North Bay has been acquired through land trusts and open-space districts (see chapter 8). Marin has a couple of small county parks but has secured huge amounts of parkland by other means (see chapter 4).

Communing with Nature

Alameda and Contra Costa proceeded differently from other counties, creating a bi-county East Bay Regional Parks District in the 1930s. The district today includes nearly 100,000 acres in sixty parks and reserves and is the largest urban parks district in the United States. It serves as the bulwark of the greenbelt in the East Bay, protecting public open space from the bay front to the western delta, along the Berkeley-Oakland hills, around Mount Diablo, east of Altamont Pass, and south along the Diablo range.[21]

The creation of the regional parks district was an important departure in the history of Bay Area open space. It brought together a coalition across two counties and several cities in pursuit of a common goal, borrowing from water developers the innovative tactic of a special district. And it combined the idea of big-city recreation with a greenbelt that could block urban expansion over the hills. It served as a model for Bay Area open-space preservation far into the future.

The East Bay Regional Parks District had less in common with the state parks than with earlier civic models, such as Olmsted's chain of parks and parkways in Buffalo, New York; the Metropolitan Park Commission of Boston; Chicago's greenbelt reaching out to the Indiana Dunes; and Horace Cleveland's plans for Minneapolis.[22] Certainly, it recalled the hopes of Olmsted, Robinson, and Oakland Progressives for a park and parkways system in the East Bay hills, as when Werner Hegemann declared that "beautiful sites like Wildcat Canyon must be held forever in a natural state." Moreover, it lent Oakland, Berkeley, and the East Bay their own identities after a generation of growth had raised their populations close to that of San Francisco.

The idea of public open space in the greater East Bay had been nurtured for some time before 1930. The Sierra Club had been organizing hiking trips in the East Bay hills since 1904; the Contra Costa Hills Club (founded by Harold French, a friend of Muir), since 1920. Mount Diablo, the Tamalpais of the East Bay, was an early target for hiking enthusiasts. Even more than Mount Tam, Diablo dominates the local scene—a landmark visible from the Golden Gate to the Sierra, from Shasta to Fresno. It is said that the unimpeded view from the top of Mount Diablo (3,849 feet) is the farthest in all directions in North America. But Diablo was also a prime target for developers. A hotel and stage road went in by the 1870s; in the 1910s, Robert Burgess and W. R. Hearst cooked up a new hotel and subdivision scheme, putting in the first auto road; and in the 1920s, investor Walter Frick took over the mantle of speculation. A group of locals pushed through legislation for a small state park in 1921, but it took them another fifteen years to acquire the land from Frick. The park was dedicated in 1931.[23]

Hikers and conservationists had long sought public access to the watershed lands of the private water companies on the eastern slope of the Berkeley-Oakland hills, but they had been repeatedly rebuffed. When the East Bay Municipal Utility District (EBMUD) announced that it was putting 10,000 acres of surplus watershed lands on the market on the completion of the Mokelumne aqueduct, the opportunity was seized. Robert Sibley, avid hiker and manager of the University of California Alumni Association, and Hollis Thompson, Berkeley city manager, jumped to the occasion. They organized an East Bay Metropolitan Park Association to petition EBMUD not to sell to developers and gained the backing of the Sierra Club, Oakland Park League, East Bay Planning Association, and Contra Costa Hills Club.[24]

When EBMUD proved intransigent, a new campaign was drawn up. This time, Samuel May, director of the Bureau of Public Administration at the university and a longtime advocate of parks, took the initiative. He got Robert Kahn, heir to an Oakland department-store fortune, to pay for a park study and contracted with the Olmsted Brothers, who had just completed the state parks survey. Their study was authored by Ansel Hall of the National Park Service, a resident of Berkeley. The Olmsted-Hall report of 1930 gave the park movement a guiding vision of a most expansive and eloquent nature. It not only outlined parks for the East Bay but proposed a ring of parks all around the hilltops and shoreline of the Bay Area—a vision that would not be fulfilled for another half century.[25]

May put a student, Harland Frederick, to work running the political campaign in support of the Olmsted-Hall proposal for a chain of parks along the ridge tops. They created a new East Bay Regional Parks Association with the backing of such prominent burghers as Duncan McDuffie, Ralph Fisher of American Trust, and John Miller of Richmond, and they carefully cultivated the support of all the East Bay municipalities. Park advocates rallied their troops at a mass meeting in 1931 at the Hotel Oakland, and when EBMUD (led by former Oakland mayor and governor George Pardee) again refused to budge, they simply swept by the water agency and went directly to the state legislature and the voters.[26]

They had hit upon an innovative strategy: a special bi-county park district—the first of its kind in the country. It followed the model of irrigation districts, pioneered by California in the nineteenth century and expanded in the early twentieth century to water districts for cities, such as Marin Municipal Water District, EBMUD, and the Metropolitan Water District of Southern California. The East Bay Regional Parks District was authorized by a special act of the state Legislature in 1933, on a bill authored by Assemblyman and former Oakland mayor Frank Mott. Before financing could be secured, however, Contra Costa County pulled out and Alameda County had to go it alone. The voters still overwhelmingly approved a bond issue and property tax to finance land purchases, in the midst of the Depression.

The district, led by Charles Lee Tilden, began purchasing land from EBMUD but was stymied by the high price demanded by their old nemesis, George Pardee. The district could buy only three parks: Tilden (Wildcat Canyon), Sibley (Round Top), and Temescal (reservoir). Redwood Park was added in the mid-1940s and Chabot Park in the early 1950s, for a total of 5,400 acres—barely half the original surplus lands. Like the state parks, the district got a real boost from New Deal workers, who built trails, picnic areas, an amphitheater, roads, and bridges. Also like the state parks, the district operated under the dominant rubric of scenic drives, hiking trails, and general recreation. It was wildly popular from the outset and brilliant at bringing the pleasures of the country to city dwellers. Ever since, hardly a kid in the East Bay (including my own) has not grown up with such delights as Tilden Park's carousel, trains, animal farm, and Lake Anza.

Even more than the state parks, which had been engineered from above by old-line conservationists, the East Bay parks were a product of public mobilization. The campaign brought together university students, city officials, hikers, and women's groups. It had the avid support of leading Oakland realtor Fred Reed, who emphasized the parks' benefits for home values, and of Tommy Roberts, a labor leader from the Central Labor Council of Alameda County, who campaigned tirelessly for the district among the unions and was elected to the first board of directors. In short, the East Bay Regional Parks District brought together people across the class divide to make common cause for public recreation. The district has held to this vision of mass recreation to this day, and its parks are some of the best used in the Bay Area greenbelt.[27]

The Greater East Bay

After the Second World War, the East Bay park system expanded rapidly, doubling to more than 10,000 acres in the 1950s, doubling again to more than 20,000 acres in the 1960s, and redoubling in the following decade to around 60,000 acres. Three dozen parks were added to the system in the process. They came into being in various ways: purchases of private land, as at Sunol; donations of ranches, such as Garin; and transfers from public agencies, as at Briones (carved from EBMUD watershed lands).

When William Penn Mott took the reins in 1962, he revitalized district finances and energy. A crucial move within his first year of office was to secure the reentry of Contra Costa County, whose only major open space up to then was Mount Diablo State Park. The district quickly made major new acquisitions, such as Black Diamond Mines and Las Trampas, trying to keep ahead of the wave of urbanization engulfing the valleys east of the Berkeley-Oakland hills. Mott made a major policy shift in 1967 by moving out of the hills to acquire shoreline properties. The first bayside park was Coyote Hills in southern Alameda County, followed by Point Pinole in west Contra Costa County and Crown Beach (a former private amusement park in the city of Alameda). The shift toward the bay was spurred by the rise of the movement to save the bay, as different branches of open-space activism reverberated off one another. Kay Kerr, friend of Mott and a founder of the Save San Francisco Bay Association, was an active supporter and honorary chair of the campaign to bring Contra Costa County back into the regional parks district.[28]

This golden age of the East Bay park district is sometimes attributed to the vision of Mott, who was something of a latter-day Stephen Mather. He ascended from superintendent of the Oakland Park and Recreation Department (1945–62) to general manager of the East Bay Regional Parks District (1962–68) to chief of California State Department of Parks and Recreation (1968–85) and thence to director of the National Park Service (1985–89). He enthusiastically pursued the mission of the parks, moved fluidly through the corridors of power, and mustered broad support for their projects from philanthropists such as Laurance Rockefeller. He was part of the network of upper-class conservationists who carried so much weight around the Bay Area in the middle of the twentieth century. He was, naturally, on the board of the Save the Redwoods League, and he enjoyed a close association with such men as William Lane from the peninsula, Norman Livermore of Marin, and Robert Sproul and Clark Kerr, presidents of the University of California. Ronald Reagan brought Mott to Sacramento and then to Washington, D.C., yet Democratic governors Pat Brown and Jerry Brown both thought highly of him.[29]

This kind of Great Man Theory of history leaves much to be desired. It ignores the contributions of such key players in park district management as Richard Trudeau, successor to Mott as general manager from 1968 to 1985, and Hulet Hornbeck, who served as acquisitions chief under Trudeau. They added twice as much acreage to the district's parks as Mott had. Following that, Pat O'Brien, general manager, and Bob Doyle, assistant general manager, provided excellent leadership and made large additions to regional open spaces in the 1990s and 2000s.[30]

On its left flank, the park district had the formidable assistance of public-spirited citizens such as Jean Siri, Barbara and Jay Vincent, and Lucretia Edwards in west Contra Costa County. For fifty years, they were tireless advocates for the environment in their corner of the region, especially access to San Francisco Bay's waterfront. It took them ten years to secure the whole of Point Pinole from Bethlehem Steel Corporation, after Mott bought only a sliver. Siri, who called herself "a professional troublemaker," declared, "I did every dirty thing I ever did to get that park"—including setting up a false-front West Contra Costa County Conservation League with a sheaf of letterhead. She was the wife of Sierra Club president Will Siri but never got involved with the club; her interest in the bay grew out of Save Francisco Bay Association and her daughter's school science project. Jean Siri later secured Brooks Island off Richmond for the parks district after going out to dig native artifacts with her children and finding that Richmond wanted to level it for an airport. The Vincents did so much work along the Richmond shore that the city dedicated a park to them (now a part of the Rosie the Riveter National Historic Park).[31]

Save Mount Diablo has played the most important supporting role in the eastward expansion of the parks and open-space system. It was founded in 1971 by Dr. Mary Bowerman and Art Bonwell, residents of central Contra Costa County. Bowerman had done her doctorate at UC Berkeley in the 1930s on the ecology of Mount Diablo, whereupon she realized that the state park was not sufficient to protect the mountain's unique, endemic wildlife. Bonwell was chair of the Mount Diablo Regional Group of the Sierra Club's Bay Chapter. He wisely recruited young blood to the cause, such as Bob Doyle, Bob Walker, and Seth Adams.[32]

The group's initial mission was to expand Mount Diablo State Park, which was only 6,700 acres on the summit. They mobilized the conservation-minded elite of Orinda and Lafayette, including an older generation of activists from the parks district expansion of 1963 and the Contra Costa Parks Council, including Hulet Hornbeck and Susan Watson. In the process, they built a public constituency for open space in a notoriously conservative part of the Bay Area. These efforts were instrumental in getting the park expanded to 19,000 acres, with the help of State Senator John Nejedly. Save Mount Diablo pushed the East Bay Regional Parks District to acquire Black Diamond Mines and Morgan Territory, and pushed local cities to establish Lime Ridge, Shell Ridge, and Crystyl Ranch parks, bringing the total parklands around Mount Diablo to nearly 80,000 acres.

A key to the success of the regional parks district has been robust financial support from the people of the East Bay. Postwar expansion was funded by a special legislative tax override passed in 1963 and a bi-county tax increase in 1971, as well as injections from the 1964 state parks bond issue. After the tax revolt broke out in California, everything changed. Proposition 13 in 1978 brought park acquisitions to a halt. For a while, the district had to depend on private philanthropy, such as the Adopt-a-Parks program.

But the green movement and the local citizenry showed their resolve by voting through a large district bond measure in 1988. That put $225 million in the hands of district managers, allowing them to go on a shopping spree in the early 1990s when property values were down. The result was that, once again, the parks district was able to follow the eastward march of urbanization and to set aside precious open spaces before they could be

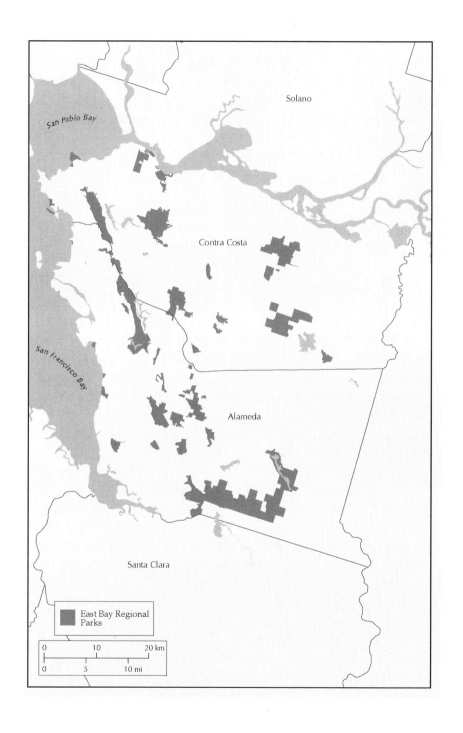

MAP 7. East Bay Regional Parks, 2005. Drawn by Darin Jensen. Courtesy of Bay
Area Open Space Council, GreenInfo Network, Greenbelt Alliance.

swallowed by the advancing wave of pavement, pools, and parking lots. Some 30,000 more acres were added, including Pleasanton Ridge, Round Valley, and Crockett Hills (Carquinez Strait). Big state bond measures in 2000 and 2002 had the district buying again, farther east at Clayton Ranch, Vasco Caves, and Big Break Regional Shoreline near Antioch, where the bay meets the delta.[33]

The East Bay Regional Parks District's holdings today dwarf the original parks in the hills behind Berkeley and Oakland. But it has kept to the original goal of making its parks accessible and inviting to the public. Today that means more hiking trails than before—user surveys show that walking is the public's first choice—and also more picnic areas, education programs, swimming areas, and youth camps, all within ten to thirty minutes of residential areas. All this is possible even though most of the district's money must go to land acquisition just to keep up with the rate of urbanization, and amenities like the Tilden Park Carousel are too expensive to include today.[34]

The district has established an admirable record on several counts, including good site selection, recreation, and park management. But in the end it all comes back to a solid foundation of public support. There is no mystery in that, because the district always made serving the public its primary goal, even as it has set aside great swaths of open areas for hiking, ridge tops for unencumbered views, and habitat for wildlife. It turns out that picnic areas and cougars can coexist if enough green space exists for both people and predators to catch their breath.

PEOPLE FOR THE PARKS

The recreational parks are an essential component of the Bay Area greensward, around 400,000 acres altogether. They began with the great urban landscape parks, expanded with city playgrounds and neighborhood parks, and became more ambitious in the state parks system and county and regional parks. The parks—city, county, regional, and state—benefited enormously from the booming economy and tax revenues of the postwar economy, but just as much from the political climate of the New Deal–Great Society era in which taxing the rich and spending for the public weal was something Californians—even liberal Republicans such as Governor Earl Warren—did as a matter of course.

Just as the landscape parks became busy playgrounds for the people in due time, outlying state and county parks quickly became popular with the growth of outdoor recreation between the world wars. Growing affluence, the model-T Ford, and asphalted roads paved the way in the 1920s as the middle class lit out for the great outdoors. Across the country, the early fishing, hiking, and hunting clubs of the well-to-do, such as the Boone and Crockett Club, gave way to the Izaak Walton League and the National Wildlife Federation, the first national organizations of outdoor recreationists, mostly fishermen and hunters. Even elite organizations such as the Sierra Club promoted wider public access to the wonders of the American West and worked hand in hand with the National Park Service.[35]

Mass recreation exploded after the Second World War, when the lower middle and working classes were able to join the flood of city dwellers to the countryside, thanks to better wages, more leisure time, and, of course, more cars. So great was the demand, parks could not be added fast enough. There was always too little money when parkland had to be bought back from private owners. Land costs were magnified by dilatory action in acquiring parcels before the city caught up to the prime places. As a result, landowners, often from the oldest and wealthiest families, were paid handsomely by the time the public interest prevailed. As Herbert Jones, longtime Republican state senator from San Jose and president of the Sempervirens Club, observed, the state had paid "an appalling price" for its delayed acquisition of parklands, allowing massive payoffs to landowners (which, sadly, remains true to this day).[36]

The recreational parks were soon taken to task for catering to mass tourism and the almighty automobile. A new generation of nature preservationists arose to challenge park management and the very idea of mass access to wildlands. The Wilderness Society, which led the return to preservation, was founded in the 1930s by three veterans of federal land management: Aldo Leopold, Bob Marshall, and Robert Sterling Yard. They were disillusioned by the way the Park Service, Forest Service, and Bureau of Land Management pandered to resource extraction and popular recreation. Soon they were joined by a new generation of Sierra Club leaders, for whom the National Park Service's Mission 66 program (launched in 1956) was the fatal step. The club would be instrumental in generating support for a new part of public parklands—wilderness parks—through its biennial Wilderness Conferences (begun in 1947) and publishing program.[37]

The legislative achievements of the new era of preservation were spectacular: the Wilderness Act of 1964, the Wild and Scenic Rivers Act of 1968, the classification of huge areas of national forest land as wilderness under Rare I and II in the 1970s, and the Alaska Lands Act of 1980. California gained millions of acres of protected land in wilderness areas. While most of these places were far away from the Bay Area, some effects were felt close at hand. Portions of Point Reyes National Seashore were declared wilderness, as were parts of state parks (such as Henry Coe) and certain of the East Bay regional parks (such as Sunol and Los Trampas).[38]

Nevertheless, the number of backpackers feeding the backcountry boom would never have grown so large by the 1960s without the mass appeal of the recreational parks to the previous generation. The people came to enjoy their pleasuring grounds around the Bay Area and learned to savor the easy joys of nature near at hand. This is where most people came into contact with forest and stream, trails and wildlife— certainly, that was true of my family's outings to Memorial and Portola parks or my teenage ramblings to Pescadero and San Gregorio beaches. Many such ordinary folk would become stalwarts of the environmental movement to come.

While the recreational parks set an important precedent of saving precious places and outlying open space, a more radical vision of land conservation came into play as the pressures on natural resources multiplied. All around California and the far West, the postwar boom meant clear-cutting forests to build houses, turning rivers into staircases to power the cities, and building giant ski resorts for fun and profit. The havoc wreaked on the countryside would compel the Sierra Club, for one, to break with its genteel past of working inside the system and take a more militant, oppositional stance to government and capital. Modern environmentalism would be born out of the ashes of elite conservation. So the story goes.

But closer to the city, a parallel story was unfolding. Upper-class conservation was becoming less genteel, more modernist, and more oppositional. It was facing down the forces of progress at home, in the shape of postwar suburbanization. Out of that collision, a new kind of movement for open space would emerge in the Bay Area, fighting freeways and industry, putting tighter restrictions on growth, and thinking up new kinds of national parks in the region's backyards. From this would be born a mass movement to save the San Francisco Bay and the California coast that would reverberate off national trends and lead the way to a new age of environmental consciousness.

4

THE UPPER WEST SIDE

Suburbia and Conservation

The open vistas across the bay and the invigorating views from the encircling hills make San Francisco a self-conscious city, prone to a good deal of preening as well as self-preservation. This happy sense of being Nature's Children has been especially pronounced in the upper-class suburbs along the west side of the bay, in the green redoubts of Marin, San Mateo, and northern Santa Clara counties. Conservationists in these areas have been the most intense in their affection for the greensward and the most generous in setting aside public open space.[1] West-side residents came to understand that what wealth and nature provide is all too easily squandered. Without the intervention of these elite conservationists over the years, the hills and shores, woodlands and marshes would have disappeared underneath asphalt and condos long ago. It may be self-interested, but the fierce anti-development stance of the well-heeled has served the rest of us better than might have been expected.

THE BULLDOZER IN THE COUNTRYSIDE

The postwar boom set the trip wires that turned the guerrilla skirmishes of the Bay Area conservationists into a full-scale insurgency against the forces of urban expansion and property capital. The metropolis had ballooned during World War II as the launching pad for the war in the Pacific, and it kept on the same upward course in the decade after the war. By 1955 the population had already doubled that of 1930, to more than three million. Suburbanization blew outward even faster.

The sprawling city exploded across hill and dale as never before, producing a wide spectrum of environmental effects: disappearing farmlands, bay fill, water pollution, energy waste, loss of open space. This ratcheted up the confrontation of the city with the country, exposed the ambivalence of urbanites about city life and rural landscapes, and raised the stakes in land-use decisions. More generally, it generated a sense of deep unease that urban growth was out of control and that the California landscape would be forever altered.

Modern environmentalism was born from the belly of the city in the confrontation with the bulldozer in the countryside, as Adam Rome has shown. The long-standard version of American environmental history, endlessly reproduced by popular reference, is that modern environmental protest came from the union of two great ideas: John Muir's clarion call for wilderness and Rachel Carson's call to arms against pesticides. But their voices floated above the tumult of postwar urbanization, whereas a critique of growth arose spontaneously from a thousand voices and a broad mobilization to protect the land and waters bubbled up from below.[2]

We need histories of the ordinary people who turned the ideas of the notables into a mass environmental movement. Furthermore, the Bay Area deserves greater recognition for its place in the national reaction against postwar suburbia. This role derives not from some natural prescience of the local, but from the intersection of a budding green political culture and rampant suburbanization. Californians had a clearer picture of the environmental impacts of midcentury suburbia because the state was well ahead of the country in feeling the full force of urbanization based on autos, highways, electricity, cheap gas, tract housing, and mass affluence by the 1920s. The postwar explosion of suburbia brought the greatest environmental assault on the state since the mining era.[3]

Close Encounters of the Suburban Kind

Long a backwater of urbanization, Marin County eventually came under intense development pressure. Completion of the Golden Gate Bridge in 1937 and U.S. Highway 101 north caused its population to double in fifteen years, and that was just the appetizer. In the heady days of the postwar boom, the State Highway Commission was in an ecstasy of freeway planning, which would be finalized in a statewide Freeway and Expressway Plan in 1959. The freeways were fed by the Collier-Burns Act of 1947, which committed California to gas taxes and metro freeway systems, and the Interstate and Defense Highway Act of 1956, which provided 90 percent federal funding. Marin was no exception. State highway 1 was to be made into a freeway up the coast, with two cross-country freeways connecting up with U.S. Highway 101, opening up the coast to development. A second bridge was built, from Richmond to San Rafael, in 1958 and a third was proposed, from San Francisco to Marin via Angel Island.[4]

In the step-down from war, old Army and Navy bases at the Golden Gate headlands, Sausalito, and Angel Island became available for privatization. The State Lands Commission gleefully sold off bays and wetlands to speculators, triggering massive projects from Richardson Bay to Point Reyes. Meanwhile, the U.S. Army Corps of Engineers was damming Lagunitas and Nicosio creeks and the headwaters of the Eel River to make water available for thousands of new homes in Marin. Capitalists from far and near set their sights on the valleys, bays, and coastline, buying land in anticipation of the roads and water to come.

All this was marked out in the West Marin General Plan of 1964, a nightmare vision for Marin conservationists. It anticipated a quarter million new residents along the coast to join the quarter million along the U.S. Highway 101 corridor. County politics were in the thrall of the forces of urban expansion. The Board of Supervisors had been taken over by a prodevelopment crew, led by Ernie Kettenhofen, while Jack McCarthy, the son of a contractor-developer, blew the trumpets of progress as the local state senator.[5]

The San Francisco Peninsula was already the principal avenue of suburbanization out of San Francisco in the mid-twentieth century. Local politicians played their assigned roles. State Senator Richard Dolwig was known as the developers' best friend in Sacramento, along with his close ally, Assemblyman Carl Britschgi. The supervisors welcomed the State Highway Commission's plan to crisscross San Mateo County with freeways, including Highway 1 down the coast to Santa Cruz, Interstate 280 along the foothills, and the Bayshore Freeway (U.S. Highway 101 south). These would have been connected by three freeways over the mountains to the coast, at Pacifica, Half Moon Bay, and Woodside, plus expressways down Skyline Drive, Junipero Serra Way, and El Camino. The supervisors also invited in the Corps of Engineers to build a dam on Pescadero Creek to supply the up-and-coming coast.

The 1930s had ushered in a new breed of merchant builder who accelerated the pace of mass suburban development. These developers combined subdivision, construction, and sales under one company, filling tracts with several hundred homes at a time. The new builders were masters of modern design, marketing by means of model homes, and financing through local savings and loan associations and federal New Deal programs, which made the thirty-year, low-interest mortgage the standard.[6] By the 1960s, an even more ambitious generation of community builders combined thousands of homes and apartments with shopping centers, schools, and industrial parks.

As a result, the suburbs became notoriously orderly and repetitious, because they were being built out quickly by a few large developers. Typical house styles were quite limited, minor variations (such as flipping the blueprint) on a handful of designs for flat-roof Moderns, ranch houses, or, by the 1960s, split-levels. Compared with the subdivisions for the rich, middle-class and working-class tracts featured smaller houses, greater monotony, and less regard for landscape amenities. By the 1940s, curvilinear streets and cul-de-sacs had displaced the grid almost completely from the new suburban metropolis.

The Bay Area had some of the pioneer merchant builders, even before the famous Levitt of New York and Philadelphia, such as Henry Doelger, David Bohannon, Joseph Eichler, and Henry Kaiser. The flatlands of the peninsula counties were the proving grounds of these builders in the 1940s and early '50s. Doelger built in Daly City and South City; Bohannon did his work from Redwood City to San Mateo; Eichler's first tracts were in Sunnyvale and Palo Alto. Similar developments occurred a decade later in the string of Marin towns along U.S. Highway 101: Greenbrae, Larkspur, Corte Madera, and Terra Linda. The Bay Area also had its share of the even larger community builders of the 1960s, such as Jack Foster, who created Foster City near the San Mateo bridge. Bohannon's Hillsdale was a pioneer mixed-use development, which maximized the return on a large parcel through high density, commercial rents, and a mix of classes.[7]

The New Metropole

Meanwhile, Silicon Valley was beginning its explosive growth, creating the three-headed monster that is the San Francisco–Oakland–San Jose metropolitan area. In the postwar era, the electronics industry exploded across the old agrarian landscape, creating an entirely new Edge City on the southern flank of the bay. In the process, the face of the Valley of the Heart's Delight would be entirely reengineered over the course of a mere quarter century. The acres in orchards shrank by three-quarters (from 101,000 in 1940 to 23,500 in 1973), and cities expanded to more than 100,000 acres by 1968. Of that newly urbanized space, residences occupied just under 50 percent, industry 10 percent, and commercial strips 5 percent; the rest lay under asphalt in streets and parking lots.

San Jose's sprawl would become an indelible part of the American imagery of postwar suburbia, and Stanford's Industrial Park would become the symbol of the new high-tech industrial landscape. Both the city and Stanford University played pivotal roles in the Silicon Valley growth machine. San Jose was run by a notorious growth coalition led by city manager Dutch Hamann—the man who wanted his city to be the Los Angeles of the north. But working behind the scenes were real estate developers such as John Sobrato and Carl Berg, who would have as much to do with the success of the valley as the city managers, university planners, and electronics moguls who got most of the press.

Housing carpeted the valley. The dominant postwar fashion was the single-family home, and the norm was the detached house on a quarter-acre lot, with wide street frontage and a prominent garage and driveway. It was a habit that devoured land by the hundreds of square miles on the South Bay flatlands. Larger homes for the rich went up one by one or in small groups, on acre or larger lots in the western foothills in Los Altos Hills and Portola Valley. Meanwhile, industrial properties were built up at a great rate on both sides of U.S. Highway 101, creating a central employment corridor running from Palo Alto to Santa Clara.

Where and when development occurred was left to developers and the market. Speculators would buy up large parcels from farmers in a helter-skelter fashion, seeking out cheap land that could be converted to urban uses at vastly increased property values. The farther out the better, as it meant a larger payoff. This made developers the enemies of urban concentration, egged on by the immense spatial freedom provided by the automobile. By leapfrogging outward, developers could advertise their housing tracts as "country living" or provide spacious industrial sites with ample parking.[8]

In order to make the jump to urban uses, developers had to secure zoning approval and urban services—roads, water, sewerage—from the relevant local government: county, special district, or city. Because existing jurisdictions were small, most of the land lay outside any city limit and under the suzerainty of the county. As new tracts and industrial parks were dreamed up, they were annexed willy-nilly by ambitious cities. Zoning was unrestrained by any significant advanced planning, except in Palo Alto. Towns set aside vast tracts of empty space for industry and approved every housing subdivision that came forward. Elite suburbs such as Los Altos Hills and Saratoga established large lot sizes to enforce their social character and keep out the subdividers. No one looked to the collective effects.

San Jose was the worst offender. From 1950 to 1960, it made more than 500 annexations, from 1960 to 1970 more than 900. San Jose ballooned in size, grabbing almost half the population of the county and passing San Francisco as the most populous city in the Bay Area. It also became the third-largest city in the state in land area, hitting 174 square miles by 1975. The redrawn map of the city became the poster child for the suburban sickness of the age. It seemed the very embodiment of what one writer called "the geography of nowhere."[9]

MARVELOUS MARIN

Marin County is synonymous with environmental ethics and open-space protection. Marin was the scene of some of the earliest conservation efforts in the region, and after World War II the county led the charge against urban encroachment. Its record on land protection has been extraordinary: more than half the county is off-limits to developers today. From the Golden Gate to Point Reyes, over 100,000 acres on the western slope are in national and state parks; nearly 40,000 acres of ranchland in the north county are protected under the Marin Agricultural Land Trust; 14,000 acres of islands of open space are maintained on the urbanized east side by the Marin County Open Space District; and the flanks of Mount Tamalpais are protected by the Marin Municipal Water District's 20,000 acres of watershed. Nowhere else in the United States is there so much superlative open space within an arm's reach of a major urban center.

The Great Ladies

The pivot of green Marin has long been the Marin Conservation League, founded in 1934 by Caroline Livermore, Sepha Evers, Helen van Pelt, and Portia Forbes. The league was a mix of grandes dames of the San Francisco aristocracy, of which the Livermores were a fixture, and pioneering professional women, such as van Pelt, one of the first female landscape architects in the United States, and Mary Summers, head county planner and one of the earliest women in her profession. They stood in the tradition of twentieth-century garden clubwomen promoting conservation and public improvement (Caroline Livermore established the Marin Art and Garden League in the 1940s). While often self-effacing and known by their husbands' names, they were formidable characters and keen political operatives. They worked behind the scenes through their social contacts at the same time as they did the public legwork of going to city councils, boards of supervisors, and the state Legislature. Because many members were too shy to speak up, league stalwart Grace Wellman created a troop she called the Nodders and Frowners as a Greek chorus for league arguments at public meetings. As Marin conservationist Martin Griffin says, "[Caroline] and her 'ladies,' as she called them, had the vision, connections, and clout to be effective."[10]

First among the great ladies was Caroline Sealy Livermore, who served as Marin Conservation League president from 1941 to 1961. Daughter of a leading capitalist of Galveston, Texas, she married San Francisco engineer-businessman Norman Livermore, a fixture on city corporate boards and the Marin Republican Party central committee. They moved to the leafy glades of Ross, north of Mill Valley, in 1930, where she raised five sons. Caroline was a powerful presence in the county and an indefatigable warrior for the green cause. "She was much more remarkable than people realize," observes Boyd Stewart, Point Reyes rancher. "A very intelligent, sharp woman, she grew up before women were generally accepted in business, but she was one of those women who made her own place in the world—without making a lot of commotion about it. . . . She had grown up with wealth and didn't have any interest in it. She fitted in any place with any group." Dorothy Erskine of People for Open Space put it succinctly: "Caroline was a master."[11]

The league's first concern was how to cope with the impact of the Golden Gate Bridge. They put out a publicity pamphlet entitled *Marvelous Marin County* and sponsored a surrogate county master plan in 1935, which was drawn up by Hugh Pomeroy and draftsmen hired with New Deal funds. Pomeroy had helped write the ambitious Olmsted-Bartholomew regional plan for Los Angeles that was nixed in 1930, but in Marin his ideas would serve as the county's open-space vision for decades. At the league's behest, the Marin Board of Supervisors adopted the first county zoning ordinance in the state, in 1937; hired a permanent planner four years later; and produced a recreation plan. This was an amazing achievement, given that Marin politics were ruled by rural landowners, ambitious developers, and a conservative board. It would serve as an example for planning all around the Bay Area in the postwar era (see chapter 6).[12]

Hard on the heels of the planning study, the league started campaigning for the first parks in Marin. Led by Sepha Evers, they secured Drake's Beach in 1938, Stinson Beach in 1939, and Shell Beach in 1943. They got the state to purchase Samuel P. Taylor State Park (formerly a popular private campground) in 1940 and Tomales Bay State Park in 1948, and the county granted the state Shell and Stinson beaches. League member Verna Dunshee, who was president of the Tamalpais Conservation Club, was instrumental in the expansion of Tamalpais State Park to more than 6,000 acres.[13]

As the postwar boom hit, the Marin conservationists were ready. Every highway, bay-fill, and army-base project triggered a response from the warriors of the Marin Conservation League and a burgeoning group of allies, including the Sierra Club, Audubon Society, and Nature Conservancy. The counterattack was led by Caroline Livermore. She saved Angel Island from being bulldozed into Raccoon Strait to build a bridge from Telegraph Hill to Tiburon and stopped the filling of Richardson Bay for a New Town development (Reeds Port). Livermore's tactic was to create associations on the spot—not exactly what political scientists have in mind in referring to the American genius for joining clubs and building "social capital."[14] She formed the Angel Island Foundation to buy the island for a state park and the Richardson Bay Foundation to purchase tidelands out from under the Utah Construction Company. She started a Marin branch of the Audubon Society for the sole purpose of getting the bird-watchers on her side. She secured Angel Island State Park in 1954 (completed in 1962) and killed off Reeds Port in 1958. Fittingly, Angel Island's highest peak is named Mount Caroline Livermore.[15]

The Livermores are second only to the Kents as Marin's most famous family. Norman Sr. was a director of the Save the Redwoods League, chair of the California Academy of Sciences, and a founding director of the Marin Municipal Water District. Norman ("Ike") Livermore Jr. was appointed Secretary of Natural Resources under Governor Ronald Reagan in the 1960s, in which office he helped to stop the Dos Rios Dam on the Eel River. Putnam Livermore served as attorney to the Nature Conservancy and cofounder of Trust for Public Land, as well as chair of the California Republican Party, and helped organize opposition to San Francisco's freeways. George Livermore was an architect and pillar of San Francisco society. John Livermore, a mining geologist, ran the family's 10,000-acre ranch on Mount St. Helena in Napa County. Robert Livermore became a rancher in the East Bay.[16]

The Evers family also made a large mark on Bay Area environmentalism. Sepha's husband, Albert, an architect, was the first president of the California Roadside Council, a group created in 1929 by San Franciscan Cora Felton to fight the proliferation of billboards—the advertising medium of choice in the new automobile age. The Evers' son, Bill, a San Francisco attorney, cofounded the Planning and Conservation League, lobbying arm of conservation groups in Sacramento, in 1965. PCL grew directly out of the Roadside Council's success and the work of Helen Baker Reynolds of San Francisco. PCL's earliest campaigns were to save Lake Tahoe and the coast. Twenty years later, Evers would help launch the Greenbelt Alliance (see chapter 6).[17]

Save Our Seashore

In the early 1960s, the fight jumped from east to west Marin. Point Reyes was the first point of attack. Back in the 1930s, the days of the Olmsted state park survey and the New Deal, Point Reyes had been listed as a possible park, but the idea languished until revived in 1958 by former National Park Service West Coast planning director George Collins. Collins was aware of the surging development pressures and of hearings held that year on West Marin freeways. He got the Park Service to verify the merit of the site and then picked up the eager support of newly elected Congressman Clem Miller, who introduced the legislation for a national seashore. In 1961 a group of activists led by Bill Grader, Barbara Eastman, and Margaret Azevedo formed the Point Reyes National Seashore Foundation, with Caroline Livermore as honorary chair and Doris Leonard, wife of past Sierra Club president Richard Leonard, as vice-president. Sadly, Miller died three weeks after the park bill was signed in 1962.[18]

The battle for Point Reyes was not over, however. Point Reyes was not only one of the first two national seashores, along with Cape Cod; these were the first national parks to be purchased entirely from private landowners. It was an expensive proposition, unlike the usual dedication of lands already under federal jurisdiction. The funds ran out halfway through acquisition of the original 53,000 acres. The Park Service proposed selling off part of the park to raise the money for the rest, after the Nixon administration refused to release more funds. Meanwhile, loggers were clearing timber on private in-holdings, speculators were driving up land values, and the Park Service was building a four-lane expressway across the peninsula (stopped only by lack of funds).

Marin activists jumped into the breach. Their leader this time was Peter Behr, who had been elected to the Marin Board of Supervisors in early 1962 after a recall of a man who had tried to reduce the size of the national seashore by half. In 1969 Behr, George Collins, and Barbara Eastman created a new group, Save Our Seashore (sos), which was backed by a wide coalition that included Harold Gregg, then president of Marin Conservation League; Sylvia McLaughlin of Save San Francisco Bay; and Edgar Wayburn of the Sierra Club. They gathered a half million signatures from around the Bay Area on a petition to President Nixon, who freed up the money before the midterm election in 1970. Point Reyes National Seashore was eventually expanded to more than 70,000 acres.[19]

The next conservation targets in west Marin were Bolinas Lagoon and Tomales Bay, flanking the Point Reyes peninsula. In 1961 Martin Griffin and the newly minted Marin Audubon Society started to look for a way to save the bird rookeries of the lagoon. The Board of Supervisors had created a Bolinas Harbor District to build a marina-centered development. Audubon headed them off by quietly buying up (Audubon) Canyon Ranch and other ranches all along the coast, which they later turned over to the county, state, and federal governments for parklands. They were backed financially by

Alice (Mrs. Roger) Kent, Caroline Livermore, Gwin Follis (wife of the president of Standard Oil of California), Grace and Theodore Wellman of the Marin Conservation League, and the Nature Conservancy. "Our first Audubon Canyon Ranch letterhead looked like a Who's Who of San Francisco and Marin County," Griffin recalls.[20]

The harbor district was dissolved in 1969, and Bolinas Lagoon ended up under the jurisdiction of the Marin County Open Space District. Unfortunately, a housing development called Seadrift had already gone in, which blocks winter storms from sweeping over the barrier beach to clear out sediment from the lagoon. By 1970 the contest had moved up to Tomales Bay, where Audubon repeated its strategy of selective purchases to block coastal access and stop the in-filling, which thwarted the grand plans of the local syndicate, Land Investors' Research Group (builders of Rohnert Park in Sonoma County).

At the same time, the southern flank of Marin was in jeopardy from the Marincello project just back of the headlands of the Golden Gate. Marincello was the brainchild of construction magnate Tom Frouge, backed by Gulf Oil Corporation. It was a typical New Town, a mixed-use development for 30,000 people. Following quick approval by the county supervisors in 1965, construction began—only to be halted by a falling out between Frouge and Gulf Oil. Attorney Marty Rosen, who was converted to the conservation cause by the case, began tying up the project with lawsuits, helped by Doug Ferguson. This gave time for public opposition to mount. When Frouge died unexpectedly in 1969, the project was thrown into limbo and the conservationists had their opening. Huey Johnson, who had just set up an office of the Nature Conservancy in San Francisco, stepped in and bought the land, with the financial help of heiress Martha Alexander Gerbode.[21]

Turnabout Is Fair Play

By this time, the tide had turned in Marin. The long-standing progrowth majority on the Board of Supervisors was overtaken by events. In 1968 Peter Behr was joined by two other conservation-minded members, Peter Arrigoni and Michael Wornum, following a campaign spearheaded by Roger Kent, a power in the state Democratic Party as well as in Marin County. The new green majority nixed support for Marincello, asked Cal-Trans (the former State Highway Commission) to put the kibosh on any new freeways or bridges, and threw out the West Marin General Plan. Alas, Caroline Livermore died in 1968, just at this moment of triumph.[22]

The planning commission ordered a new, ecologically informed study of the county, to be undertaken by an expanded planning staff, and a new plan for controlled growth. A revised general plan was adopted in 1973 and is still in effect. It is a brilliantly simple plan that won widespread favor. It divides the county into three parts for three different purposes: urban development on the east side along the U.S. Highway 101 corridor; recreational open space on the west side; and agriculture in the center and north. The astounding success of the green machine meant the county virtually halted urban development: the growth rate fell from 113 percent in the 1940s and '50s and 42 percent in the 1960s to less than 7 percent in the 1970s and 4 percent in the 1980s. County population has stabilized at around 250,000.[23]

A key figure in the replanning of Marin was Margaret Azevedo, longtime member of the county planning commission and resident of Tiburon. She had been a co-founder of the Point Reyes and Richardson Bay foundations in the early 1960s and would serve later on the Citizens Advisory Panel for Transit of the Golden Gate Bridge district and the board of the California Coastal Conservancy. Azevedo played a central part in shaping the county general plan to confine development to the eastern corridor. Her early work was with Oakland's recreation department and the federal Works Progress Administration.[24]

Control over water was another device by which conservationists stanched the blood supply to the growth machine. In 1971 Marin voters rejected a project to pipe in Russian River water; in 1973 the Marin Municipal Water District declared a moratorium on further hookups and halted all development on watershed lands. The county refused to sign on to water projects farther north, even as its supplies hit rock bottom in the drought of 1976–77. Developers were apoplectic when, soon thereafter, the county shut off the water to all but a limited number of new homes each year and adopted the most draconian water conservation measures in the country.[25]

Such a turnabout by a county to an antigrowth regime was unprecedented in the annals of California government and virtually unheard of in American federalism. Exclusive suburban municipalities, such as Hillsborough on the peninsula, have sometimes halted growth (usually by means of large-lot zoning) once they filled in, but no political unit of this scale had ever done so. It goes against the grain of American boosterism and local growth machines.[26]

Out of the open-space struggles of Marin came the most outrageous move in the history of the Bay Area greenbelt: creation of the Golden Gate National Recreation Area. GGRNA extends from Point Reyes south through most of western Marin County to the headlands and Fort Baker; then crosses to the Presidio and down the beachfront of San Francisco; takes in Crissy Field, Alcatraz, Fort Mason, and the Maritime Park on the city's northern waterfront; and continues south into San Mateo County, including the northernmost beaches, Sweeney Ridge, and the Phleger estate. It covers 75,000 acres in all.

The idea for a national park flanking the Golden Gate had been a dream of Edgar Wayburn's since he had come west in the 1940s. Wayburn, who moved to Bolinas, had been active in the mobilizations for Mount Tamalpais State Park expansion, Point Reyes National Seashore, and saving the Marin Headlands. He was a believer in the principle that "wilderness begins in your own backyard." Wayburn was five-time president of the Sierra Club, beginning in 1961. A conservationist of the old guard, Wayburn volunteered his time while keeping up his medical practice in San Francisco well into his eighties. Like so many greens who came of age before the 1960s, he stayed a lifelong Republican and preferred to hobnob with friends in high places rather than engage in public confrontations. And, like most men of his era, he formed a lifelong partnership in conservation with his wife, Peggy Wayburn.[27]

If Wayburn had the vision and the pedigree, Amy Meyer had the moxie for grassroots organizing. Like many activists who went on to do great things, Meyer fell into the struggle for the park by accident; it was at first a sideline to neighborhood organizing by this local artist, teacher, and housewife from San Francisco's Richmond District. Meyer and Wayburn formed People for a Golden Gate National Recreation Area in 1971, and they proved to be a formidable team. Meyer herded scores of environmental organizations, neighborhood groups, and San Francisco politicians into the fold. Wayburn cooked up a grandiose scheme that included the Pacific beaches and northern waterfront of San Francisco, the Presidio, Alcatraz and Angel

Island, and the west side of Marin County all the way to Point Reyes. He also wanted the San Mateo coast but would have to wait.[28]

The GGNRA proposal was made possible, in good part, by a second post-war wave of decommissioned military bases around the Golden Gate: Forts Funston, Miley, and Mason in San Francisco and Forts Cronkite, Berry, and Baker in Marin. Public interest was piqued by the heated debate over the disposition of the federal prison on Alcatraz and the Indian occupation in 1969–70. Furthermore, a federal study had recommended the creation of more national parks near urban centers, leading to a proposal by Secretary of the Interior Walter Hickel called Parks for the People. Yet the National Park Service came up with a disappointing proposal for a national recreation area of only 4,000 acres. So Wayburn cultivated Hickel's successor, Rogers Morton, who bucked his own department to support a more ambitious park of 34,000 acres.[29]

A key to success was the intervention of Democratic Representative Phillip Burton of San Francisco, a power in Congress. In 1972 Burton would push through a bill for the GGNRA, gaining leverage over Republicans because the Nixon administration wanted to win California in the upcoming election. Burton and the park seemed an odd coupling, however. He was a New Deal liberal, a champion of the working class, and in no way an outdoorsman. That Burton became a convert to the green cause is a testament to the broad appeal of environmentalism in the Bay Area in the 1960s. Burton was particularly close to Wayburn, whom he called "my guru," but he had seen the force of the environmental movement at work and wanted to expand his liberal coalition.[30]

At the time, bridges were being built between labor and greens in the Bay Area through the efforts of the Sierra Club's Mike McCloskey, Will Siri, and Dwight Steele (a former labor lawyer) and labor leaders such as Dave Jenkins of the International Longshore and Warehousemen's Union (ILWU) and Jack Henning of the San Francisco Central Labor Council. Jenkins was put on the board of the Bay Conservation and Development Commission and Henning on the board of the Nature Conservancy. Many unionists had joined the coalition opposing the Panhandle Freeway across Golden Gate Park. A 1971 Conservation and Jobs conference had brought the two movements together to talk, and in 1973 the Sierra Club took the unprecedented steps of creating a Labor Liaison Committee and supporting the oil workers' national strike against Shell Oil.[31]

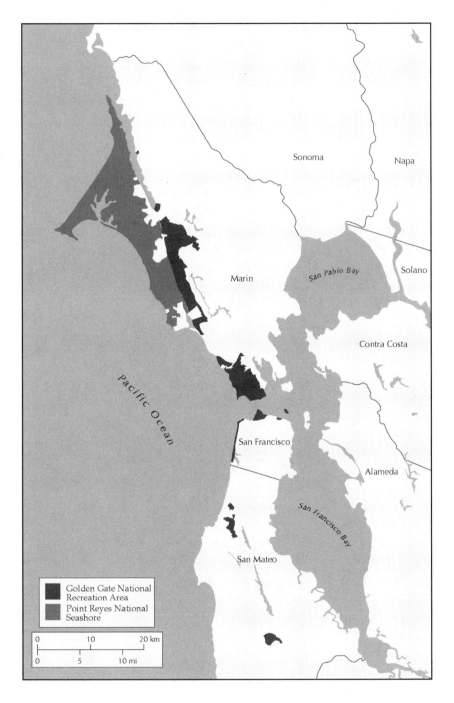

Golden Gate National
Recreation Area
Point Reyes National
Seashore

| 0 | 10 | 20 km |
| 0 | 5 | 10 mi |

MAP 8. Golden Gate National Recreation Area and Point Reyes National Seashore, 2005. Drawn by Darin Jensen. Courtesy of Bay Area Open Space Council, GreenInfo Network, Greenbelt Alliance.

Burton would take the alliance farther when he expanded Redwood National Park in 1978 and provided millions in compensation for lost timber jobs. That move was part of his finest legislative achievement: the 1978 Omnibus Parks bill. With his usual legislative genius and brute political strength, Burton cobbled together every possible park proposal around the country and won overwhelming support from his colleagues for one of the greatest land conservation measures in American history. Included in the bill was money to expand GGNRA all the way north to Point Reyes and south to Sweeney Ridge in San Mateo. Amy Meyer said of Burton that "he has done more to turn land into parks than any congressman ever in the history of the United States." Burton died in 1983, at the height of his powers, just as the Reagan revolution was sweeping the New Dealers away.[32]

Assembling the GGNRA turned out to be more difficult than anyone had expected. The military bases were transferred smoothly to the National Park Service, along with Alcatraz. But the State Department of Parks and Recreation under William Penn Mott refused to turn over several parks for nothing, and the Marin Conservation League and Tamalpais Conservation Club backed the department up. San Francisco was worried about loss of tax revenues from granting land within its domain to the national park, and the Nixon administration refused to release funds to buy the 17,000 acres of private land within the park boundaries. It took a decade to work out all the kinks. In the end, California handed over Stinson and Muir beaches and the Maritime Historic Park in San Francisco but held onto Angel Island and Mount Tamalpais; San Francisco ceded its beaches and aquatic park; and Phil Burton got the money released to buy the rest of the private holdings. As a result, GGNRA actually controls and manages 34,500 acres, while another 40,000 acres remain under state parks and other government agencies.[33]

The Presidio was the last and most controversial piece of the GGNRA. Burton had wisely included a clause in the original legislation that granted the Presidio to the park when the Army was finished with it. This did not take place until 1994. But with the Republican takeover of Congress that year, the Presidio transfer came with strings attached: it would have to fit the neoliberal premise that parks pay for themselves, or it would be sold off. A Presidio Trust, created to oversee its financial affairs, was chaired by San Francisco's leading capitalist of the day, Donald Fisher of the Gap. Disposition of the Presidio's buildings and grounds has been embroiled in controversy ever since, with strong public opposition to commercialization. The first director of the trust, Jim Meadows, was forced to resign over questionable deals. But first he granted a long-term lease to filmmaker George Lucas to build a major special-effects studio center on the site of Letterman Hospital.

The Golden Gate National Recreation Area has proven to be a showcase for open and flexible management of parks in the midst of a metropolitan area. The intensive demands on these parklands and the variety of recreational, scenic, and habitat uses require a lively balancing act by the Park Service. Brush fires of popular outrage appear recurrently. A broad-based group, the Golden Gate National Parks Association (now Conservancy), was formed to mobilize citizen input and financial support, and it has been so successful that the National Park Service is using it as a model for its parks around the country.[34]

PROTECTING THE PENINSULA

Although Marin got most of the attention, conservationists on the San Francisco Peninsula were quietly engaging in their own battles against runaway development. San Mateo and northern Santa Clara counties have a huge amount of open space to show for it, some 265,000 acres in parks, open-space reserves, watersheds, and beaches (divided about evenly between the two counties).[35] This achievement outdoes Marin and is all the more remarkable when we consider that the peninsula and Silicon Valley suffered the greatest onslaught of postwar suburbanization, leaving environmentalists less time to save anything.

The social makeup of the peninsula differs from Marin's. In the nineteenth century, it was the primary locus of suburban estates of San Francisco's elite, in Atherton, Millbrae, Hillsborough, and Menlo Park. Unlike Marin's, the old-money class of the peninsula has not been active in conservation and even today is invisible in the green movement. After the Second World War, a new layer of upper and upper-middle classes developed farther south, as the growth of Silicon Valley and Stanford University gave the area one of the greatest concentrations of electronic companies and scientific and technical workers in the world. Recruits to the conservation and greenbelt movements have come mostly from this new bourgeoisie. Yet here, too, it took on a militant cast that would confound many of the fondest profit-making schemes of capital.

Redwood Revival

The Santa Cruz Mountains have retained their allure and sense of remoteness to this day. So it is not surprising that renewed logging in the 1940s rekindled conservationist fires on the peninsula. Two prized sites were Butano Forest and the Pescadero Creek watershed, just north of Big Basin in San Mateo County. In a 1946 report to the State Parks Commission, Frederick Law Olmsted Jr. recommended Butano and Pescadero for inclusion, saying that it was remarkable that some groves of virgin redwoods had survived there. The state acquired Portola Park in the Pescadero Creek watershed in 1945, but Butano fell into the hands of the Santa Cruz Lumber Company. Ironically, this was land Timothy Hopkins had given to Stanford University when he was a trustee fifty years earlier, and the university had sold it off for cash. A campaign to save Butano was launched through a new group, Butano Forest Associates, formed in 1946 with the participation of the Loma Prieta Chapter of the Sierra Club, Save the Redwoods, Tamalpais Conservation Club, and Marin Conservation League. After initial setbacks, Butano State Park was secured a decade later.[36]

In the late 1960s, Pescadero Creek was threatened by logging by Santa Cruz Lumber and another company, Bayside Timber. San Mateo conservationists leapt into action, led by Claire Dedrick of Menlo Park and Grace Radwell of La Honda. San Mateo County purchased almost 5,000 acres of the Pescadero watershed for a park in 1968, followed by Sam McDonald Park in 1970 and Heritage Grove Park in 1974. This time Stanford University did the right thing, handing over McDonald's bequest of redwoods to the public. As a result, the Pescadero watershed has been almost wholly protected, although the terms of sale allowed cutting to continue for several more years. An uproar over clear-cutting pressured the county to deny Bayside Timber a permit, and the ensuing legal fracas forced the Legislature to rewrite the state Forest Practices Act (the state and the counties have been squabbling ever since over who has the right to regulate timber cutting).[37]

Meanwhile, a new organization called Conservation Associates bought Castle Rock east of Big Basin, as well as the Forest of Nisene Marks in Santa

Cruz County (with the aid of the Nature Conservancy), and gave both of them to the state park system. Conservation Associates were George Collins and Doris Leonard, veterans of the National Park Service and the Sierra Club, and Dorothy Varian, widow of electronics pioneer Russell Varian. They came together as a tribute to Russell Varian, who had died on a hiking trip in Alaska with the Leonards in 1959.[38] In 1968 a plan was hatched in Dorothy Varian's home in Cupertino to resuscitate the Sempervirens Club as a trust fund to complete Big Basin and Castle Rock state parks and to save the Waddell Creek basin, west of Big Basin.

The Sempervirens Fund became the project of Los Altos pharmacist and Sierra Clubber Tony Look, who had just spearheaded the purchase of Berry Creek Falls for Save the Redwoods League, and his partner Howard King, a retired Hewlett Packard engineer. The fund became independent in 1972, with an impressive list of sponsors, including William Penn Mott, Newton Drury, David Packard, and Ansel Adams. The fund initiated a Skyline-to-the-Sea trail from Castle Rock through Big Basin, a distant dream of Andrew Hill. It then worked closely with state parks director Mott to define a more complete Big Basin State Park and purchase the missing pieces, doubling the park's size to 18,000 acres by the 1990s.[39]

Guardians of the Foothills

Silicon Valley's meteoric growth gave birth to a new environmental mobilization. The green wave rose up around Stanford University, where it took on the advancing tide of industry, highways, and cold-war science. Stanford played midwife to the birth of microelectronics and the digital age in the postwar era, and it was the epicenter of the industrial earthquake that produced Silicon Valley—the world center of electronics.[40] Ironically, Stanford has been the prime target for environmental rebellion among the favored classes it drew to the area and nourished over the last fifty years.

The midpeninsula foothills—a golden land of oaks, chaparral, grass, and grazing—separate the forested mountains from the bay flatlands where most commercial development is concentrated. They provide a summer-sweet spot of openness between the country and the city, a scenic scrim running the length of the peninsula. They are, moreover, the public doorstep to the residential hideaways of the upper classes that stretch from

Woodside to Saratoga. As Silicon Valley grew, engineers, businessmen, professors, and their ilk moved into the little towns, such as Portola Valley, Los Altos Hills, and Los Gatos, tucked into the oak savannas and woodlands nestled at the base of the Santa Cruz Mountains.

The leading force for open land on the peninsula for almost fifty years has been the Committee for Green Foothills. The committee was born at a 1962 meeting in the living room of Ruth Spangenberg (living rooms are to Bay Area greens what garages are to the electronics wizards of Silicon Valley). The group was made up of Stanford faculty and neighbors trying to protect their homes from the expansion of the Stanford Industrial Park, which had opened in 1959 on the south side of campus. Among the twenty-six founding members were Lois and George Hogle, Martin Litton, Jack and Eleanor Fowle, and Morgan and Katy Stedman (and my childhood pediatrician, Richard Cutter). Author and Stanford professor Wallace Stegner was designated its first president.

Two years later, Stanford announced plans for a second unit of the industrial park, to be located across Junipero Serra Way in the foothills, with Ampex as the anchor tenant. The newly formed committee went to work and forced a city referendum in Palo Alto. They lost the vote by a hair, but Alex Pontiaeff of Ampex heard their appeal and pulled his company out, so the project died. When the same proposal resurfaced in 1969 as the Coyote Hill plan, the Committee for Green Foothills sued. Again Stanford won, but the committee was able to negotiate strict limitations on further expansion beyond Junipero Serra Way (later Boulevard)—their "line in the sand" to protect the foothills. Stanford next moved to develop land along the northern edge of campus, creating the office park on Sand Hill Road that is today the venture-capital capitol of Silicon Valley and the world. The university won again, but, in a delicious irony, the committee was later befriended by Sand Hill developer Tom Ford, who became one of the great benefactors of peninsula open space.[41]

Another front in the 1960s land-use war broke out around Woodside when Pacific Gas and Electric (PG&E) sought to route power lines across the Santa Cruz Mountains to the Stanford Linear Accelerator. Forming a group called Save Our Skyline, committee founders Don Aitken and Lois Hogle succeeded in having the power poles dropped in by helicopter instead of an ugly swath being cut up the mountainside. The committee's young lawyer,

Pete McCloskey, took the case to the U.S. Supreme Court and won; he then helped rewrite the law authorizing the Atomic Energy Commission. McCloskey, a liberal Republican, went on to Congress, where he became a visible critic of the Vietnam War.[42]

The committee had strong links to the wider Bay Area conservation crowd. Stegner and Martin Litton were Sierra Club directors in the 1960s. Stegner had been lassoed by Bernard DeVoto and David Brower into becoming the literary voice of the American West in the 1950s. Litton was a lion in defense of the redwoods, Grand Canyon, and Mineral King, and Brower called him "my conservation conscience." Morgan Stedman was on the Santa Clara County Planning Commission, fighting sprawl. Barbara Eastman, an active member and peninsula resident, was a key organizer of Save Our Seashore. Dorothy Varian of Conservation Associates was a friend. Bill and Mel Lane, owners of *Sunset* Magazine, were longtime, discreet allies; Mel would direct both the Bay and Coastal commissions.[43]

The members of the committee also connected to national currents of the time. Eleanor Fowle was the sister of liberal Democratic Senator Alan Cranston, a former real estate broker in Los Altos. Lewis Mumford, who had advised Stanford President Donald Tresidder to keep the foothills open back in 1947, wrote to Ruth Spangenberg at the founding of the committee that "since Dr. Tresidder's death the university has gone for more 'fashionable' and more expensive advice, promising immediate profits and ultimate debacle. Keep up the fight. The weight of good sense and public decency is on your side!"[44]

Although the early presidents of the Committee for Green Foothills were all men, the key activists have largely been women, such as Lois Crozier-Hogle, Katy Stedman, Mary Davey, and Lennie Roberts. They got involved as young housewives married to professionals, raising children but looking for outlets for their energies, and most have stuck with the organization for decades. Many were already involved in women's clubs, civic associations, and social reform. They were all educated, worldly, well connected, and ready to brave a good deal of hard opposition to their views at a time when environmentalism was still a strange and radical idea. But they never took themselves too seriously and enjoyed referring to themselves as "the Green Feet." Most of the first generation of Committee for Green Foothills activists were ready converts to feminism and the first women to serve on any number of local public bodies.[45]

Another defining moment in the peninsula open-space movement was the defeat of Palo Alto's massive foothills housing scheme in the late 1960s. Interstate 280 was going to make the hills readily accessible for commuters, a golden opportunity for developers and their backers in city government. Mayor George Morgan and planner Charles Luckman wished to make Palo Alto "the Wall Street of the West," and a large parcel of land in the hills above I-280 had fallen into their hands. Its development would have allowed them to increase their domain by a third. But a handful of rebel women from the Palo Alto Civic League, led by music teacher and Committee for Green Foothills member Nonette Hanko, intervened to question the city's unbridled enthusiasm.[46]

In particular, the opponents questioned an expensive planning report, which justified the project, that the city had solicited from planning consultants Livingston-Blayney Associates. The opposition egged the City Council into demanding an additional cost-benefit analysis of the foothills development. What they got back was unexpected: a pathbreaking statement of the *costs* of growth and not just its benefits (see chapter 6). In the aftermath, the city dropped the project. More than that, Hanko and her allies got the City Council to set aside the land as Palo Alto Foothills Park, which, at 1,400 acres, was the largest municipal park created in a half century.[47]

Stanford's plans for its vast domain of more than 8,000 acres continue to vex the peninsula's defenders of open space. From 1960 to 1985, the campus doubled its building space, from four million to eight million square feet. A major fight broke out in 1985 over a proposal to put the Reagan Presidential Library in the foothills; Stanford lost, and the library went to Southern California. That same year, for the first time the university submitted to a review of its development plans, signing an agreement with Palo Alto and Santa Clara County. Four years later, Stanford easily won approval from the county for two million square feet of build-out, with the promise that it would have to return for further review as this quota was filled. By the late 1990s, Stanford was back at the table asking for a General Use Permit on almost five million more square feet over ten years.

The university ran into concerted opposition from the Committee for Green Foothills, a new group called the Stanford Open Space Alliance led by Peter Drekmeier (son of two Stanford professors), and county supervisor Joe Simitian, a Berkeley planning-school graduate. The greens wanted an end to further intrusion into the foothills across Junipero Serra Boulevard. When the dust settled in 2000, the county approved Stanford's plan but extracted a twenty-five-year moratorium on foothills expansion. Yet within a year, Stanford was back, asking for a variance for a Carnegie Foundation Center that sneaks over the growth boundary.[48]

SECURING SAN MATEO

Highway plans unveiled in the 1950s were the vanguard of development on the peninsula and at the forefront of conflict, as in Marin. The first trigger was Highway 84, the Willow Freeway, slated to go through the middle of Portola Valley and over to the coast; it was stopped in the early 1960s, thanks in part to unstable geology. The next highway plan to earn the wrath of the budding green movement was straightening and widening Skyline Boulevard (now Drive). Thanks to the efforts of Eleanor Boushey of the Committee for Green Foothills and mayor of Portola Valley, Skyline was left to its curvy ways and named California's second scenic highway in 1968. Interstate 280 could not be halted, but the San Francisco Water Department had it rerouted farther away from the reservoirs, and billboards were banned along it.[49]

The Dynamic Duo and the Island of Sanity

Another highway fight, against extending the Junipero Serra expressway into Menlo Park, brought Claire and Kent Dedrick into the antidevelopment fold in the mid-1960s, when they came on the board of the Committee for Green Foothills. A research scientist in Stanford's medical school, Claire ultimately quit her job to work full-time on environmental causes and local elections. She teamed up with Janet Adams to put out San Mateo County's first environmental newsletter, *The Black Mountain Gazette.* They proved to be a formidable team. Dedrick and Adams managed Pete McCloskey's campaigns for Congress, as well as those of several local mayoral and county-supervisoral candidates. They helped start up the Peninsula Regional Group of the Sierra Club Bay Chapter in 1967 and founded the Peninsula Conservation Center two years later. The Peninsula Conservation Center (now called Acterra) has served as a vital gathering spot for environmental groups for more than three decades.

From there, Adams and Dedrick went statewide. Janet Adams directed the popular mobilization to protect the California coast in 1971–72 (see chapter 5). Activists from that campaign went on to found the California League of Conservation Voters, under the leadership of Adams and Marion Edey of Los Angeles. The California league is the oldest and most potent of the state-level green political action committees.[50] Claire Dedrick became a vice president of the Sierra Club and then Secretary of the California Resources Agency under Governor Jerry Brown in 1973. She was instrumental (along with Phil Burton) in the battle to expand Redwood National Park. Neither woman gets the credit she deserves. Dedrick came under intense fire from all sides in the redwoods fight and ended up embittered by her experience (though she continued in other posts in state government). Adams moved to Florida and dropped out of sight.[51]

At the northern end of the peninsula, on the San Francisco–San Mateo border, stands Mount San Bruno. This island of open space, mostly grasslands, is one of the Bay Area's most striking features and the first thing seen by anyone arriving from San Francisco airport. It is 1,300 feet high and 6 miles long, covering 3,300 acres, and is precious habitat to a dozen rare and endangered species. It is also a looming target for real-estate promoters. In the mid-1960s, Crocker Land Company, Ideal Cement, and Chase Bank proposed to hack off the top of the mountain and dump it in the bay for the airport and 30 square miles of housing developments. A decade later, the Rockefeller interests were proposing a New Town of 25,000 people to go on the mountain's saddleback. Bette Higgins of South San Francisco formed the Save San Bruno Mountain Committee to defeat these schemes. They won and then secured a county park covering 2,400 acres.[52]

David Schooley, a veteran of those battles, founded San Bruno Mountain Watch and became the mountain's chief defender. He has watched the unprotected edges be chipped away. A variance to the Endangered Species Act in 1982 (passed with Mount San Bruno specifically in mind) allowed developers to cut a deal in which 10 percent of the mountain could be built on in exchange for habitat mitigation on another 20 percent. As a result, housing projects such as Terrabay and Pacific Pointe have pushed up the sides of the mountain, reducing open habitat and introducing invasive species. The eighteen-story Mandalay Tower is a true blot on the landscape. As the old conservation saying goes, "Victories are always temporary; only the defeats are permanent." More encouraging is the way local residents have rallied behind groups such as Friends of San Bruno Mountain to volunteer in habitat restoration.[53]

Calling Their Bluff

By the 1970s, land-use battles were breaking out all over northern San Mateo County, and freeways were once again the flash points. First, the greens blocked the continuation of Interstate 380 from the San Francisco airport to Pacifica. Then the Save Sugarloaf Committee killed a housing development and kept the last remaining promontory of hills north of Highway 92 as open space. The Highway 92 freeway from the San Mateo Bridge to Half Moon Bay was never completed west of Crystal Springs Reservoir.

An old sore with the highwaymen was the extremely unstable Devil's Slide south of Pacifica on Highway 1—the main route south from San Francisco along the coast. CalTrans proposed to bypass Devil's Slide by routing a freeway over Montara Mountain. In 1971 a coalition led by the Committee for Green Foothills and Sierra Club went to court to have the National Environmental Policy Act applied to the project, and CalTrans backed off. When Republican Governor George Deukmejian took office in the 1980s, the engineers had to be stopped again by lawsuits and the California Coastal Commission. In 1993 Devil's Slide took out the coast road entirely; CalTrans saw a new opening and left the road closed for months. But the county supervisors convened a panel of engineers and geologists who proposed a tunnel through Montara Mountain. The greens, realizing they could not hold off CalTrans forever, shifted tactics to a county initiative to force a tunnel route, which passed handily in 1996. The tunnel finally got under way a decade later.[54]

The San Mateo coast is 50 miles of Pacific shore every bit as marvelous as Big Sur and Point Reyes. Less rocky, it has some of the finest beaches in California, plus long bluffs and shallow river valleys that have been intensively farmed for more than a century. But it has had scant protection, except for a narrow strip of state beaches, and at midcentury its future was looking grim. The tide began to turn, slowly but surely, after 1970. Sweeney Ridge was snatched from the developers' hands and included in the Golden Gate National Recreation Area. Then the action turned to the urbanizing midcoast area around Half Moon Bay. The initial trigger was the ambitions of the developers Deane and Deane (backed by Westinghouse), who owned 8,000 acres nearby. The opening shot from conservationists was to save Montara Beach, north of Half Moon Bay, which was made a state park in 1972. The Committee for Green Foothills found its energies increasingly directed westward, as local activists contested one large-scale project after another at such coastal properties as Cascade Ranch, Corral de Tierra, Pigeon Point, and Mirada Surf. There were some defeats, but they held the line.

The California Coastal Commission was installed during the 1970s with the help of San Francisco Peninsula greens, but its protections depended on local planning and regulation. Soon, a fierce debate broke out over San Mateo County's coastal plan. One was prepared in 1980, but the Board of Supervisors kept amending it to accommodate developers. So Lennie Roberts and the Committee for Green Foothills leapt into action with the Save Our Coast initiative of 1986. This county proposition would make the coastal plan whole again, including prohibition of onshore oil operations, an urban growth boundary around Half Moon Bay, and farmland protections. Despite bitter opposition from real-estate interests, the county farm bureau, and the Board of Supervisors, the measure carried by a two-thirds majority.[55]

None of this would have been possible without a change in the political coloration of the county. It began in the mid-1960s through the intensive campaigning of Dedrick, Adams, Roberts, and their compatriots, who started pushing green candidates such as Eleanor Boushey and Olive Mayer onto city councils. By the end of the decade, they had ousted Senator Dolwig (later convicted of mail fraud) and Assemblyman Britschgi.[56] In the 1970s, Arlen Gregorio went to the State Senate, and Dixon Arnett replaced Britschgi in the Assembly. By 1976, the greens had a majority on the Board of Supervisors (John Ward, Fred Lyon, and Ed Bacciocco). This was critical to saving Mount San Bruno. Yet things reversed again in the early 1980s, putting off for another decade a stable green hegemony of the kind achieved in Marin after 1968.

By the end of the twentieth century, the San Mateo coast was falling under the sway of the greenbelt forces, and developers were steadily losing ground. With a larger populace around Half Moon Bay, a new watchdog group, the Coastal Protection League, spun off from the Committee for Green Foothills. Many of the key coastal properties had been secured by the Peninsula Open Space Trust, and the Midpeninsula Regional Open Space District had been extended to the coast (see chapter 7). The last frontier of land conservation is the tenacious farm area of the lower coast around Pescadero, where it has proved hard to put together a farmer-environmentalist coalition of the kind achieved in Marin.

San Jose and the Santa Clara Valley lagged a decade or more behind the west side in spawning a green rebellion. Unlike the peninsula, here almost

TABLE B. County Size, Urban Area and Protected Land

	County	Urbanized	Protected
Alameda	477,800	144,000	104,700
Contra Costa	462,400	145,200	111,000
Marin	336,300	41,400	178,000
Napa	483,400	21,400	99,300
San Francisco	30,600	24,900	5,100
San Mateo	289,500	71,100	107,800
Santa Clara	839,800	185,100	201,800
Solano	530,600	55,400	66,000
Sonoma	1,013,900	72,800	133,600
Total	4,464,400	761,400	1,007,200

SOURCE: Greenbelt Alliance, 2006.
NOTES: Urbanized land means density greater than 1.5 people per acre; protected land includes parks, reserves, and easements, but excludes city parks.

all the planned highways were constructed, a process spearheaded by county transportation planner Jim Potts. Opposition to untrammeled growth finally broke out around 1970, as the hidden costs of the city's Los Angelization came due, and the struggle would take the form of service limits and growth boundaries (see chapter 6). Santa Clara County joined in the Mid-peninsula Regional Open Space District and began funding county parks again in the 1970s, ending up with a fine complement of open space in the surrounding mountains by the end of the twentieth century (see chapter 7).

The west Bay Area, from Marin to Santa Clara County, illustrates land conservation at its strongest. The swaths of open space preserved by 1975 are truly impressive in scale and beauty, from Point Reyes in northern Marin County to the national park straddling the Golden Gate to the open foothills of the peninsula. Later, the west-side greenbelt would be further expanded by open space districts and land trusts. More than 40 percent of protected open space in the Bay Area lies on the west side, even though it contains the three smallest counties, along with less than half of Santa Clara County. (See Table B.)

The west side certainly reveals the upper-class element in land conservation. The rich and near-rich nestle into the mountain slopes and charming foothills, then fight to keep their capacious front and backyards unspoilt. Along with its famous greensward, the west side has the highest house prices in the western United States. Marin is well known for the green cape of rectitude it pulls tightly around itself, and it has become even *more* exclusive over time, ironically, because the greenbelt is something worth paying for—and guess who has the money?[57]

Still, the rich can occasionally confound their critics, becoming truly public spirited. David Packard, Sigurd Varian, Gordon Moore, and Bill Hewlett were all pioneer capitalists of Silicon Valley. Nonetheless, the Varians were liberals and environmentalists. Lucille Packard was friendly with the Green Foothills faction, and the four Packard children became ardent environmentalists; the Packard Foundation funds green causes far and wide. Gordon Moore spent many happy childhood days on family lands near Pescadero; now the Moore Foundation helps pay for open-space acquisition. Even the reticent Hewlett gave money to protect his beloved Lake Tahoe.

More significant than the largess of bourgeois philanthropists, however, has been the hard work of green activists of the West Bay, from the great ladies of the Marin Conservation League to the leaders of the Committee for a Golden Gate National Recreation Area to the professionals and their spouses in the Committee for Green Foothills. At a minimum, their existence reveals a contradiction between the consumption and production interests of the upper classes. But they have shown much more than that; they have thrown many a monkey wrench in the works of land development and promoted a generous zeal in saving public spaces from the absolute rule of profit.

Nonetheless, the elite's achievements would be inconceivable without the environmental consciousness that coalesced into a mass popular movement around the bay. The pioneer greens of the West Bay had a wide base of support. Claire Dedrick remembers how well educated the public was by the late 1960s and how the conservationists drew in hippie kids, factory workers, and dignified Republican ladies, all at once.[58] Something was afoot—the metamorphosis of upper-class conservation into a mass environmental movement—and nowhere more so than in the Bay Area, especially in the fights to save the bay and the coast.

5

THE GREEN AND THE BLUE

Saving the Bay and the Coast

San Francisco Bay is the defining feature of the metropolis. It saves the Bay Area from being an endless urban plain like Los Angeles. The bay is not merely an arm of the Pacific, but one of the world's great estuaries, a place of encounter between the sea and freshwater. However, this centerpiece of the metropolis has been treated in anything but a benign fashion. Fouled, filled, and strangled, the bay came to bear little resemblance to the paradise that once provided feasts of oysters and clams, salmon and steelhead, ducks and geese to the Ohlone and Miwok people.

San Francisco Bay was intensively used and sorely abused after the 1849 Gold Rush. The state of California wasted no time in turning over to private owners the entire shoreline and intertidal zone, which thereafter became the exclusive province of docks, factories, railroads, and warehouses. The deeper waters were plied by thousands of seagoing ships, local ferries, tugboats, and barges. The bay became the commercial heart of the region, but was reduced to a functional place of production and circulation of goods and people—of no more aesthetic or spiritual import than today's freeways. The city turned its back on the workaday bay, leaving it to the depradations of utilitarian plunder and profit.

Yet as the bay reached its nadir in the post–World War II era, a political movement broke out to save it in the early 1960s. Saving the bay was one of the first mass, popular mobilizations on behalf of the natural environment, here or anywhere in the world. This went well beyond the interventions of early conservationists, by putting in place a regulatory commission more extreme in its powers than the most radical of New Deal projects. Neither red nor black—the colors of earlier mass movements—this new environmentalism was green and blue.

The environmental movement to save the bay reversed a century of degradation. Since the Bay Conservation and Development Commission was established in 1965, the bay has shrunk no further and has had hundreds of acres of wetlands restored. Its waters are no longer rank, and aquatic life is abundant, with shorebirds in large number feeding along the mudflats and marshes. Most important, the bay is now seen as a vast scenic, recreational, and ecological open space instead of a dour transportation hub and industrial landscape. Some 180 of 275 miles of bay shoreline are now open to the public, compared to only 5 miles when the blue-green revolution began. The bay has come back as the visual centerpiece of the metropolis, a watery commons for the region, and a source of pride to Bay Area residents.[1]

BYE-BYE BAY

True to California's mining legacy, the biological and mineral wealth of San Francisco Bay have been extracted without mercy. Early observers marveled at the abundance of wildlife, fish, and shellfish. The beavers were the first to go, followed by coastal fur-bearing mammals—fur seals, otters, elephant seals, and sea lions—by the end of the nineteenth century. Commercial fisheries flourished in a wide variety of species—salmon, striped bass, sturgeon, sardine, and shad—as California became the nation's leading commercial fishery by the 1920s; then they collapsed one by one, due to overexploitation, pollution, and climate cycles.[2]

Oysters, farmed intensively off South San Francisco, were a staple of local restaurants until pollution fouled the beds. The Morgan Oyster Company, which had come to dominate the industry, sold its beds in the 1920s to the Pacific Portland Cement Company, which dredged reefs of oyster shells, used to make cement, from the bay bottom until the supply ran out. Meanwhile, the southern bay (and some of San Pablo Bay) was crisscrossed with dikes to create 34,000 acres of salt evaporation ponds. Millions of dollars' worth of salt were extracted. Many small companies worked the southern bay over the years, but only one company, Leslie, survived after the Great Depression; it owned almost 100,000 acres in and around the bay by the 1940s. Most postwar fill projects would be built on former Leslie and Pacific Coast Cement lands.[3]

For a century no one thought twice about throwing anything and everything into the bay. Canneries, rendering plants, and tanneries discharged a river of effluvium; smelters, steel mills, and coal-gas plants spilled toxic ooze; and thousands of ships quietly voided their wastes into the waters. City sewers discharged along the waterfront, and every rainfall brought a foul load off the streets into creeks and storm sewers. Water quality was already bad by the 1870s, and the bay was a virtual cesspool by the early 1900s, choking off the shellfish beds and driving out the schools of herring, sardine, and shrimp supporting the whole food chain (all of which effects were blamed on Chinese fishermen, according to the racial lunacy of the time).[4]

Water pollution only became worse as the coastal plain urbanized beyond San Francisco. With the tremendous growth of the East Bay in the early twentieth century, the Oakland estuary became so rank that a channel had to be cut through to improve circulation, turning Alameda into an island. After the Second World War, rapid growth south down the peninsula and into southern Alameda County put the southern bay into a parlous state, with the mudflats putrefying underneath the public's noses.[5]

The shoreline of the bay has been fair game for land developers, farmers, and public spectacle, powered by the hunger for cheap land and accumulation of capital. Because the shoreline is shallow (only 6 feet on average), it is easy to turn it into solid ground by diking and filling. Some 50,000 acres of bay fill have reduced considerably the original 500 square miles of open water. Before 1950, the central bay was the area most altered by filling. Railroads and piers were built out into the bay, and San Francisco's waterfront was completely reconfigured. Oakland estuary and West Oakland were extensively filled in the early twentieth century for railyards, docks, and industry. Treasure Island was built out in the middle of the bay for the 1939 World's Fair. San Francisco and Oakland built airports with runways out into the bay to handle wartime traffic.

The military has treated the bay as its pond. The assault reached high pitch by World War II, with the full-scale militarization of the Bay Area. The Army and Navy expanded their Oakland and Alameda bases, moving the East Bay's shoreline a mile closer to San Francisco. The Navy built

Hunter's Point shipyard, expanded Mare Island, and took over Treasure Island for its arsenal. The wartime shipyards in Richmond and Sausalito filled the shorelines. And none of the military gave a hoot about sewage pouring into the bay from bases or ships.

The highway builders also had their eyes on the bay. Having helped to kill off the ferries, which once plied the waters by the score, they began dreaming up new bridges and freeways. One bridge, the Northern Crossing, was to go from Telegraph Hill in San Francisco over to Tiburon in Marin County, while another, the Southern Crossing, would have cut a diagonal from Hunter's Point to Bay Farm Island south of the city of Alameda. As soon as the Bayshore Freeway was completed up the peninsula, the engineers envisioned a parallel freeway out in the bay itself.

The military-industrial vision of the bay reached its apotheosis in the postwar Reber Plan, which proposed damming the northern and southern ends of the bay, converting them to freshwater reservoirs, and filling the shallows all around the bay for industrial use; a couple of submarine bases were thrown in for good measure. Reber's scheme was only one of many wild-eyed proposals for multiple bridges, dams, and freeways at war's end. A separate plan to bridge and dam the southern bay was promoted by transportation engineer Jim Potts of San Jose, for example. The Army Corps of Engineers built a model of the bay as big as a barn to test the effects of these ideas on the estuarine hydrology—solely out of concern with sedimentation of shipping channels.[6]

By the 1950s, the cities around the San Francisco Bay were engaged in extravagant fill projects to accommodate their garbage. Sanitary landfills—a nice name for city dumps—were creating bulbous protrusions along the eastern bay shoreline at Berkeley, Albany, and Emeryville. San Jose's rising tide of garbage was smothering the old port of Alviso, San Francisco's was filling up Candlestick Cove (leaving Brisbane landlocked), and flat-topped mountains of waste were building up on the wetlands by Palo Alto, Mountain View, and Hayward.

Housing developments had entered a new phase, too, scaled up to mammoth proportions by the early 1960s. Bay Farm Island, a former diked pasture by the Oakland airport, was built out as Harbor Bay Isle by Ron Cowan. Jack Foster built his eponymous city on another former marsh south of the San Francisco airport, Brewer's Island (a Leslie Salt property), with the help of a municipal improvement district provided by special legislation from State Senator Richard Dolwig. Leslie began its own project, Redwood Shores, on salt ponds north of the port of Redwood City. The grandest of them all was the Rockefeller-Crocker $3 billion plan to cut the top off Mount San Bruno and dump 200 million cubic yards of it on 10,000 acres of former oyster beds (owned by Ideal Cement) between Redwood Shores and Foster City—an area bigger than Manhattan.[7]

The postwar splurge of filling and abuse began to send shivers through many residents of the comfortable hillsides around the region, who could see the bay clearly from their windows or on their daily commutes. They were particularly shocked when the Corps of Engineers published a map of all potential "reclaimable" lands in the bay, which showed nothing remaining but the Pleistocene river channels.[8]

THE SOUL OF A REGION

The public would have to fight to reclaim the San Francisco Bay it had lost long before. Opposition to the bay's destruction was set in motion by Kay Kerr, Sylvia McLaughlin, and Esther Gulick of Berkeley, over lunch at the Town and Gown Club. They had been riveted by bulldozers ripping away at Point Isabel and garbage dumped into the harbor behind it, as well as by Berkeley's plan to fill another 2,000 acres (doubling the size of the city). Unschooled in politics, the trio invited thirteen conservationists to Gulick's living room for a strategy session in 1961, including representatives from the Sierra Club, Save the Redwoods League, East Bay Regional Parks District, and Audubon Society. No one thought it was a winnable battle, so David Brower suggested they start a new organization and get everyone else's mailing lists. Thus was born the Save San Francisco Bay Association (now Save the Bay).[9]

The movement to save the bay originated with the East Bay elite. One of the triumvirate was the wife of university president Clark Kerr, another was the wife of Board of Regents' and Homestake Mining Company president Donald McLaughlin, and the third was married to an economics professor. They were well connected and used those contacts well, whether to lobby the governor at university functions or bring friends into the association, such as Newton Drury's wife, photographer Ansel Adams, and William Penn Mott—who was made the group's president. They quickly mustered support from all over, including Caroline Livermore's Marin Conservation

League and the Save Our Bay Action Committee on the peninsula, led by Claire and Kent Dedrick, Janet Adams, and Eleanor Boushey.

With surprising rapidity, the movement to save the bay became a mass political uprising. The association grew to 18,000 members by 1970 (wisely, they charged only $1 for membership). They got a boost from the Bay Area's most popular radio personality of the day, Don Sherwood, who joined at the urging of Kerr. Books were written and television programs made about the threatened bay. The bay became a wildly popular cause, and hundreds of thousands of people (including me) were converted to environmentalism in the process. Nothing was more essential to the foundation of the Bay Area's green culture. It all goes through Save the Bay.[10]

Save the Bay wanted a single agency to plan for and regulate use of the bay and its shoreline. State Senator Eugene McAteer of San Francisco and Richmond Assemblyman Nicholas Petris were recruited to lead the fight in Sacramento, and Governor Pat Brown called a special session of the Legislature to deal with the bay. In 1965 the McAteer-Petris Act passed, placing a moratorium on fill and creating the Bay Conservation and Development Commission (BCDC). The commission showed its teeth immediately, stopping Redwood Shores, further expansion of sanitary landfills, and all talk of damming parts of the bay. It was backed up by several lawsuits brought by Save the Bay against local cities, the State Lands Commission, and the Army Corps of Engineers, as well as by local citizens including Jean Siri and Jay and Barbara Vincent badgering the agencies to enforce the law. BCDC was made permanent four years later, despite a fierce last-ditch effort by developers and Senator Dolwig to gut it. Melvin Lane, peninsula environmentalist, Portola Valley city councilman, and co-owner of *Sunset* magazine, was named the first chair of BCDC and Joe Bodovitz, former journalist and director of San Francisco Planning and Urban Renewal (SPUR), the first executive director.

BCDC has been spectacularly successful, a model of environmental regulation for the public good, as well as of efficient, effective government. It has minimized bay fill and kept the developers at bay, while insisting on marsh restoration and shoreline planning. It has increased public access by promoting bayside parks, trails, and views. Most of these places are on former landfills, industrial sites, and working waterfronts. BCDC's fame spread far and wide, and it was cited by the first United Nations Conference on the Human Environment as a model of citizens' initiative.[11]

Yet the threats to the bay did not disappear. Some properties were grandfathered in, allowing expansion and new proposals. No sooner was Red-wood Shores shot down, for example, than another scheme, South Shores, rose in its stead. Mobil Oil, which bought Bair Island (a diked and drained tidal marsh) from Leslie Salt, wanted to put in 4,000 housing units. Red-wood City's council happily accommodated them in 1982. With the help of the Committee for Green Foothills, conservationists created Friends of Red-wood City and put a referendum on the ballot. Since Redwood City was still a solidly blue-collar town, the case was argued on fear of taxes and traf-fic. Mobil spent $100,000 and the friends $15,000, but the latter won by a razor-thin margin.[12]

Over the next two decades, the biggest challenges faced by BCDC came from the expansion of the Bay Area's ports and airports. BCDC has repeat-edly confronted the Port of Oakland's dredge and fill plans, as the ship channel goes lower and lower to accommodate bigger and bigger con-tainer vessels. In the latest round, to get a 50-foot channel, BCDC extracted Middle Harbor Park, a beach and wetlands, from the port. In 2000 San Francisco International Airport (SFO) announced a plan to fill 2 square miles for new runways, in order to accommodate increasing traffic (SFO was the fifth-busiest airport in the country in the 1990s) and reduce delays on days with poor visibility. Then the economy and air traffic plummeted in the recession of 2001–02, and the airport put its expansion on hold. Save the Bay and other greens are vigilant, but the heavy artillery of busi-ness and government keep pounding away to soften up public opinion for raiding the region's most precious open space. This is one more affir-mation of Kay Kerr's view that "What we have learned is that the bay is never saved."[13]

A CRABBED ESTUARY

The watershed of San Francisco Bay covers 60,000 square miles, or almost two-thirds of the land surface of California. So the bay has been altered profoundly by developments upriver. Hydraulic gold mining washed mountains of sediment out of the Sierra Nevada, clogging the rivers with gravel and raising the bed of San Pablo Bay by a yard. Giant dams at Shasta and Oroville, which can store two years' runoff, have reduced peak flows, affecting bay circulation and salinity. Most importantly, the waters flowing into the Sacramento–San Joaquin Delta are diverted into the world's largest plumbing system. Almost half the outflow to the bay is redirected, some going as far as San Diego. The rivers are worth their weight in tomatoes, rice, almonds, tract homes, and lobbyists.[14]

Estuarine fish and wildlife have suffered disproportionately. Dozens of dams have radically obstructed the salmon and steelhead runs up the Sacramento and San Joaquin river systems. Striped bass, which breed in the delta, plummeted once the State Water Project started pumping in 1972. Water withdrawals, agricultural pesticides, and prolonged drought in the 1980s combined to devastate native fish populations, which had declined by 80 percent by 1990; a decade later they had recovered to only half the total of thirty years earlier. The only commercial fishing left in the bay is for herring and anchovies. Sturgeon have recovered, only to become the target of caviar poachers.[15]

Up to 1970, the state's water development had raged unchecked in the era of concrete dams, whether for agriculture, hydropower, or city water supplies. After the water seekers had finished off John Muir and flooded Hetch Hetchy Valley, the Sierra Club had lost interest in the damming of California's rivers, leaving no meaningful opposition for decades. Up went the gigantic federal Central Valley Project in the 1940s and the State Water Project in the 1960s. When the club launched its famous battles on the Colorado in the 1950s, it had to start from scratch to learn about the damage done to rivers by these massive dams and diversions. A battle over the Stanislaus River in the early 1970s brought the problem home to California again and gave birth to Friends of the River among a new generation of river runners and backpackers. Unfortunately, the cause was lost on the Stanislaus and New Melones reservoir filled.[16]

The cumulative effects of the upstream projects were felt in the fishery and water quality of San Francisco Bay. So when a third megaproject was proposed (which included the Peripheral Canal), Bay Area greens were in a fighting mood. They had come to realize that the bay and the Sacramento–San Joaquin Delta constitute a single estuarine system—the largest in North America south of Alaska. Bill Davoren, a former assistant to Stuart Udall at the Department of the Interior and deputy to the director of the California Coastal Commission, founded the Bay Institute in 1981 to promote this vision of a unified system. As he puts it, "Save the Bay was working on the bay as a bathtub, but no one was looking at the faucet."[17]

When the Legislature approved the Peripheral Canal package in 1982, all hell broke loose. A group led by Mark Dubois of Friends of the River, Michael Storper of Friends of the Earth, Nick Arguimbau of Citizens for a Better Environment, and Helen Burke of the Sierra Club went to work. The Planning and Conservation League backed the canal, then reversed course. The Sierra Club went along with the idea because former club activist Gerald Meral had been elevated to deputy director at the California Department of Water Resources, then found its membership in rebellion (I quit the club and never rejoined). Arguimbau dug up a little-used provision of the California Constitution: the popular referendum. A statewide coalition formed, and the referendum passed handily in 1982, backed by 90 percent pluralities in some northern counties and a majority in Los Angeles.[18]

State water development had been stopped in its tracks. Environmentalists then went on the offensive, bringing lawsuits against the State Water Resources Control Board to reduce diversions in order to maintain bay water quality. They won a key decision in 1986 when Judge John Racanelli required review of the effects of diversions.[19] The water board launched the bay-delta hearings to get public input, and after years of testimony, it issued a draft decision in 1992 that would have required greater outflows. But Governor Deukmejian spiked that at the behest of the farm lobby. In turn, the U.S. Environmental Protection Agency rejected the state's weakened water-quality standards. At the same time, Congress passed the Central Valley Project Improvement Act, which, among other things, dedicated a proportion of Sacramento–San Joaquin outflow to environmental amelioration.

California's water supply system was thrown into crisis. Congressman George Miller (D–Contra Costa) forced the state and federal governments to broker a deal to jointly manage water supply and water quality. The result was the Bay-Delta Accord of 1994, out of which came a long-term planning process under the rubric CALFED.[20]

For five years, 1995–2000, CALFED negotiators tried to build consensus between water diverters and conservationists, on the theory that win-win solutions are possible and objective science can mediate political disputes. CALFED is a classic case of trying to square the circle; there is not enough water to go around. It has mostly served as a facade behind which agriculture and Southern California squeeze more out of the bay and delta, using state water bonds approved in the mid-1990s. Some improvements of fish habitat have been achieved upstream using some of the same bond revenues, but little mitigation money has gone to the bay itself. In fact, water releases for fish runs have come chiefly because of suits over listing chinook salmon sub-species under the Endangered Species Act. As one former Save the Bay staffer put it, "Frankly, the only thing that's worked is litigation."[21]

THE GREENING OF THE BLUE

San Francisco Bay is the most altered aquatic ecosystem in the United States. More than 250 nonnative species of plants and animals have taken hold. The intruders, like the bubonic plague, have followed the trade routes. They have come as fry in the ballast of ships, in shipments of oysters, and as well-meaning introductions. In days past, most came from the Atlantic, such as the oyster drill or the Atlantic shipworm, but recent invaders come mainly from Asia. A saltwater Asian clam has taken over the upper estuary and a freshwater Asian clam the delta, and Chinese mitten crabs are pandemic in brackish waters.[22] In 1999 the Legislature instituted a ban on oceangoing vessels discharging untreated water in California. Five years later, San Francisco BayKeeper won a suit to make EPA regulate the discharge of ship ballast along the coasts. Alas, the crabs were already out the barn door.

Estuarine ecology has also been radically altered by the loss of tidal marshes and mudflats. Only 20 percent of 300 square miles of salt and brackish marshes around the bay's fringe remain, along with 100 square miles of freshwater wetlands diked for pasture and, until recently, 60 square miles in salt ponds.[23] Shorebirds and migratory waterfowl still use the bay and delta in large numbers, but endemic species have suffered mightily. The California clapper rail, salt marsh harvest mouse, and gray fox are threatened with extinction. No one knows what the future of the delta will be, since the islands are sinking and the levees cannot be sustained indefinitely—but subsidence prevents most areas from ever returning to marsh.

Salt of the Earth

Bay Area greens have been trying to restore bay ecology by bringing back the tidal wetlands. In the past, Save San Francisco Bay Association focused on bay fill, the role of wetlands for wildlife was poorly understood, and BCDC did not have jurisdiction over salt ponds or other previously converted tidelands. So a separate movement arose down in the South Bay. It started in the

1960s when Palo Alto was pondering a Utah Construction Company development plan; instead, the city set aside a Bayside Park (now the Baylands Nature Preserve) in 1963, thanks to the determination of Harriat Mundy, Lucy Evans, and Emily Renzel, local Audubon Society members. In the process, Palo Alto made the first breach in the dikes in a century, restoring some of the area to tidal marsh.

Buoyed by this success, local citizens began agitating for restoration of the rest of the southern bay, which had once held some 60,000 acres of tidal marsh. The leaders of this effort were Florence LaRiviere and her husband, Phillip, who were loosely affiliated with Save San Francisco Bay, Peninsula Conservation Center, and Audubon Society. In the mid-1960s, they began to put together a coalition with other advocates around the southern rim of the bay, such as Ralph Nobles of Redwood City and Frank and Janice Delfino of Alameda. The idea of federal protection came from Arthur Ogilvy, a Santa Clara County planner.[24]

Democratic Congressman Don Edwards of San Jose introduced legislation, and Congress authorized the San Francisco Bay National Wildlife Refuge in 1972 (later named for Edwards). The refuge lies mostly south of the San Mateo Bridge, carved out of salt evaporation ponds. Among the original acquisitions was a purchase from Cargill, which sold 12,000 acres of former Leslie Salt ponds to the U.S. Fish and Wildlife Service in 1979, but with the kicker that the company could continue to make salt in perpetuity on two-thirds of that. The LaRivieres were not satisfied with the size of the refuge or the operating salt ponds; they and their allies wanted the whole southern bay reserved and the salt ponds breached to restore tidal marshes. So they constituted a Citizens Committee to Complete the Refuge in 1985. Public mobilization proved easier this time, with dozens of groups and cities signing on, and Edwards secured a doubling of the refuge to 40,000 acres in 1988.

A third victory came in 1997, following a Citizens Committee and Audubon Society campaign to recover Bair Island, the site of the South Shores project nullified by referendum in Redwood City. The giant Japanese construction firm Kumagai Gumi had bought the 1,700-acre island from Mobil Oil in the go-go days of overseas investment in the 1980s. Working with activists in Japan, Bay Area greens managed to shame Mr. Kumagai into selling by publishing his picture in a full-page ad in the *New York Times*.[25]

A last major addition came after Cargill decided to sell almost all its remaining salt ponds in the late 1990s. A deal was reached in 2004 for 16,500 acres, but land values had risen so much that Cargill walked off with $100 million plus major tax breaks. The money was raised from federal and state governments and four green-friendly foundations (Hewlett, Packard, Goldman, and Moore). The first breaches in the dikes were made in 2005, bringing the return of thousands of waterfowl. But Cargill retains its old salt production rights and some 3,000 acres of prime development land near Redwood City and San Jose.[26]

Dikes and Ducks

In the North Bay, the picture is muddier. Suisun and San Pablo bays once had extensive tidal wetlands along their northern shores, 59,000 acres and 55,000 acres, respectively. By 1890, almost all these wetlands had been diked for hay farms, duck puddles, or salt ponds. Most of the vast Suisun Marsh was bought up by hunting clubs a century ago. Since hunters prefer dabbling ducks, they turned the marsh into seasonal wetlands (making it the largest brackish marsh in North America). The State Department of Fish and Game secured the Grizzly Island Wildlife Area for public-access hunting on about a quarter of the marsh. With all the hoopla over bay fill in the 1960s, the clubs—more than 150 of them—sought protection of their own through passage of the 1974 Suisun Marsh Protection Act and creation of the Suisun Resource Conservation District thereafter. Current talk of tidal restoration has the clubs nervous.[27]

Around the northern shore of San Pablo Bay, most of the tidal marsh was converted to hay farms, with some evaporation ponds near the mouth of the Napa River. The very first segment of the San Francisco Bay National Wildlife Refuge was a thin band of tidal marshes and mudflats south of Highway 37 (mostly created by hydraulic mine tailings), which already fell under the purview of the federal government. In 1989 Congresswoman Barbara Boxer (D-Marin), pushed by Save San Francisco Bay Association (which began to focus on marshes in the mid-1980s), secured funds to buy the first hay farm, at Tubbs Island. The 2004 Cargill deal brought 3,000 acres of salt ponds into the northern refuge. Hay farmers have become willing sellers, and the U.S. Fish and Wildlife Service has acquired large tracts near the mouth of the Petaluma River. Skaggs Island, formerly part of Mare Island Naval Shipyard (decommissioned), is slated to be transferred to the refuge, as well.[28]

The California Department of Fish and Game has about 10,000 acres around the mouth of the Napa River in the Napa-Sonoma Marsh Wildlife Area. These are former Cargill ponds purchased in 1994 with penalty money from a Shell oil spill. Another 1,600 acres were added a decade later. The state also manages the Petaluma Marsh, the largest remnant tidal marsh anywhere on the bay, and Bel Marin Keys and Bahia wetlands on the western flank of the Petaluma River. It took fierce battles, led by Barbara Salzman of Marin Audubon against developers and the city of Novato, to secure these areas. Bahia is a good example of how land-acquisition financing has to be cobbled together from many sources; the Coastal Conservancy, Wildlife Conservation Board, CALFED, North American Wetlands Conservation Council, Marin County Open Space District, Marin Community Foundation, and Marin County all contributed funds.

The most poignant wildlife refuge is Antioch Dunes on the Contra Costa shoreline. Sand dunes there were once 120 feet high, driven by the constant winds funneling from the bay to the delta. They were, predictably, hacked down for building materials, mostly bricks for San Francisco. Later, two paper mills were built over them, one of which still stands. Finally, a small refuge was created in 1980 on the remaining dunes, the first in the country established to save endangered plants and insects: Lange's metalmark butterfly, Antioch Dunes evening primrose, and Contra Costa wallflower.[29]

The Tides of Marsh

From an afterthought in the early days of saving San Francisco Bay, marsh restoration had become the leading edge of bay conservation by the 1990s. The first halting steps at restoring tidal wetlands came in the 1970s, but it took twenty years' experience to find out what worked and what did not. Now, more restoration work goes on here than anywhere in North America, and the region pioneered many of the techniques used elsewhere.

One of the first restoration practitioners was Philip Williams, a hydraulic engineer from England who got involved in the campaigns against dams and early surveys by BCDC and the Coastal Conservancy in the 1970s. His best-known bay work is the wetland created de novo at Crissy Field in San Francisco for the National Park Service. Some other marsh restorations underway in the early 2000s were Eden Landing (state), Bahia Lagoon (Marin Audubon), San Leandro Bay (East Bay Regional Parks District), and Alviso Slough (federal).

Perversely, restoration work includes destroying stands of Atlantic cord-grass that have colonized the southern bay after an ill-advised introduction by the U.S. Army Corps of Engineers on a restoration project thirty years ago. Once the Atlantic and Pacific varieties cross-pollinated, hybrid clusters began spreading like wildfire, crowding out more-diverse native marsh plants. This proliferation of cordgrass threatens the entire ecology of bay tidal wetlands if it is not stopped.[30]

As enthusiasm for restoration built up, a broad coalition of scientists, government officials, and environmentalists came together as the Baylands Wetland Ecosystem Goals Project, under the auspices of the San Francisco Estuary Institute, founded in 1994. The goals report, issued in 1999, recommended restoring 100,000 acres of the original 190,000 acres of tidal wetlands around the bay—of which barely 40,000 exist today. A coalition called San Francisco Bay Joint Venture, consisting of some thirty nonprofits and government agencies, has been assembled under the aegis of the Coastal Conservancy to promote habitat restoration. What is most significant, however, is not the government imprimatur but the way restoration of the bay tidelands has come from below, out of the swirling tides of Bay Area green activism.[31]

GOING COASTAL

For a century after the Gold Rush, the foggy Northern California coast was left to lumber schooners and fishermen plying their hard trade. The favored environments of the Bay Area were traditionally the sunny climes behind the Coast Ranges, not those on the shoreline itself. The best one could expect of a coastal retreat was a beachside cottage in Santa Cruz, where the coastline turns inward from the fog's chilly reach. Not until the 1960s did the Pacific coast become a major conservation cause. This was partly because recreationists finally discovered it, but mostly because of the development of offshore oil drilling and onshore power plants from San Diego to Humboldt County.

Nuclear Power Disarmed

The bay metropolis's thirst for electricity has long been slaked by the Pacific Gas and Electric Company, whose power lines are like so many intravenous tubes feeding the region. In the early twentieth century, PG&E became the new "octopus," replacing the Southern Pacific. When the utility lost its monopoly over hydropower to the federal government in the 1930s, it then turned to natural gas and oil-fired electric plants. These required huge volumes of water, so gigantic installations went up around San Francisco Bay at Antioch, Pittsburg, and Hunter's and Protrero Points in San Francisco; at Moss Landing on Monterey Bay; and at Morro Bay farther south.[32]

After World War II, PG&E's ambitions dovetailed with the fantasies of the atomic scientists, with nuclear energy as the panacea for a voracious national energy appetite. So the company began drawing up plans under the watchful eye of the Atomic Energy Commission (AEC). An experimental reactor was built with General Electric on Vallecitos Creek near Pleasanton in 1958, another at Humboldt Bay. PG&E targeted the coast for commercial reactors because of the unlimited supply of cooling water and the distance from urban habitation: Point Arena, Bodega Head, Davenport, and Nipomo Dunes. Since no place bears greater responsibility than the Bay Area for loosing the atomic beast on the world, it is only fitting that some of the first electric reactors were to be built here and that the antinuclear movement should have taken root here in opposition to them.[33]

It all began at Bodega Head, a little peninsula north of Point Reyes and part of the Pacific plate shearing off the mainland at the San Andreas Fault. A power plant was proposed in 1957 and secretly approved by Sonoma County. But a few obstreperous people from around the Bay Area, led by David Pesonen and Karl and Jean Kortum and backed up by Joe Nielands, Joel Hedgpethe, Doris Sloan, and Harold Gilliam, decided to fight. This was a typical Bay Area political mélange: militant conservationists, committed scientists, liberal politicians, and the odd leftist. This motley crew took on PG&E, the AEC, Sonoma County, the Army Corps of Engineers, and the state Public Utilities Commission—and won against all odds. Along the way, they formed a group under the clunky title of the Northern California Association to Preserve Bodega Head and Harbor and recruited Secretary of the Interior Stuart Udall to their cause. Seismic concerns, sharpened by the great Alaska earthquake, rang down the curtain on this piece of regulatory theater in 1964.[34]

Diablo Canyon, on the central coast, was another story. There, PG&E

won the battle to build a plant but lost the war over nuclear power. Things had started off smoothly enough when the company sat down with leaders of Conservation Associates and the Sierra Club, who did not oppose nuclear power but wanted to save Nipomo Dunes (now a state park) in San Luis Obispo County. The conservationists had proposed an alternative site, at Diablo Canyon, farther north, to which PG&E acceded in 1966. But the younger and more militant members on the Sierra Club board—Martin Litton, Fred Eissler, and David Pesonen—were opposed to nuclear power on principle, and a fierce controversy blew up between them and the older generation, led by Richard Leonard, Will Siri, and Ansel Adams.

Within a year, David Brower shifted position and backed the Young Turks. He had become thoroughly impatient with the "Bohemian Club diplomacy" of the Old Guard. Relations became increasingly strained between the two factions, as well as over Brower's high-handed style, mounting costs of publications, and rebelliousness of the staff. In 1969 Brower was ousted as executive director. He then established Friends of the Earth, with antinuclear politics at the top of its agenda. Modern environmentalism had its first international organization. The turning tide would catch the Sierra Club, which soon rewrote its bylaws and did an about-face on energy policy and internationalism.[35]

Meanwhile, protests against Diablo Canyon were mounting from other quarters. They began locally with Mothers for Peace in San Luis Obispo, a carryover from the anti–Vietnam War movement. They were joined by peace activists of the American Friends Service Committee and Santa Cruz Resource Center for Nonviolence, forming the Abalone Alliance in 1976, which was organized around affinity groups, little hierarchy, and consensus-building. Feminist, communitarian, anarchist, and environmentalist ideals were closely joined, with civil disobedience the main tool of action. Blockades of the construction site were undertaken in 1977–78, and major demonstrations followed the Three Mile Island nuclear disaster in 1979.[36]

PG&E and the AEC pushed forward with the licensing of Diablo Canyon, triggering a huge encampment in 1981 at which 1,900 people were arrested. Then PG&E's invincibility suffered a meltdown when a company engineer announced that pipes in the cooling system had been installed backward. The octopus had met its Captain Nemo. As one historian has observed, "Far from being pesky but harmless gnats, the antinuclear movement halted nuclear construction."[37] With state reforms after the energy crunch of 1973, PG&E became a promoter of conservation and alternative energy sources, creating the world's largest geothermal field, at the Geysers (Napa County), and what was once the world's biggest wind farm, at Altamont Pass (east of Livermore).[38]

Shoring Up the Coast

By 1960 the whole California coast was coming under assault from the forces of development. The California Highway Commission, which had turned U.S. Highway 101 into a freeway from Los Angeles to Morro Bay, had big plans to continue north up Big Sur and all the way to Sonoma. Behind the highways would come the housing tracts; in Southern California, from Laguna Beach to Ventura, they were already rushing to the sea like lemmings. In Northern California, where the coast was as yet little touched, big plans were afoot. Resistance began to mount, and soon environmentalists up and down the California coast would forge an alliance to change the rules of the game.

In Northern California, the trigger point was the Sea Ranch in northern Sonoma County. In 1965 Oceanic Properties, a subsidiary of Castle and Cooke, announced that a swath of Sonoma's coastal terraces would become an elite retreat of 5,000 homes. Despite the beautiful artifice of Sea Ranch's design, it quickly drew fire from Sonoma County residents for blocking public shoreline access. Opponents failed to stop county approval, so they looked to state intervention (see chapter 8).

They were not alone for long. In 1969 the great oil blowout of the Santa Barbara channel took place. The spill became national news and a touchstone of environmental awareness. Overnight, coastal protection became popular in Southern California, especially in its main conservation redoubt, wealthy Santa Barbara. Offshore drilling, it should be recalled, was a technology not much older than nuclear power then, and just as prone to error, and petroleum was the largest source of energy in the state.[39]

A statewide movement was formed, with the goal of installing coastal controls along the lines of the San Francisco Bay Conservation and Development Commission. Bill Kortum of Sonoma joined in 1969 with Phyllis Faber of Marin, Janet Adams from the peninsula, and Ellen Stern Harris of Los Angeles to put together the California Coastal Coalition (later Alliance), which ultimately grew to more than 100 organizations. Adams took the lead in the north, Harris in the south. Of Adams, one participant admired, "She's just absolutely wonderful—a dynamic, energetic, nonstop, ethical, principled dynamo. She's like a bulldozer with heart."

Coastal protection bills were put before the Legislature in 1970–71 by Alan Sieroty, John Dunlap, and Pete Wilson but faced stern opposition. When the Legislature refused to act, backers decided to go directly to the voters with an initiative. They overcame a well-funded campaign by such corporations as PG&E, Chevron, and Bechtel, using an all-volunteer force operating on a shoestring. In 1972 California voters approved Proposition 20.

Ironically, the new law was stronger than the bill that failed in the Legislature in 1970–71. Proposition 20 created the California Coastal Commission to oversee all development in the coastal zone (defined as 1,000 feet back from the sea in cities to 5 miles back in rural areas), with five regional commissions to review projects and issue permits. It also mandated long-term city and county master plans for the coastline. Mel Lane and Joe Bodovitz moved from BCDC to the California Coastal Commission, and the new body functioned with the same effectiveness. After an interim period, it was confirmed by the Legislature in 1976, with the help of Claire Dedrick and Phil Fradkin of the California Resources Agency.[40] The coastal commission continued to hold development at bay under subsequent directors Michael Fischer and Peter Douglas.[41]

Giving Sanctuary

Red flags had been raised in the 1960s over the decimation of marine mammals, such as whales and sea lions, by uncontrolled hunting. This concern grew with the devastation of shorebirds, such as brown pelicans and peregrine falcons, by concentrations of DDT and other pesticides in the food chain. Scientists and citizens began to mobilize around the Bay Area under such groups as the Friends of the Sea Otter, the Point Reyes Bird Observatory, and the Oceanic Society. Then the Santa Barbara oil spill brought California's ocean resources into the national spotlight.

The Marine Protection, Research, and Sanctuaries Act, passed by Congress in 1972, has proved critical to protecting the blue Pacific Waters off the coast of California. Under this act, the federal government has created thirteen marine sanctuaries, four in California: Channel Islands (1980), Gulf of the Farallones (1981), Cordell Bank (1989), and Monterey Bay (1992). The latter three abut the Bay Area and cover roughly 7,500 square miles from Point Reyes to Monterey. The sanctuaries prohibit oil drilling, as well as mining, dumping, and sewage disposal. Their principal goal has become to protect coastal wildlife. The Central California coast, one of only four regions of near-shore oceanic upwelling in the world, is particularly rich in marine life. The United Nations has designated it a World Biosphere Reserve.[42]

With protection, marine bird and mammal populations have revived outside the Golden Gate. Nonetheless, sea otters continue to die because of fecal parasites from cats and opossums arriving in storm runoff, a distinct adverse effect of urbanization along the coast. Meanwhile, no one knows the effects of the thousands of barrels of radiation waste dumped by the Navy in the Gulf of the Farallones or the continuous low-level outflow of pesticides from the Central Valley down the Sacramento River and through the bay.[43]

Despite the sanctuaries, California's coastal waters have suffered the same overfishing as the rest of the world's—thanks to unrestricted takes by industrial trawlers. The sanctuaries do not limit fishing, which is regulated under the 1976 Fisheries Conservation Act; it has not proved up to the task. Rockfish have been annihilated by bottom trawling, which was finally interdicted along much of the West Coast in 2002. California was forced to take action with its own Marine Life Protection and Management act in 1999, which gave the state Department of Fish and Game stronger control through coastal marine reserves and fishing bans. Fish and Game followed a year later with No Take Zones all through the Channel Islands. An even stronger California Ocean Protection Act passed in 2004, and the state established a system of fiteen reserves that have funding in 2006.[44]

Offshore oil drilling remains a sensitive issue all along the coast. When the Reagan administration sought to revive exploration, it was fiercely resisted up and down the coast. A federal moratorium was imposed in 1982 but must be renewed by Congress every year. In the second Bush administration, oil was again king and the moratorium was lifted, leading to proposals to double the size of the Gulf of the Farallones reserve in order to protect the Sonoma County coast.

San Francisco Bay is the heart of the Bay Area's landscape and the most critical of all its open spaces and public resources. The Pacific coast is the defining feature of California. In size, the protected blue zones of the bay and coastal waters are larger than all the green areas on land put together, and they establish a welcome buffer against flagrant urbanization. Everyone knows that the juncture of bay and ocean makes San Francisco one of the most remarkable sites for a big city on the face of the earth. But it would not be the glorious place it is if destruction of bay and coast had not triggered a mass mobilization of the citizenry on the side of the environment, changing the region's politics forever.

6

ENCOUNTERS WITH THE ARCH-MODERN

Regional Planning and Growth Control

The Bay Area uprising for parks, open space, and bay protection has been a confrontation with the forces of urbanization, at times a mortal combat against the paved and the dammed. This has all too frequently been defensive, piecemeal actions by local groups feeling the hot breath of the bulldozers on their neck of the woods. In the efforts to save the bay, protect the coast, and create a national park at the Golden Gate, one finds only hints of a larger vision of containment of the monster metropolis. Nonetheless, an overarching vision has been nurtured all along by one organization above all: People for Open Space/Greenbelt Alliance. It has provided a necessary utopian and regionalist outlook on the greenbelt, urban planning, and metropolitan government for fifty years.

This has been more than a struggle to conserve green and pleasant lands. Local efforts to stop the development juggernaut have shown, time and again, the need to grasp the nettle of urban growth, including suburban sprawl, housing supply, and local government. Planners and conservationists have therefore tried to rewrite the rules of the game through strong planning measures, government reform, and urban-growth controls. This has added up to a movement of considerable force, even though the utopian ideals of the most radical planners and visionaries have not been realized. As a result, urban development in the Bay Area has not been an uncontested

TABLE C. Bay Area Population Growth (nine counties, by decade)

	Total Population	Increase	Percent
1940	1,734,308	156,299	9.9
1950	2,681,322	947,014	54.6
1960	3,638,939	957,617	35.7
1970	4,630,576	991,637	27.3
1980	5,179,759	549,183	11.9
1990	6,023,577	843,818	16.2
2000	6,783,760	760,183	12.6

SOURCE: www.bayareacensus.ca.gov/historical/copop18602000.

process, as in so much of the United States; it is questioned, fought over, reconfigured, and frequently stopped in its tracks.

Here again we are witnesses to the two-headed nature of the greening of the Bay Area. On the one hand, the leadership has typically come from the elite—both upper class and intellectuals from the university communities—and its agenda has been both utopian and conservative with regard to the city. On the other hand, the growth-control movement has called up a kind of mass participation in and cry for local democracy, and its success in making the city more livable is inextricable from popular conviction and mobilization.

CRY CALIFORNIA

The greenbelt movement was born in opposition to the great postwar growth boom, when the Bay Area was adding a million people a decade. Future prospects must have seemed appalling to people accustomed to a compact city like San Francisco and a still-modest regional population of a couple million. Developers and government agencies had delirious visions of doubling and redoubling the scale of the metropolis. One such document, promulgated by the U.S. Army Corps of Engineers and U.S. Department of Commerce in 1960, envisioned more than fourteen million people in the Bay Area by the next century. The Association of Bay Area Governments predicted a more modest seven million by 1990—a figure only a decade off the mark.[1] As it was, in 1980 the Bay Area would have more people than forty other states of the union.

Midcentury Modernists

In counterflow to the prevailing Babbitry of local growth boosters, a broad critique of unchecked urbanization emerged among the region's intelligentsia. Nothing quite like this had happened before, even in the age of Muir. The Bay Area in the 1960s was a hotbed of radical, green ideas among planners, political scientists, biologists, and journalists. The usual accounts of postwar environmental thought, running from Fairfield Osborn to Rachel Carson, give too little credit to these Californians.[2]

The first voices in the wilderness were a group of planning students from Berkeley who came together in the late 1930s as a group calling themselves Telesis. Telesis was a typical avant-garde formation, complete with manifesto. They mounted an influential exhibit called "Space for Living" in 1939 that awakened San Franciscans to modern planning and architecture and to the idea of an urban greenbelt. The group included future Berkeley professors T. J. (Jack) Kent, Francis Violich, and Mel Scott and architectural stars Vernon DeMars and Garrett Eckbo. They had imbibed the classic teachings of Raymond Unwin and Le Corbusier on the need for rationalized urbanism and of Patrick Geddes and the London greenbelt planners. They came of age during the New Deal, with its ambitious National Resources Planning Board and greenbelt cities under Rexford Tugwell. They were influenced by Catherine Bauer, the leading housing reformer of the day, who had moved from New York to the Berkeley faculty and was married to architect and professor William Wurster. They admired Lewis Mumford, the great American civic prophet and frequent visitor to Berkeley. They were, in short, Modernists and social democrats of the age of social planning, humanist faith, and utopian hopes.[3]

A second group came out of the Bureau of Public Administration at UC Berkeley, under Samuel May. The bureau was steeped in the Progressive tradition of municipal reform and had long been an incubator of regionalism and good government. May, who had been instrumental in securing the East Bay Regional Parks District, later served as chair of the State Planning Commission under New Deal Governor Culbert Olson and tried to establish a nine-county Bay Area Regional Planning Commission in 1940–41 (Jack Kent served as May's assistant at the time). The bureau (later the Institute for Governmental Studies) continued to advocate for metropolitan governance in the 1950s under Victor Jones, and in the 1960s and '70s under

Eugene Lee, publishing a raft of monographs on the balkanization of local government and advantages of regionalism. Institute publications circulated widely in the national debates over metropolitanism. Under the institute's aegis, Mel Scott produced the first urban history of the Bay Area, which was at the same time a plea for regional planning and government.[4]

A third branch of critical intellectuals was made up of scientists, reformers, and journalists who provided a scathing indictment of environmental ruin. The prologue was a 1956 conference at the University of California, Berkeley, convened by Carl Sauer, Lewis Mumford, and Marston Bates entitled "Man's Role in Changing the Face of the Earth." The themes were picked up in the 1960s by Raymond Dasmann, San Francisco–born ecologist and UC Santa Cruz professor, especially in his panoramic survey, *The Destruction of California*. Alf Heller, a Wells Fargo heir, and Sam Wood, a planner, formed the think-tank California Tomorrow in 1961 and wrote *California, Going, Going . . .* ; they also launched the journal *Cry California*, whose editor William Bronson compiled a photojournalist account, *How to Kill a Golden State*. Harold Gilliam wrote timely books on San Francisco Bay, Point Reyes, and regional ecology, used his post at the *San Francisco Chronicle* to popularize environmental issues, and helped ghost-write Stuart Udall's *The Quiet Crisis*—a nationally influential environmental tract. Later, Berkeley writer Ernest Callenbach hit a nerve with his tale of *Ecotopia*.[5]

The Bay Area's critics of the exploding metropolis were not simply Romantics recoiling against modernity nor Progressives going about the business of good works. They drank from the fount of midcentury Modernism, in which the world could be made better by the enlightened application of foresight, science, and good government. Those who promoted regional planning, land conservation, and rational urbanism were mostly liberals—whether Democrat or Republican—who had imbibed the spirit of the New Deal. They tilted toward enlightened management and the social good, over and above the pursuit of private profit. This renders untenable the view that environmentalism is a *post*modern phenomenon, a new social movement unleashed by the heat-death of the universe of social management. Postmodern sensibilities lay far in the future.[6]

Where this Green Society ideology diverged from mainstream Modernism and New Deal liberalism was its deep attachment to nature and the California landscape. It was more Mumford than Le Corbusier, just as it had been more Muir than Pinchot—in both cases, wrestling with modernity's devils.[7] In terms of environmental history, members of this society stand closest to Bob Marshall, "the people's forester" and founder of the Wilderness Society in the 1930s. Marshall is the one national figure usually pointed to as embodying a joint social and environmental vision despite his wealthy background. In the Bay Area, this kind of green ideology with a rose-colored hue persisted into the postwar era, while Marshall is a tragic figure who was marginalized by red-baiting and the Wilderness Society's conservative turn—even before his early death. Bay Area heroes of the greenbelt, such as Dorothy Erskine, Jack Kent, and Harold Gilliam, lived long and influential lives, keeping the green flame alive.[8]

Guardian of the Greenbelt

A necessary complement to the ruminations of intellectuals was the practicality of citizens who took up the cause of planning, from Marin to San Jose. When it came to regional planning, the key Bay Area activist was Dorothy Ward Erskine, founder of People for Open Space (later Greenbelt Alliance). Erskine was another of the grand ensemble of women movers and shakers who put in place the Bay Area greensward. Born in San Francisco in 1896, she was the granddaughter of Quakers and daughter of one of the pioneer woman doctors of the city. She greatly admired Progressive stalwart Alice Griffith, who had brought Jane Addams' ideas to San Francisco, and Erskine volunteered in Griffith's Neighborhood House in her youth. Dorothy attended UC Berkeley and married a liberal lawyer, Morris Erskine. They lived in an apartment on Telegraph Hill that Dorothy designed in 1940 in the new Modern style, with sweeping bay views through walls of glass.

Dorothy Erskine was no genteel liberal. In the 1920s, she became friends with Lincoln Steffans, Sara Bard Field, and the Carmel bohemians. In 1931 and 1937, she traveled to Russia to see if what she'd heard about the socialist experiment was true (it almost cost her husband his contract with Bank of America). In the 1940s, she gave talks on Russia at the California labor school and was friendly with the labor movement.

Along with Josephine Duveneck and Martha Gerbode, Dorothy joined Griffith's Housing Association in a study of slum conditions in Chinatown (prompted by the federal Housing Act of 1937). In 1939 she met the young idealists of the Telesis group from Berkeley and became an advocate of city planning. In the early 1940s, she organized the San Francisco Housing

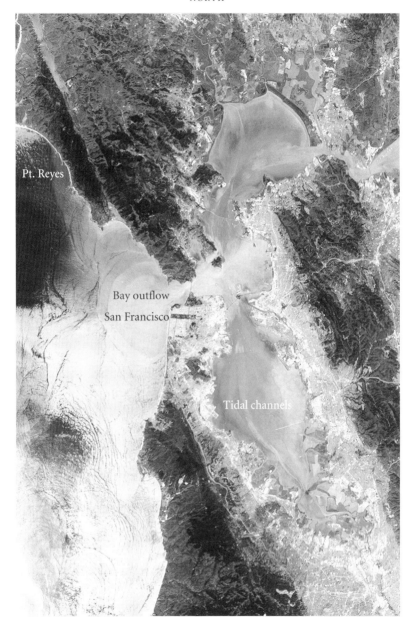

The Bay Area from the International Space Station, Expedition 4, April 21, 2002.
Photo by astronaut Ed Lu; courtesy of NASA.

John Muir in his vineyard, at his home in Martinez, c. 1910. Photographer unknown; Marion Randall Russell Papers, Bancroft Library, University of California, Berkeley.

Mount Tamalpais through the painter's optic, c. 1850s. Watercolor by William McMurtrie; Robert Honeyman Jr. Collection, Bancroft Library, University of California, Berkeley.

(Facing page, top) Redwood drag line in the Santa Cruz Mountains, c. 1880s. Photographer unknown; courtesy of San Mateo History Room.

(Facing page, bottom) Founders of the Sempervirens Club at Big Basin, 1900. *From left*: Louise Jones, Carrie Stephens Walter, J. G. Packard, W. W. Richards, Mr. Hamilton (guide), R. S. Kooser, J. F. Cooper, Charles Wesley Reed. Photo by Andrew Hill, courtesy of History San Jose.

San Francisco watershed and Crystal Springs reservoir, looking north
up the Peninsula, 2005. Photo by Jitze Couperus.

(Facing page, top) Above Route 580, Cronin Ranch, East Bay hills, 1994.
Photo by Matt O'Brien, from the series "Back to the Ranch."

(Facing page, bottom) Urban sprawl, east side of the Santa Clara Valley,
looking north, 1996. Photo by Bob Dawson.

Shadow Cliffs Regional Park, east of Mount Diablo, site of the former Kaiser quarry, 1987. Photo by Bob Walker; courtesy of Oakland Museum of California.

Golden Gate Park's western end emerges from the sand dunes, c. 1884. Photo probably by Isaiah Taber; located by Greg Gaar.

Mount Diablo, Mt. Diablo State Park, looking from the east, 1984.
Photo by Bob Walker; courtesy of Oakland Museum of California.

Tilden Regional Park, above Berkeley, looking east from Inspiration Point
trail, 2004. Photo by Richard Walker.

The Golden Gate from the Marin Headlands, 2004. Photo by Richard Walker.

Board of the Committee for Green Foothills, 1972. *Front row, from left*: Pat Barrentine, George Norton, Ruth Spangenberg, Lois Crozier-Hogle, Kent Dedrick, and Tom Jordan; *back row*: Kirke Comstock, Claire Dedrick, Larry Dawson, Mary Moffat, Mary Gordon, Paul Smith, Eleanor Boushey, Norman KcKee, Kathryn Stedman, and Morgan Stedman. Photo by Lowell Johnson; courtesy of the Committee for Green Foothills.

Caroline Livermore, c. 1960. Courtesy of the Marin Conservation League.

Foothills in summer, Santa Clara County, 2004.
Photo by Cait Hutnik, "Light of Morn."

Mount San Bruno, with South San Francisco lapping at its heels, 2005.
Photo by Brian Hauk; courtesy of Skyhawk Photography.

South Bay salt ponds at winter flood tide, Coyote Hills Regional Park, c. 2000.
Photo by Alvin Docktor.

Florence and Phillip LaRiviere at their Palo Alto home, 2005. Photo by Richard Walker.

Berkeley's landfill and marina, 1987. Photo by Bob Walker; courtesy of Oakland Museum of California.

Dorothy Ward Erskine, c. 1950. Photographer unknown; courtesy of John Erskine.

San Mateo County coast north of Point Año Nuevo, looking north, 1985. Photo by Bob Walker; courtesy of Oakland Museum of California.

Dougherty Valley, Contra Costa County, looking south, 1996. Photo by Matt O'Brien, from the series "Back to the Ranch."

Coyote Valley, south of San Jose, looking east from Coyote Ridge, 2004. Photo by Cait Hutnik, "Light of Morn."

Mission Peak Regional Park, looking east over Fremont, c. 2004.
Photo by Alvin Docktor.

Monte Bello Ridge, looking southeast at Black Mountain and the Santa Clara Valley, 2004. Photo by Frank Crossman.

Christo's twenty-two-mile "Running Fence" under construction, Marin County, 1976. Photo by Phillip Hofstetter.

Napa Valley vineyards in winter, 2004. Photo by Richard Walker.

Office development in Sonoma, near Petaluma, 2005. Photo by Richard Walker.

Clearing forest for vineyards, Sonoma County, 2005. Photo by Richard Walker.

Bill Kortum, longtime Sonoma County environmental activist, 2005. Photo by Richard Walker.

Industry on the Contra Costa Shore, near Pittsburg, looking east, 1996. Photo by
Bob Dawson.

New EBMUD Sewage Plant, at San Lorenzo, looking east, 1987.
Photo by Bob Walker; courtesy of Oakland Museum of California.

Carl Anthony, founder of Urban Habitat,
2005. Photo by Margaret Paloma Pavel;
courtesy of Earth House, Oakland.

Aging power plant at Potrero Point, San Francisco, looking north, 2005.
Photo by Richard Walker.

Warning sign, hazardous substance area, industrial East Oakland, 2002.
Photo by Scott Braley.

Brownfield site near Hunter's Point, San Francisco, 2002.
Photo by Scott Braley.

Young community gardeners, East Oakland, 2002. Photo by Scott Braley.

Tract houses going up at Discovery Bay, east Contra Costa County, 1987.
Photo by Bob Dawson.

Authority to build public housing and was an advocate of housing integration. Then she formed the San Francisco Planning and Housing Association to force the city to create a professional planning department (it was one of the last holdouts in the country). The latter group morphed into the San Francisco Planning and Urban Renewal (now Research) Association (SPUR), which pushed the city into the federal Urban Redevelopment program in the 1950s.[9]

In the 1960s, Dorothy Erskine was active in several of the key battles to save the character of San Francisco, such as stopping the freeways and a proposed United States Steel tower next to the Ferry Building. She worked closely with Frida Klusmann in saving the city's cable cars. And in the early 1970s, she organized the campaign for a funding initiative to expand the city's parks.[10]

By the mid-1950s, however, Erskine had broadened the scope of her interests to a regional scale. Her first foray into greenbelt planning was the fight for Butano Forest (see chapter 4). In 1958, Erskine put together Citizens for Regional Recreation and Parks, in concert with Jack Kent and Barbara Eastman, drawing in the Sierra Club, Marin Conservation League, and conservationists around the Bay Area. This was a happy marriage of citizen volunteers (mostly women) and professionals (mostly men), which was common at the time. It was also a joining of the Modernist city planners with environmentalists and urban reformers of a social bent, a rare alliance in American politics.

Formation of the citizens group was prompted by a national debate over the growth of outdoor recreation, which led Congress to create the Outdoor Recreation Resources Review Commission in 1958. One of the Erskine group's first endeavors was to compile an inventory of open space and parks, which was incorporated into the California Public Outdoor Recreation Plan of 1960.[11] They made a public splash with three conferences, on open space, parks, and the bay. The third one, in particular, became an "exposé of environmental destruction," in Erskine's words. "People suddenly woke up. This publicity led directly to the movement to save San Francisco Bay." When Kay Kerr, Sylvia McLaughlin, and Esther Gulick first convened, they invited Erskine, Kent, and Scott to give them advice, and they continued to meet in Erskine's living room for years thereafter. Her son calls her "the fourth lady" of Save San Francisco Bay. She served as vice-chair of the first board of directors of the Bay Conservation and Development Commission.[12]

Erskine was, typical of women of her time, self-effacing and happy to shift credit onto others, but she was effective in getting people to do the right thing. Her guiding principles could have come from Karl Polanyi, the great British critic of unleashing the naked market on society: as she put it, "Land is a resource, not a commodity subject to speculation and mindless use. Nothing could be more formidable [than our struggle]!" And her view of social-movement politics could have come from Saul Alinsky, Chicago's prophet of popular organizing, when she declared: "It's only the crises that educate us, one after another. Fighting these battles you begin to develop a sense of a team with different new people willing to join. And the effect upon the individual in doing this is really extraordinary. . . . People completely change under the effort to do these things."[13]

METRO-TOPIAN VISIONS

The battle for urban containment and a regional greenbelt was fought on two fronts from the 1950s to the 1970s. One was the reform of urban government. If government could be rationalized, good land use would follow. The other was visionary planning. If idealized, rational landscapes could be drawn up to instruct public and civic leaders, a better city would be born. This was the planner's mantra going back to the Progressive era.[14] But the visionaries faced a situation in which postwar metropolitan areas were flying apart into thousands of little suburban pieces. Governmental fragmentation and lack of planning came under fire from planners, political scientists, and conservationists.[15]

Modernizing Government

Metropolitan government reform was in the air in the 1950s.[16] Explosive suburbanization had led to a proliferation of jurisdictions outside the sway of the old core cities of San Francisco and Oakland, creating a Bay Area of nine counties, 100 cities, twenty-four transit districts, and 108 special districts for parks, water, sewage, and the rest. Studies on government consolidation and regional planning started circulating early in the decade, and bills were being introduced in the California Legislature by the late 1950s.

A state Office of Planning (later the Governor's Office of Planning and Research) was created in 1959, along with a special Governor's Council on Metropolitan Area Problems. The council recommended the creation of metropolitan districts with responsibility for regional planning, transportation, parks, and other functions. In the face of opposition from local governments, this was scaled back to a proposal for a Metropolitan Area Commission that would review all changes in local jurisdictions in the state. In the end, the Legislature passed a county-by-county version of this, the Knox-Nisbet Act of 1963.[17]

Some semblance of control on suburban sprawl was imposed by the new act, which required every county to create a Local Area Formation Commission (LAFCO) to approve municipal annexations and incorporations. The act further called for general plans to be drawn up by all counties and cities. Although these were not as yet mandatory, local governments had to begin thinking about their future in a systematic and public way for the first time. The Legislature also thought about parks when it passed the Quimby Act of 1965, under which local governments can require developers to dedicate space for parks or open space or pay fees for park acquisition. LAFCOS, general plans, and developer fees are still the primary tools for the geographic rationalization of urban growth in California. Nonetheless, the planners were far from satisfied with these half-measures. They wanted regional government and planning.[18]

Regionwide infrastructure programs were getting under way, as well, and these called for new governing bodies. A Bay Area Rapid Transit (BART) district had been authorized in the mid-1950s to build a regional light-rail system and bonds approved in 1962. Planning for new freeway systems was being undertaken by the State Highway Commission in the late 1950s, and when the freeways ran into opposition, the state launched an immense Bay Area Transportation Study (BATS) in the early 1960s. At the same time, San Francisco was pushing the Legislature to create a Golden Gate Authority, which would manage all port facilities and bridges around the bay in a unified manner. All these proposals found avid support from the powerful Bay Area Council, a front for the biggest banks and corporations, headed by liberal capitalist Edgar Kaiser, and from a wide spectrum of businesses, governments, newspapers, and labor unions.[19]

Such initiatives lit a fire under local officials from smaller cities, which wanted no part of metropolitan government. In 1960 administrators and politicians from a majority of the Bay Area's cities and counties came together in a forum to consider their options. The conclave of suburban officials, jealously guarding their prerogatives, pasted together an alternative to regional government. Their solution was a weak council of governments with no regulatory powers, and so the Association of Bay Area Governments (ABAG) was conceived in 1961. The Golden Gate Authority was killed off soon thereafter. San Francisco and Oakland did not join ABAG until years later.[20]

Thinking Like a Region

But the winds of land-use reform were still blowing, so ABAG was given the job of preparing a regional growth plan. ABAG's chief planner, Jim Hickey, and Jack Kent were closely aligned, and Citizens for Regional Recreation and Parks worked hand in hand with Hickey's staff in preparing the open space part of the plan (which appeared as a draft in 1966 and in final form in 1970). The planners were emboldened by the times, and they produced a regional blueprint that was nothing less than a revolutionary document for an American city. It called for city-centered development and an astounding 1.8 million acres of greenbelt, or about one-third of the region. As Kent later observed, "There is no other metropolitan region in the USA that has ever said there ought to be a big permanent greenbelt around its existing system of central cities."[21]

As enthusiasm for greenbelt planning built up, the Citizens for Regional Recreation and Parks became People for Open Space in 1968. Erskine and Kent pulled together an impressive board and got a Ford Foundation grant to bolster the case for the benefits of open space. This unprecedented study, which came out in two reports, triggered a national debate over the costs of sprawl. People for Open Space launched their campaign for the Bay Area greenbelt by using an aerial view of Los Angeles captioned "On a clear day, you can see (sprawl) forever." Alf Heller and California Tomorrow weighed in for state land planning and regional government, catching the tenor of a national debate over stronger control by the states.[22]

The ABAG plan built up an enormous impetus, but the association's General Assembly refused to be bound by it. Local governments wanted nothing to do with rational planning. So the developers continued on their merry way, and ABAG remained a debating society.[23] East Bay Assembly-

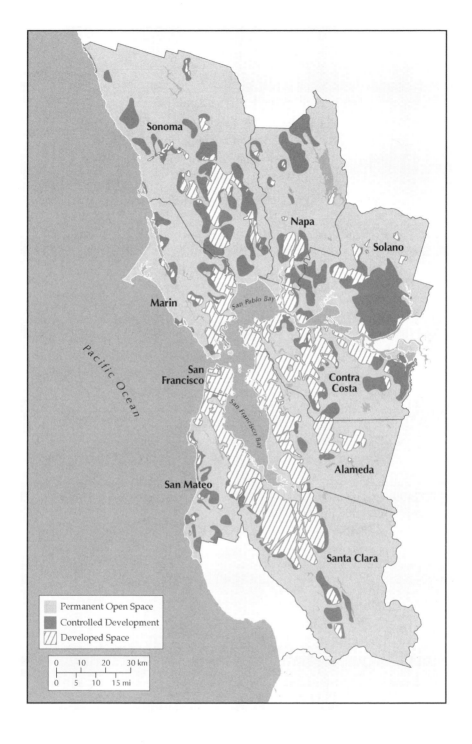

Legend:
- Permanent Open Space
- Controlled Development
- Developed Space

0   10   20   30 km
0   5   10   15 mi

MAP 9. Proposed Bay Area Greenbelt, ABAG Plan of 1970. Drawn by Darin Jensen. Source: Association of Bay Area Governments, 1970.

man John Knox carried bills for regional government for years before giving up in 1975. The utopian moment had passed. The planning movement got a couple of consolation prizes out of the Legislature in 1970 and 1972. One was a requirement that city and county general plans include open-space elements, and the other was a mandate that zoning conform to general plans.[24]

Another consolation prize was authorization of the Metropolitan Transportation Commission (MTC) in 1970. The MTC supplanted CalTrans as the overseer of highway and transit planning for the Bay Area, including the flow of state and federal transportation funds to the region. MTC grew out of the BATS initiative, after ABAG refused to take on the responsibility for regional transportation. As a result, ABAG ended up with no money and MTC with no regional vision. Worse, MTC became a lord unto itself, immune to citizen input and urban planning. To this day, it refuses to accept responsibility for land-use development patterns as a consequence of its transportation policies, which favor highways and suburban rail over city streets and buses.[25]

Back to the Future Greenbelt

Soon after the defeat of regional planning, Jack Kent restructured People for Open Space. In the mid-1970s, it became a membership organization with a professional staff. Larry Orman, one of Kent's former Berkeley planning students, took the helm as executive director. Kent and Orman started rounding up grants to do research into land-use dynamics, farmland conservation, and housing policy.[26]

Above all, they envisioned a sweeping Greenbelt Action Network that would marry Kent's comprehensive ideas about the metropolis as a unity of housing, employment, transportation, and open space with popular mobilization along the lines of the Save San Francisco Bay Association or the California Coastal Alliance. The ambitious greenbelt network idea had to be abandoned, however, after Hewlett Foundation got cold feet and pulled its support. People for Open Space scaled back its plans and focused on farmland loss (see chapter 2), backed by Packard and other foundations. It followed that successful campaign with another on housing, after Allen Jacobs took the helm from Jack Kent.[27] In the mid-1980s, it would undergo a major transformation in strategy and organization and change its name to Greenbelt Alliance.

Greenbelt Alliance revisited the planners' utopian vision one last time in the late 1980s. President Bob Mang and executive director Larry Orman approached Angelo Siracusa of the Bay Area Council to try to build a coalition. While the two groups were leery of each other, they were the only ones with a regionalist perspective. They put together the Regional Issues Forum in 1988, a discussion group among leading business and nonprofit players, including financier Mortimer Fleishhacker III (long a backer of People for Open Space), Edie Dorison of the Silicon Valley Manufacturers' Group, Bob Kirkwood (later vice president of Hewlett Packard), Rob Elder of the *San Jose Mercury*, and Martin Paley of San Francisco Foundation.

Out of their discussions came a major initiative, the Bay Vision 2020 Commission. The commission, chaired by UC Berkeley Chancellor Ira Michael Heyman (a liberal professor of planning law) and directed by Joe Bodovitz (former director of BCDC), put out a report that echoed all the classic themes of the planning visionaries: the greenbelt, housing justice, regional governance, city-centered development, higher-density housing, and better transit. But Heyman and Bodovitz did most of the work by themselves, and the commission failed to mobilize broad support. Greenbelt Alliance and Bay Area Council followed up with an effort to get the Legislature to authorize a study commission for regional planning but lost by one vote.[28]

The greenbelters have been able to rally some of the leading corporations behind their vision of the Bay Area because of practical considerations of maintaining the region's economic vibrancy and attractiveness for skilled workers, given the green tinge to the popular imagination. As the Bay Area Council puts it: "In the Bay Area, the links between well-managed growth and a healthy economy are stronger than in most places. This region has prospered in large part because it is a wonderful place to live. To the extent that growth problems threaten our quality of life, they threaten our prosperity, too."[29] The commitment goes beyond mere lip service, because the blue blood of the civic burghers runs a bit greener here, too.

With the demise of the Bay Vision 2020 Commission in the early 1990s, Greenbelt Alliance dropped the utopian quest for regionwide solutions. It would, instead, refashion itself as an umbrella group for local battles (see below). After almost twenty years as executive director, Larry Orman moved on to found GreenInfo Network. Jack Kent died soon after. A new executive director, Jim Sayer, took over in mid-decade, followed by Tom Steinbach in the 2000s. Funding and staff have grown, and the organization is more active than ever. The years of metro-topian futility are well forgotten, but one can still mourn the end of hard-edged idealism and the passing of the era of Modernist faith in large-scale social engineering and betterment.[30]

CITY LIMITS

Local governments oversee the process of urban growth in the United States, and they do a haphazard job of it. The lure of money is strong, and local officials are normally in the thrall of developers. In established cities, developer power may be checked to some degree by well-organized popular opposition, leaving urban politicians to broker agreements. Peripheral suburbs, on the other hand, are normally ruled by a class alliance of landowners, realtors, lawyers, and merchants looking to profit from growth, and the machinery of government rests in the hands of city managers looking to expand their domains.[31]

The result is suburban sprawl, cookie-cutter houses, and chintzy commercial strips. By the time enough people move in, settle down, and create opposition to unregulated expansion and disappearing amenities, the machinery of growth has moved on to the next willing jurisdiction. Things were worse before the regional-planning and good-government reforms. Prior to that, according to one legal scholar, there was only "the unstoppable building industry." But then "the development abuses of the 1950s and 1960s brought about the modern era of expanded regulatory controls."[32] But the abuses did not stop entirely, and the abused began to rise up in protest.

Because local governments are the mediators of growth, local politics is the principal arena of combat between developers and their opponents. Such opposition has been widespread throughout the Bay Area since the 1970s, with the result that growth controls are in place in almost every jurisdiction in the region. These include annual quotas of new building permits, urban limit lines, hillside protections, restrictions on water and sewer hookups, and environmental impact reports. In addition, it has become much harder to break zoning designations and general plans, thanks to alert citizens and watchdog groups, and growth-control militants have been elected to city councils, water districts, and county boards of supervisors.

Every building cycle has brought a new outburst of growth-control activism. The first postwar boom peaked in 1958, the next in the early 1970s, the next in the mid-1980s, followed by the high-tech era of the 1990s, and the international housing bubble of the early 2000s. In preceding chapters we have seen the protests that broke out in the early 1960s and peaked in the early 1970s. What was going on in Marin, on the peninsula, and in the save the bay movement was certainly "growth control," but the term only came into general currency later. "No growth" and "slow growth" became the watchwords of local activists in the early 1970s. The trigger issues for local growth control varied by place and time, but water supply, sewage capacity, crowded roads, and air pollution were key points of conflict.

The Petaluma Revolution

In 1972 Petaluma shocked the civilized world of planning law with a resolve to limit growth to a fixed number of housing units per year (500). The town's new regulations quickly became the greatest cause célèbre in land-use control since *Euclid v. Ambler* (the case that affirmed the constitutionality of zoning in the 1920s). Denunciations rained down from the business press, most predicting disaster from such a radical intervention in the market. But Petaluma won in the courts in 1975. In fact, the predicted growth curve did not hold its arc when the building cycle turned down that same year, and applications for permits failed to bump up against Petaluma's ceiling until the late 1990s. The fight over growth controls split the towns-people for years, and the city council oscillated from green to progrowth majorities. Nonetheless, the city managed to reconfigure itself, keeping the highway strip along U.S. 101 under rein, making a stand to keep the historic downtown viable, and preserving a visual break between the urban and the rural on its fast-growing eastern edge.[33]

Similar battles broke out in the East Bay. The gauntlet was thrown down in 1972 by the city of Livermore, then at the edge of exurbia, when it adopted limits on sewage hookups and other services until adequate systems were in place. Shockingly, this was done by a citizen initiative led by a group of people associated with the Lawrence Livermore National Laboratory—an unheard-of bit of hubris against the usual cozy growth alliance between developers and city officials. The initiative was upheld by the courts in 1976. Livermore subsequently shifted to a city council measure limiting growth to 2 percent per year.[34]

Pleasanton, at the western end of the Livermore Valley, adopted timing controls in 1976 (2 percent per year) on the Petaluma points system, in response to edicts from the Bay Area Regional Water Quality Control Board for getting sewer construction funds. Pleasanton switched to a unit limit of 500 per year in 1986, after the massive Hacienda Business Park was approved. The latter action had triggered such an upsurge of growth-control fever among the citizenry that environmentalists nearly won a referendum on the project. To get around the sewage problem created by rapid expansion in the Pleasanton-Dublin area, developers proposed a massive Tri-Valley Super Sewer to carry wastes all the way to the bay, but the greens deep-sixed it.[35]

As in Marin, East Bay environmentalists took on the growth-fueling effects of water supply, opposing EBMUD's aggressive service-area annexations, declining block rate structures, and search for new supplies. For years, Jean Siri of El Cerrito and Helen Burke of Berkeley fought a lonely battle to reform EBMUD. When Burke was finally elected to the board in the 1970s, EBMUD's white male board of directors was the oldest elected body in the United States and utterly self-perpetuating. By the early 1980s, an environmentalist majority was swept into office, and progrowth general manager Jerry Gilbert was driven from office. The new board put the kibosh on massive plans for service extension in the Dougherty Valley, near San Ramon.[36]

Down in the South Bay, the city of San Jose—poster child of postwar sprawl—did an about-face in the early 1970s. This began with a political upheaval that dislodged City Manager Dutch Hamann and elected the first women and green sympathizers to the City Council. San Jose adopted an urban-service limit line, leaving considerable land within the city limits as an urban reserve. San Jose became one of the first cities to adopt a "pay as you grow" tax, or impact fees, on developers. These principles were incorporated into the General Plan of 1975. Growth limits were solidified with the adoption of the first green line in the General Plan a decade later. As a result, San Jose stopped expanding like crazy.[37]

Reforms were instituted as the county level as well. Santa Clara County installed a Local Area Formation Commission (LAFCO) in 1964 to check wild annexation and negotiate city limits among warring jurisdictions. Three years later, the Board of Supervisors began to refuse new services to unincorporated areas. By 1970, cities were required to provide services on land they annexed, and an open-space plan was added three years after that. With the help of local ranchers, People for Open Space pushed Santa Clara

County to adopt a strong rural zoning ordinance in 1977 and got 200,000 acres in the south county put in an agricultural reserve. But permit applications still rolled in, so the county supervisors declared a moratorium for two years while preparing a comprehensive report on the need to control growth: *Living Within Our Limits*. On the basis of that report, a new general plan was completed in 1980. A radical appeal for government control of industrial development lost at the county level, but stricter industrial regulations and fees were imposed by Sunnyvale and Palo Alto—a first in the country.[38]

Astonishingly enough, San Jose was *ahead* of Petaluma in adopting growth limits, and Santa Clara County was not far behind Marin and Napa counties in protecting rural land. But Silicon Valley had by then such a bad reputation for sprawl that neither the municipal nor county government has gotten much credit for their achievements. Yet San Jose and Silicon Valley towns have a much tighter planning system and greater density today than any comparable Sunbelt city such as San Diego or Phoenix.[39]

The initial burst of local growth controls rode in on the back of popular mobilizations of the 1960s and major environmental victories in the larger political arena. These growth controls segued well with challenges to large developers and governments that could be made under the National Environmental Policy Act of 1969, the California Environmental Quality Act of 1970, the federal Endangered Species Act of 1973, and the state's 1972 Comprehensive Planning mandate. The federal Clean Air Act of 1970 and Clean Water Act of 1972 were further points of leverage (see chapter 9). By the early 1980s, the growth-control movement was everywhere in the Bay Area and was spreading nationally.[40]

In the late 1980s, growth control swept through Southern California, which was in the throes of a huge housing bubble. But this time, growth limits rode in on the conservative horses of tax revolt and fear of immigrants, giving growth control a different cast in the southland. By taking away property taxes, Proposition 13 left many local jurisdictions near bankruptcy and made everyone take cognizance of the fiscal costs of growth. Suddenly, development fees became attractive even to progrowth city managers and politicians, as did a shift from housing and industrial land use to retail malls.[41] By 1990, more than half of California cities had growth-control measures, and a quarter had limitations strong enough that projects could easily be vetoed.[42] Bay Area controls remained the most numerous, however, and the most likely to alter development outcomes.[43]

Edge Cities

Yet even as local citizens were coming to grips with rampant urbanization, the nature of California's great metropolitan areas was changing. They were growing larger and denser, meaning that the commuter fringes were moving more than 50 miles out, the central cities were reviving as office complexes, and old suburban nodes were becoming edge cities.

Contra Costa County exemplifies the changes set in train as the Bay Area shot up from five million to six million residents during the 1980s. The urban dam once provided by the East Bay hills broke down completely. Central Contra Costa had been a genial suburban realm of 150,000 people in 1960, with the eastern part of the county beyond Mount Diablo still thoroughly agricultural. All through the hills and dales, cattle roamed on large ranches. But in the 1980s, the corridor along Interstate 680 was transformed by corporate-office decentralization from San Francisco to Concord, Walnut Creek, and San Ramon. Population exploded from 350,000 in 1980 to nudge one million by the end of the twentieth century. The I-680 corridor was written up as one of country's most dynamic edge cities. Meanwhile, housing developments spread out around the office clusters into the old ranchlands in the foothills and up Mount Diablo, while the outer fringe of sprawl moved out Highway 4 to Antioch and Brentwood, gulping up farmlands. By 1990, Contra Costa had more (and a higher proportion of) urbanized land than any Bay Area county other than San Francisco.[44]

Walnut Creek became the hub of the transformed county. Fortunately, it had hired an enlightened group of city planners, who forced developers into a high-density, nodal form of office building. Moreover, voters passed Measure H in 1986 to keep growth within the capacity of the transportation system. Suddenly the little city had "perhaps the most sweeping growth control ordinance ever adopted in a major California city." Thrown out by courts on a technicality, the measure was reinstated by the City Council in 1988. As a result, today Walnut Creek has a thriving downtown for business, culture, and shopping. Concord followed suit, going from a faceless suburb to a well-planned town with a revitalized core, and Pleasant Hill was not far behind.[45]

Elsewhere in the county, things rapidly got out of hand. Green activists began to fight developers and cities to bring some measure of control over urban sprawl. In the process, Save Mount Diablo had to retool itself from park advocate to the people's planner. The turning point was the fight to stop Blackhawk in 1973. Blackhawk was the first suburban development to lap at the foot of Mount Diablo, as well as the first of a new generation of gated communities for the nouveaux riches. It was the work of a Florida developer, Frederick Behrens, who moved into California and built a home on a nearby hilltop. Save Mount Diablo threw a monkey wrench into his plans. To stave off popular opposition, Behrens had to drastically reduce the number of homesites, surrender a huge piece of ridgeline to the state park (the largest land dedication extracted from any developer in California up to that time), and agree to pay off-site mitigation to the tune of millions of dollars.

Since then, Save Mount Diablo has monitored all large-scale developments in eastern Contra Costa County, forcing developers to provide quid pro quo to get city and county permits. Sometimes this means dedicating part of the property to open space under the Quimby Act, and sometimes it means paying for mitigation of wildlife disturbance under the Endangered Species Act. Tens of thousands of acres, including Clayton Ranch Open Space and Brushy Peak Reserve, have been saved by these tactics. Save Mount Diablo is usually involved in selecting lands for dedication or mitigation. In recent years, developers regularly approach the organization's staff to work out compromises *before* their permits come to public hearing. The group even operates as a land trust, doing pass-throughs to the state and East Bay Regional Parks District—a rare combination of conservancy and land-use advocacy.[46]

When Seth Adams began his long struggle for open space in Contra Costa County in the early 1980s, his Berkeley friends chided him for taking up a lost cause. But Adams's timing was right. Hired in 1988 as the first paid staffer at Save Mount Diablo, he began pushing for an ambitious open-space program in the eastern county, where the development wave was just hitting. That ambition has borne fruit. Today there are some twenty parks and reserves east of Mount Diablo and a total of more than 150,000 acres in public open space in Contra Costa County from Point Pinole to the delta—counting regional parks, state parks, city parks, and public watersheds.

The Thin Green Line

Urban growth boundaries, or green lines, became a popular means of local growth control in the last decade of the twentieth century. Although San

Jose had something like this in the early 1980s, it was not firm. The real inspiration came from Oregon, which in 1973 passed the country's most comprehensive state-mandated land use control act. Salem became the first city in America to adopt an urban growth boundary, in 1979, and Portland the first big city to do so, in 1980. Both have been very successful.[47] In the 1990s, Greenbelt Alliance put out guidebooks to help local activists pushing for urban growth limits, and the Association of Bay Area Governments began giving grants to cities undertaking demonstration programs for urban limit lines. Seventeen cities around the Bay Area imposed green lines in the following decade, often by popular initiatives; this was triple the rate statewide.[48]

In the North Bay, Sonoma County saw a wave of green lines imposed from Petaluma to Healdsburg; by the millennium, every town in the county except Cloverdale had a growth boundary. Marin's last progrowth bastion, Novato, put the clamps on, as did the city of Napa. In Solano County, which blossomed from 250,000 people in 1980 to more than 400,000 in 2000, bringing developers to heel was no mean trick; but Benecia—which had draped little boxes all over the Carquinez Straits—adopted a limit line, and booming Fairfield followed; Vacaville and Dixon in the north county were the only holdouts.[49]

In Contra Costa County—the worst offender for unchecked growth—the cities have proven largely immune to green lines. The worst of these—Antioch, Brentwood, and Oakley—are in the eastern county, but inner suburbs continue to approve new developments, such as Gateway in exclusive Orinda. This is why Save Mount Diablo has had to go head to head with developers. In 1990 it joined with Greenbelt Alliance and the Sierra Club Bay Chapter to try for an initiative to impose urban limit lines at the county level. It lost, but a highway tax measure passed, tying development permits to road capacity and other infrastructure, on the Walnut Creek model. The greens used it to slow annexations, working with the county LAFCO. At long last, the Board of Supervisors in 2005 began to hold the line on cities seeking to expand their nominal urban limits, but a compromise county growth boundary lost at the polls.[50]

In eastern Alameda County, things were looking bad for growth control by the end of the 1980s, thanks to spillover from Silicon Valley and the I-680 corridor. Dougherty Valley was being built out for another 30,000 residents, with more tracts planned on nearby ranches. Pleasanton and Dublin had puffed up into major towns, and Livermore politics had reverted to progrowth. Then a new revolt began in the mid-1990s. Even the Blackhawk Homeowners Association turned antigrowth. A Citizens Alliance for Public Planning formed across the Tri-Valley area in order to push for voter approval on all major new developments. When that lost, Pleasanton voters passed a green-line initiative in 1996 and a moratorium on general-plan revisions three years later.

The county Board of Supervisors installed a new general plan that included a growth-limit line, rescinded it, and then readopted it. Fearing further vacillation, Sierra Club's David Nesmith, Greenbelt Alliance's Mark Evanoff, and John Woodbury of the Bay Area Open Space Council promoted a county proposition in 1998 that locked in the general plan's limit line and installed the largest minimum lot sizes in the state in the ranching areas beyond the line. This transformed the land market in the county, because ranchers no longer expected to sell out for subdivisions. Furthermore, the city of Livermore installed an urban-growth boundary by citizen initiative in 2003, while a local land trust began reserving land south of town for vineyards. The fate of the hills north of Livermore remained very much up in the air.[51]

In the South Bay, there was a stronger history of planning controls. Over the course of the 1990s, green activists were able to secure a restrictive general plan for Santa Clara County, as well as for key cities. Seven towns installed urban-growth boundaries and open-space protections. The county plan put a buffer zone between the fast-growing south-county towns of Morgan Hill and Gilroy and firmed up controls to keep one jurisdiction from undermining the good works of another. The county also created a special zone for ranchlands, covering more than 100,000 acres in the hills. The Santa Clara Valley Manufacturers Group became a strong supporter of greenbelt planning, declaring that "These green hills are gold."[52]

Nonetheless, the best plans still have to be defended against special amendments as new developments come to light. Sergeant Ranch in southwest Santa Clara County has had two major proposals that have been turned back at the last minute. Hillside protections and ranchland subdivision controls were tightened again in 2006. And even where there are green lines, some superb open space can lie inside the line, triggering new battles, such as the one over the Almaden Valley in San Jose. The biggest of these is Coyote Valley, at San Jose's southern boundary, which became the rallying cry of South Bay growth control in the early 2000s.

In 1999 the city of San Jose approved a gigantic corporate park in Coyote Valley for the new headquarters of Cisco Systems (which briefly passed Microsoft as the world's most valuable corporation). Up to that time, the only significant development there was an IBM corporate campus, which had gone in forty years earlier. The uproar was so loud that Cisco was forced to cut a deal to pay out millions of dollars for thousands of acres of open space elsewhere in the county. The city of Salinas also successfully sued San Jose to provide more nearby housing to offset distant commuting from Monterey County. A referendum to confirm San Jose's green line passed in 2000 with an extraordinary four-fifths of the vote, thanks to a new group, People for Livable and Affordable Neighborhoods (PLAN). Then the telecommunications bubble collapsed, and Cisco cancelled its plans. Coyote Valley has not been saved, however. The fight goes on, with Greenbelt Alliance, Committee for Green Foothills, and local activists pushing to set aside more of the valley, reserve the Coyote Ridge, and concentrate development in a smaller area.[53]

THROUGH THE ROOF

The long postwar era of prosperity and rapid urbanization ended with a crash in the early 1970s. When the economy revived in mid-decade, things had changed profoundly. Beset by international pressures, the United States had suffered a sharp decline in corporate profits and a fall in the dollar. These triggered a period of rapid inflation, as companies tried to make up for lost ground. Housing prices also rose quickly, doubling across the country from 1974 to 1982—and in California they doubled that, propelled by real interest rates that dipped below zero. The era of housing bargains was over; the age of pitched battles over property and taxes had begun. A new landscape of the city and of urban politics was in the making.

Counterattack

The outburst of local growth controls led to panic among developers and their house intellectuals, and the Bay Area became a target of budding neoliberal criticism of the burden of regulation. As planner David Dowell observed at the time: "Nowhere else in the nation do communities so aggressively restrict development." A torrent of real estate and law journal articles and books rolled off the presses, turning the tide from considerations of the costs of *growth* to the costs of growth *controls*. A particularly dyspeptic attack was by Massachusetts Institute of Technology planning professor Bernard Frieden, in *The Environmental Protection Hustle.*[54]

The sirens of housing supply argued that growth controls choke off the flow of available land, driving up housing prices and excluding the poor. But their case rested on a dubious correlation of rising housing prices with the presence of more growth controls, with little rigorous testing.[55] In fact, house-price and rent increases were quite general across the United States in the 1970s, and inflation was rampant across *all* commodities, not only housing. So the argument had to rely on the fact that house prices shot even higher in California and were highest in the Bay Area. The general reason for that was not growth controls but because California was overtaking the eastern industrial centers as the growth engine of the country.

The Bay Area has had higher housing prices than any other major metropolis in the United States over the last thirty years for three main reasons. First, it is home to the world center of high-technology industries and has enjoyed robust job growth and in-migration, far in excess of housing stock during boom periods. Second, it is the richest large metropolitan area per capita, including the most billionaires, with new wealth pouring forth from stock options and instant capitalization of start-up companies. And third, it has been hit by huge influxes of finance for speculative purposes, in the savings-and-loan bubble of the 1980s and the NASDAQ frenzy of the 1990s (when the region's high-tech companies absorbed one-third of all stock increases across the country). Even so, Bay Area housing costs were passed up by Los Angeles's in the late 1980s and again in the early 2000s when Southern California basked in the limelight of a robust economy, massive immigration to jobs, and thunderous financial speculation in mortgages and housing—even though L.A. has much weaker growth controls than the Bay Area.[56]

This is not to say that supply restrictions cannot drive up housing prices; they can.[57] But the case against environmental growth-control measures is weak. Only the most stringent growth caps have slowed growth or affected house prices. More commonly, the limits are too late in coming, too capacious, poorly implemented, modified for favored developments, or evaded by clever developers.[58] Developer fees have little supply effect but add a price premium; at the same time, they provide for adequate future infrastructure to accommodate new housing starts. Permit delays, much excoriated in the business press, have not been a long-term problem; most resulted from a new level of regulation that had not been regularized by the bureaucracies (today, cities handle multiple permits in a much more competent manner) or from developers' own recalcitrance to meet environmental standards (now also more regularized).

The biggest single restriction on land supply is not growth controls but large-lot zoning, which spaces out housing and prevents multifamily housing from being built. Large-lot zoning preexisted the environmental revolt and has more to do with social exclusion than growth control. The mouthpieces of the development industry are happy to conflate two very different things—exclusion and environmental protection—and to cover their pecuniary interests with the mantle of civil rights. The trick has worked to confuse matters, especially given the class and political divide between traditional conservationists and movements of the oppressed. Many people of color, in particular, are firmly opposed to growth limits because they think all such restrictions push up costs of housing. But the exclusionary case does not apply that well to the Bay Area's greens—in contrast to New Jersey's suburban enclaves or Southern California's bastions of the "homeowners revolt."[59]

By the late 1970s, a massive counterattack was mounted by developers, property owners, and the state on the legitimacy of land-use regulation and property taxation. The great watershed is, of course, Proposition 13—written by Howard Jarvis, a lobbyist for apartment owners in Southern California. The counterattack grew stronger through the 1980s, with pressures to streamline regulations, reduce development fees, neutralize environmental impact reports, and declare regulations to be takings.[60] This resistance blunted but did not stop the revolution in land-use controls in the Bay Area.

Toward a Green Urbanism

People for Open Space took up the cause of housing more seriously in the 1980s, under the leadership of Allen Jacobs, former San Francisco city planning director and UC planning professor. House prices in the Bay Area had become a sore point by the late 1970s and were an important source of homeowner support for Proposition 13's limit on property taxes. People for Open Space countered with a 1983 report, *Room Enough*, backed by a sophisticated analysis of the causes of housing inflation and options for more responsible development.[61]

They realized, as John Woodbury of the Bay Area Open Space Council puts it, that "you have to have an answer for housing other than 'send them to Tracy.'"[62] The solutions are housing and transportation policies that emphasize density, access to mass transit, and mixed use. In the 1990s, these came under the rubrics "smart growth" and "new urbanism." These popular buzzwords refer to ways of counteracting sprawl fueled by single-family housing and auto-driven transportation by building more-concentrated housing that re-creates the urbane feel of older cities, with lots of foot traffic, street life, social mixing, and visual variety.

Some of the most eloquent practitioners of Smart Growth and New Urbanism, such as Peter Calthorpe, Christopher Alexander, and Daniel Solomon, are from the Bay Area, and they helped found the Congress for a New Urbanism.[63] Smart Growth has become the slogan of ABAG, the Metropolitan Transportation Commission, and the Bay Area Council; dozens of state legislators formed a smart-growth caucus; and the Sierra Club has a National Committee on Sprawl that is promoting smart-growth policies. Many of the older Bay Area suburbs, such as Los Altos and Walnut Creek, have tried to reconfigure their banal downtowns in a more urbane mold. Transit villages have sprung up near BART and CalTrain (commuter train) stations, such as the Crossings in Mountain View and Fruitvale Transit Village in Oakland. High-rise in-fill projects have been built by the hundreds from Vallejo to San Jose.

The Greenbelt Alliance took up the cause of Smart Growth with a passion in the 1990s, but its declarations also emphasize fair and affordable housing. The organization did a report with Bank of America and the Low-Income Housing Coalition encouraging compact development, then launched programs for affordable housing with Urban Ecology and the Silicon Valley Manufacturers Group. Greenbelt Alliance backed a billion-dollar state bond initiative in 2002 to fund affordable housing and first-time home-buyer assistance.[64]

Smart Growth is, in one sense, the natural ideology of the new, denser, postsuburban cities and a handy wish list of how to increase urbanity by people who dislike suburbia. But it can also be a cover for new forms of exclusive enclaves and new ways of tearing down working-class and minority neighborhoods, if it doesn't come with social programs and community involvement.[65] Of course, most Americans disdain density and love their cars, so it is not surprising that resistance to smart-growth planning is endemic. Calthorpe's sweeping plan for the Marin-Sonoma transit corridor lost at the polls. Hayward never adopted Solomon's dramatic downtown plan. Orinda and Lafayette refuse to consent to transit villages at their BART stations. Even Berkeley fights endlessly over high-rise redevelopment. Affordable housing almost always runs into opposition. In short, Smart Growth is still weak tea compared to the metro-topian planning ideals of the mid–twentieth century.

GREENBELT ALLIANCES

By the 1980s, the greenbelt had achieved a remarkable presence around the Bay Area, in spite of the failure of regional government, the fragmented nature of local growth protests, and fierce opposition from developers. This success was a matter of vision and strategy, institutions and funding, and the widespread acceptance of the greenbelt idea among the elite, local governments, politicians, and the general public. When the visionaries at People for Open Space assessed the defeats of their grand metro-topian schemes, they realized that the stealth pressure they had long been putting on local governments was yielding more tangible results. Something like a regional greenbelt was being built piecemeal, from the bottom up.

Think Regionally, Act Locally

In 1985, longtime POS board member Bill Evers took the lead in launching a new coalition of activists called Greenbelt Congress to develop local strategies. POS staffers Mark Evanoff and Jay Powell became mobile field organizers. In 1987, Evers merged it back into the parent organization, which changed its name from People for Open Space to Greenbelt Alliance. Evers's strategy was to engage directly with local governments and activists through decentralized organizers. The first field office went in at San Jose in 1988, under Vicky Moore, who helped ignite the green movement in the South Bay. Over the next decade, Moore assisted with launching a Santa Clara

County Open Space Authority, monitoring a new generation of general plans, and pushing for urban limit lines as well as establishing the alliance's Livable Communities Program. In the 2000s, the focus has been on protecting the hills and agricultural reserves in the south county.[66]

The alliance opened a second regional office, in the North Bay, in the early 1990s. Krista Shaw and Jessica Parsley assisted in the creation of the Sonoma County Agricultural Preservation and Open Space District, the 2020 initiative in Napa, and urban limit lines across Sonoma. Since 2000, Kelly Brown has worked with the Community Alliance with Family Farmers, bird-dogged Sonoma County's renewal of its general plan, and issued a report with the Sonoma County Farm Bureau on sprawl.[67]

A third Greenbelt Alliance office, in the East Bay, run by Evelyn Stivers, helped defeat the Tri-Valley Super Sewer and gain Pleasanton Ridge, Ohlone Wilderness, and other open spaces for the East Bay Regional Parks District. In the 2000s, David Reid worked on Contra Costa County open-space funding, developments in Antioch and Brentwood, and upholding the county green line. The greenbelt field reps have allied with the Contra Costa Central Labor and Building Trades councils, under Chuck Dalrymple and Greg Feere, respectively. The allies have found a common enemy in developers and big-box retailers that use non-union labor. The alliance narrowly lost a measure to block Wal-Mart from entering the county in 2004.

A fourth field office was established in the early 2000s to handle Solano and Napa counties, under Natalie DuMont and later Brent Schoradt. In Solano, success has been limited by the county's rampant growth ethic and conservative farm sector, but support is growing in the south-county cities of Vallejo, Benecia, and Fairfield. Those three have agreed to leave the 10,000-acre Sky Valley–Cordelia Hills triangle as open space, which the Solano Land Trust is buying up (with help from the Moore Foundation and the Coastal Conservancy). Incorrigible Vacaville in northern Solano County has been stymied by lawsuits by Friends of Lagoon Valley and threats of suits by Greenbelt Alliance.[68] (For Napa County, see chapter 8).

In short, Greenbelt Alliance has become extremely effective in curbing the outer reaches of sprawl by imitating the tactics of Save Mount Diablo, the Committee for Green Foothills, and other local activists. At the same time, the alliance has promoted better planning and redevelopment in established cities. As one field officer said, "When the field reps meet as a group, it seems like we're all just at different stages of Smart Growth. Out

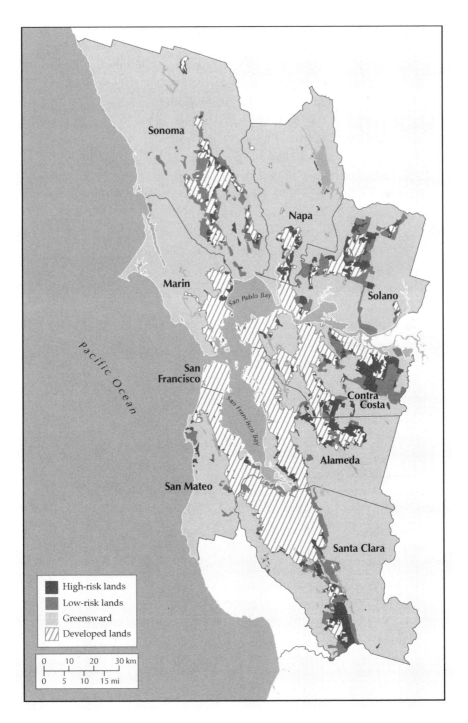

Legend:
- High-risk lands
- Low-risk lands
- Greensward
- Developed lands

0 10 20 30 km
0 5 10 15 mi

MAP 10. Bay Area Land at Risk for Development, 2001. Drawn by Darin Jensen. Courtesy of Bay Area Open Space Council, GreenInfo Network, Greenbelt Alliance.

here [in Solano County], we're still trying to get urban limit lines, while in other places they're doing livable cities."[69] A striking thing about the field reps of Greenbelt Alliance is that they are the bridge to a new generation of young activists; less surprising is that most are young women.

Another brilliant strategic stroke has been the periodic assessments of the state of open space, in Greenbelt Alliance's "Lands at Risk" reports, introduced by Larry Orman and Jim Sayer in 1988. The reports highlight the contrast between urbanized areas, secure greenbelt, and places facing imminent threat of development. The reports, which have proved to be useful media events and political tools that demand a response from local officials, include a report card on which counties make progress and which fall back between assessments.[70] By 2000, the mood was surprisingly upbeat. As one official commented in the *San Francisco Chronicle*, "Compared to other places around the country, the Bay Area is in good shape. If you go to Texas, for example, it's a total mess."[71] In the 2000s, the amount of land at risk actually started to fall.

Ribbons Around the Bay

Another unifying tactic is to create trails circumnavigating the whole Bay Area. The first is the Bay Area Ridge Trail, an idea that goes back to the Olmsted-Hall Report of 1930 (see chapter 3). In 1979 the East Bay Skyline National Recreation Trail was christened under the National Trails System Act of 1968—one of the few such trails outside of the national parks— and it rekindled the idea of a Bay Area Ridge Trail. Then Brian O'Neill, who had worked on the national trails system for the National Park Service, became superintendent of the Golden Gate National Recreation Area and took up a challenge from William Penn Mott to make the Bay Area Ridge Trail a reality. O'Neill teamed with Greenbelt Alliance and Bill Lockyer, a liberal state senator from working-class Hayward, to get the plan through the Legislature.[72] The Bay Area Ridge Trail, officially launched in 1987, is overseen by the Bay Area Ridge Trail Council, representing local governments, land trusts, and citizens' groups. The full 500-mile circuit atop the Coast Ranges flanking San Francisco Bay passed the halfway mark in 2000.

In the 1987 legislation, ABAG was given the charge of planning a Bay

Shoreline Trail to encircle the entire San Francisco Bay at the water's edge. That project, too, has been making steady headway. Just over half of its planned 500 miles is in place, thanks chiefly to the Bay Conservation and Development Commission's work over the years in gaining public access to the shoreline. But finishing the job will be challenging and expensive. Unlike the ridge trail, the shoreline trail must traverse a built-up landscape of almost fifty cities and thousands of individual landowners. In 2005, ABAG estimated that another $200 million would be required to finish the job.[73]

In short, the Bay Area has come a long way toward realizing the greenbelt visions of Frederick Law Olmsted Jr., Jack Kent, and Dorothy Erskine. It has been done in bits and pieces in one locale after another, and it has been done with an array of tools. It has been done by grassroots struggle and government, by militant stands and back-room bargaining. Behind it all, the battle for a greenbelt has been deeply influenced by the Modernist, liberal ideology of the Bay Area's metro-topian intellectuals and greenbelt planners. As a result, the movement to harness city growth has not been a grab bag of selfish defenses of local turf, but a collective project greater than the sum of its parts.

Nonetheless, for all the good work of the greenbelt movement, building has continued to squirt out into virgin territory, where new lines of defense of open space have to be drawn. The sharpest fights against sprawl in the late 1990s shifted to eastern Contra Costa, southern Santa Clara, and northern Sonoma counties (see chapter 8). The region has become far more distended now than it was when the term "metropolis" was coined in the mid-twentieth century. How can we even imagine, much less grapple with, this urban giant in the coherent way the Modernist utopians once did?

7

FASTEN YOUR GREENBELT

Triumph and Trust Funds

I n the last quarter of the twentieth century, the Bay Area grew at break-
neck speed, propelled by the microelectronic revolution taking place
in Silicon Valley. Population mounted ever higher, passing five million
by 1980, six million by 1990, and nearly seven million by 2000. Whole new
rings and flares of settlement were added to the metropolis as it spilled over
the hills encircling San Francisco Bay and rushed up and down the valleys,
reaching well into California's great Central Valley. The challenge to environmentalists was daunting, as rural landscapes were overrun at a faster pace
than ever and land values shot up astronomically.

Meanwhile, government powers, funds, and legitimacy were fast eroding
in the age of neoliberalism ushered in by the tax revolt in California and the
Reagan adminstration in Washington, D.C. Key tools of open-space protection such as state parks, land-use planning, and environmental regulation
were blunted in the 1980s, forcing the greens to lean on newer methods.
Open-space districts and land trusts became the leading edge of greenbelt
conservation efforts by the 1990s. These have proved effective, and often
more nimble, means to the same ends of acquiring parklands and public
open space.

Not only have activists and agencies been innovative in their methods
of fund-raising and land acquisition, but the Bay Area electorate has been
astonishingly willing to tax itself an extra measure for the benefit of the

greenbelt. At the same time, priorities have changed among environmental advocates over the last generation, putting more emphasis on setting aside wildlife habitat and conserving working landscapes, especially ranchlands, rather than recreational and public-access spaces.

Nevertheless, the recent turn in procuring open space in the Bay Area raises some troubling concerns. It makes the best of a bad situation, often without questioning the premises of an age of greater wealth inequality, weakened government, and diminished public spirit. It means depending on the goodwill of the rich, secretive land trusts, regressive taxes, and private property owners, rather than assuring the people's right to public space or insisting on good government rather than less government. It also favors habitat and scenery over recreation, often with an unfavorable class bias in accessibility.

SILICON VALLEY FEVER

By 1975, Silicon Valley—as the South Bay came to be known at the time—had emerged as the premier electronics cluster in the world. High-tech companies multiplied, driven by the rapidly expanding powers of microelectronics and the explosive growth of industrial and consumer markets for personal computers, business networks, and telecommunications. Venture capital and a culture of entrepreneurship provided the business stimulus, but the real force behind the electronics revolution was the creative power of thousands of the best minds in the world, flooding into the most freewheeling center of inventiveness on earth.

The Excesses of the Eighties

Santa Clara County's population hit 1.25 million people in 1980, driven by 200,000 jobs in electronics and more than 600,000 jobs overall. By 1988 employment topped 800,000 and population passed 1.5 million. San Jose expanded rapidly eastward, and housing tracts carpeted the east side of the valley up to Fremont. Commuter tracts and trophy homes began to appear to the south all the way to Gilroy. Industrial construction surged around the south end of the bay. High-tech commercial property was subject to tremendous speculation, exceeding any other part of the nation in the first half of the 1980s. When the market went bust in 1985, it took many local savings-and-loan firms with it. Among San Francisco's big three banks, only Wells Fargo escaped; Crocker Bank was ruined and Bank of America badly damaged.[1]

By the early 1980s, the contradictions of rampant urbanization were everywhere. San Jose was one of the most indebted cities in the country, thanks to years of providing new services to all comers. Schools were in double sessions, and in 1983 San Jose Unified School District went broke— the first American school system to do so since the Second World War. By 1980, Silicon Valley rents were the highest in the country, and house prices were second only to San Francisco's—well over double the national average. High-tech companies were feeling the pinch in recruiting professional workers. Roadways were overwhelmed by traffic, and commute times were getting longer. The freeways and expressways built during the paving boom of the 1960s were at capacity.[2]

No wonder, then, that growth-control rebellions hit San Jose and a host of other suburbs around the Bay Area in the 1970s and '80s. A chorus of angry voices was raised against the rampage of the developers and unchecked urban expansion. Some took the conservative route of tax revolt, while others tried the environmental route of putting up green roadblocks to developers. In both cases, popular protest had breached the walls of boosterism.[3]

The Knockout Nineties

The recession of the early 1990s brought a pause, then Silicon Valley took flight again by 1995, as high technology led the American economy out of the wilderness of global competition. Electronics became the largest manufacturing sector in the world, and electronic information processing and communications the defining technology of the day. Silicon Valley was, more than ever, the beating heart of a multilevel, global system of innovation and production, and its leading companies, such as Hewlett-Packard, Sun, Apple, Netscape, Intel, and Cisco, were the behemoths astride the New Economy, setting the agenda for the Internet age.

Silicon Valley's heartbeat was racing from 1995 to 2000. By 1999, the electronics industry was employing 230,000 people in Santa Clara County; total employment had passed one million. Another million people worked in San Mateo and San Francisco counties combined. Investors were again feeding the fires of the red-hot economy with speculative capital. The high-tech-heavy NASDAQ tripled in value from 1996 to 2000. It was the greatest stock bubble in American history, and one-third of all the value generated was invested in Bay Area companies.[4]

Silicon Valley had become the economic center of the Bay Area, stealing San Francisco's historic claim to preeminence. The technology quake sent a tidal wave all the way up the San Francisco Peninsula, where it crashed down on the city as the dot-com bubble. A good index of the geographic flip is numbers of corporate headquarters. By 1998, Santa Clara County was home to half the *San Francisco Chronicle* top 500 companies in the Bay Area. Of twenty-two Bay Area firms on the Fortune 500 in 2000, seven were in San Francisco, thirteen in Silicon Valley, and two in the East Bay.

The Wild West years of the late 1990s brought back the good times of rapid accumulation by local developers and property investors, as land values soared to new heights. Capitalists cashed in on the fevered conditions. The figures are staggering: for example, a single broker at Cornish and Carey Commercial sold six million square feet of space worth $2.8 billion in one year, highest in the nation. The outcome was an astonishing amount of new industrial space, which brought the valley total to more than 200 million square feet by the end of 2001.[5]

Exurban developments spread south all the way to San Benito and Monterey counties. To the east, commuters were flowing in from Tracy and Stockton in San Joaquin County, and beyond. Once again, traffic congestion reached appalling levels and air quality was in decline. There was a sense of urbanization careening out of control, undoing previous efforts at containment. Yet even more startling than the sprawl was the dawn of a new era of densification of the inner Bay Area. Growth brought a new flurry of in-fill and high-rise development—most spectacularly in the shining new downtown of the city of San Jose, closing in on a million people and calling itself "the capital of Silicon Valley."

Housing prices went through the ranch-house roofs. As the *LA Times* brooded, "Is it possible the Southland bubble didn't so much pop as migrate north . . . ?" Silicon Valley was ground zero of the New Economy real-estate bubble. Santa Clara County doubled San Francisco in assessed valuation by 2000, and even San Mateo County had passed up the old core city. Home prices were climbing a percentage point a month from Marin to San Jose, with median prices more than half again higher than in Los

Angeles. Meanwhile, working-class families were packing into tighter quarters or packing up to move to the Central Valley.[6]

The Roaring Nineties threw up a whole new generation of millionaires at Netscape, Yahoo, and other high-fliers of the Internet era. Venture capitalist John Doerr claimed that Silicon Valley was "the largest legal creation of wealth in the history of the planet." The heroic millionaires of the eighties bloated up into billionaires overnight; men such as Larry Ellison of Oracle, Gordon Moore of Intel, and Steve Jobs of Apple were richer than anyone's imagining. Land developers also figured among the elite, particularly John Arrillaga, Richard Peery, Carl Berg, and John Sobrato. The Bay Area led the country in the number of super-rich on the Forbes 400 list.[7]

The hype surrounding the valley and the Internet was positively giddy. This was the birthplace of a New Economy and the cultural capital of the virtual world. Its ideology was a bubbling stew of the speculative proclivities of venture financiers, technological epiphanies of engineer geeks, and bohemian counterculture of the webbies and game boys.[8] Against this roaring torrent of invention, money-making, and sheer nonsense, environmentalists bravely kept paddling in the eddies and moving upstream with remarkable speed and grace.

LANDSCAPES WITHOUT FIGURES

Open-space districts have been an essential tool in the hands of Bay Area land conservationists since the 1970s. By 2005, open-space districts had been responsible for protecting upward of 100,000 acres of greenbelt. Presently, they are found only in four counties along the west side of the bay: Marin, Sonoma, San Mateo, and Santa Clara. Traditional parks are managed by departments of city, county, and state government, whereas open-space districts are independent branches of government chartered by the state or county and have separate boards of directors. Only the East Bay runs its parks through such an instrument. More important, however, is the difference in land management.

Open-space districts see their goal as protecting land, wildlife habitat, and scenic values. They allow limited public access for hiking and do not encourage the kind of mass public recreation found at public parks. Open-space reserves are meant to keep the city and its people at bay. Part of this arm's-length approach derives from conservation biology, which had become the principal justification for land protection by the 1980s. Under this rubric, open spaces came to be seen, first and foremost, as habitat reserves for ecological sustainability. Maintaining larger pieces of land and leaving wildlife corridors between reserves are intended to keep species from becoming isolated and vulnerable to extinction. A second justification is to provide breathing space between towns and bolster green lines.

The pioneers of the genre were the Marin County Open Space District and the Midpeninsula Regional Open Space District. Marin's district was launched in 1972 in concert with the pivotal county general plan and complements the planning function (see chapter 4). Because of the way development is concentrated along the U.S. Highway 101 corridor, the district's mission has been to acquire land in the interstices to prevent towns from running together. Its holdings amount to some 14,000 acres in thirty-two parcels.[9] The Marin County Open Space District is a subordinate part of the overall Marin greenbelt, but important around the town of Novato in the northeast county.

By contrast, the Midpeninsula Regional Open Space District (MROSD) is a major player in the greenbelt of the San Francisco Peninsula. Given the modest number of parks on the peninsula in 1970, the Committee for Green Foothills and their allies saw the need for a new vehicle of land protection. Activists began meeting in the living rooms of Nonette Hanko and Mary Davey, where they revived an idea that had lain dormant since the creation of the East Bay parks: a bi-county district to protect open space.[10] MROSD required no new state enabling legislation, just the approval of San Mateo and Santa Clara counties.

The creation of MROSD was approved by the voters in Santa Clara County in 1972. San Mateo County was more reluctant and only entered the district four years later. To get past opposition from developers and farmers, MROSD's boundaries had to be restricted to the east side of Skyline Drive on the ridge top. The northern part of Santa Cruz County was added in 1992. A modest property tax, passed before Proposition 13, has yielded a good bounty, up to $20 million per year, so the district has been able to move aggressively to purchase land. By 2005, it owned almost 50,000 acres in twenty-six open-space preserves, an extraordinary record of expansion (see Appendix, table 4).[11]

The MROSD was long hamstrung by the geographic limits placed on it by San Mateo County, where the leading edge of development occurs on the

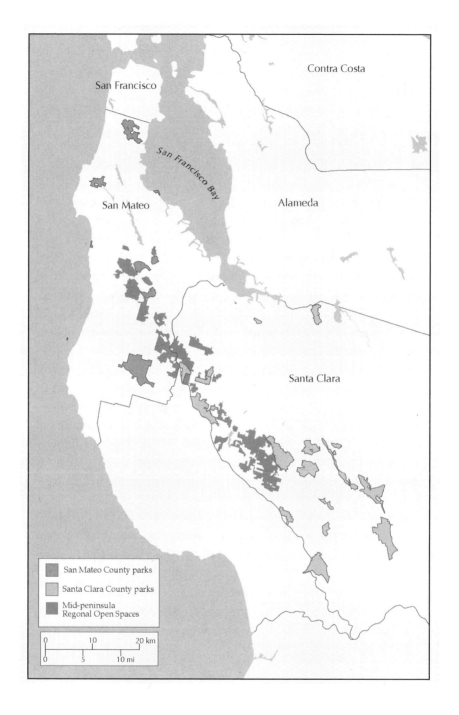

San Francisco

Contra Costa

San Francisco Bay

San Mateo

Alameda

Santa Clara

San Mateo County parks

Santa Clara County parks

Mid-peninsula
Regonal Open Spaces

| 0 | 10 | 20 km |

| 0 | 5 | 10 mi |

MAP 11. Open Space on the San Francisco Peninsula, 2005. Drawn by Darin Jensen. Courtesy of Bay Area Open Space Council, GreenInfo Network, Greenbelt Alliance.

coast. After years of trying to extend its jurisdiction to the western slope, the district gained voter approval in 1999 and final go-ahead from the courts in 2004 to extend its coverage to almost all of San Mateo County. Critical support came from the newly urbanized Half Moon Bay area, while farmers farther south still regard MROSD with suspicion.

Santa Clara County Open Space Authority, covering the southern and eastern portions of the county, was not created until 1993. Starved for funds over its first decade, it was able to protect only around 10,000 acres. The authority was established as a benefits assessment district, a conservative device created in the wake of Proposition 13, in which only property owners vote, on a per-acre basis, but measures pass by a simple majority, not two-thirds. A lawsuit by an antitax group held up the original $4 million assessment for eight years. In 2001, the greens got the Silicon Valley Manufacturers' Group to back a successful campaign for another $8 million assessment; another suit was filed and in the courts for another four years (when Contra Costa County greens tried the same trick of a benefits assessment district in 2004, they lost).

The Sonoma Agricultural Preservation and Open Space District was created in 1990 by popular vote. It is funded through a sales tax, rather than a property tax, and as of 2005 had an annual budget of about $18 million. It has accumulated holdings of nearly 30,000 acres and has transferred another almost 40,000 to public agencies for parks and refuges. Its purposes are more varied that those of the MROSD, including agricultural preservation and urban separators, and it readily acquires easements as well as fee simple properties (see chapter 8).

Elsewhere in the North Bay, Solano County greens, led by the Solano Land Trust, have been campaigning for an open-space–park district in their county. Friends of the County Regional Parks District have mustered a long list of supportive officials and citizens, but the Solano County Board of Supervisors voted down the idea because of farm opposition. Despite further negotiating between the Friends and farmers and ranchers, the proposal failed again in 2005, in large part over fears of a new sales tax on the Sonoma model.

The East Bay Regional Parks District has not been unmoved by the demand for habitat conservation and wildlands. Las Trampas, Ohlone, Sunol (formerly a park), Mission Peak, and Morgan Territory have been set aside as wilderness areas. The parks district is also keen on protecting wildlife corridors. The district is working with Save Mount Diablo and other activist groups to create a chain of parks from Mount Diablo to the southern Diablo range across the Altamont Pass.[12] Even San Francisco got in the swing of things with a proposition in the 1970s to fund a Natural Areas Program in the city's parks, from small sites such as Hawk Hill in the Sunset district to large areas such as McLaren Park on the southern flank of the city.

Limited access to open-space reserves may be justified by wildlife ecology, but it can also smack of elitism. It is not surprising that the chief proponents of open-space districts have been in the West Bay, with their more upper-class constituencies. West-side greens favor scenery and hiking over the more motorized, organized, and well-pottied recreational uses of the masses. West Bay open spaces are for light use; on the peninsula, this continues the tradition of San Francisco's closed watershed. Moreover, the open-space reserves are far away from the poor and people of color.[13] The Midpeninsula Regional Open Space District has tried to ameliorate the problem by not charging use fees, providing an educational program for schoolchildren, and asking county bus lines to serve its reserves. The East Bay Regional Parks District, by contrast, continues to be attentive to popular demand and to maintaining easy public access to most of its parks. Regrettably, the shift away from popular recreation to nature reserves goes against the spirit of public enjoyment of parklands that was so prominent in the middle of the twentieth century.[14]

REVOLTING TIMES

Just as the open-space districts were getting started, public funding shrank dramatically in the wake of California's tax revolt. The antitax, antigovernment mood swept California in 1978 with the passage of Proposition 13, which capped local property taxes, and in 1979 with Proposition 4, which trimmed state taxes (Southern California's favorite son, Ronald Reagan, cut federal taxes after his election to the presidency in 1980). Just as the rich were getting richer in the 1980s and '90s because of the New Economy and the stock market run-up, they got an extra fillip from tax cuts at every level of government.

Meanwhile, governments got relatively poorer. The loss of revenues was amplified by spending restrictions at the state and federal levels under the neoconservative Republican regimes of Reagan and Bush Senior in Washington, D.C., and Governors George Deukmejian and Pete Wilson in California. The Reagan administration turned off the tap on the Land and Water Conservation Fund, for example. Reality hit hard in the recession of the early 1980s and harder yet in the recession of the early 1990s, leaving California governments in a nearly permanent state of fiscal crisis.[15]

Park and open-space systems up and down the state suffered a blow from the fiscal clampdown. At the local level, the Midpeninsula Regional Open Space District lost one-third of its revenues, and the East Bay Regional Parks District halted all further land purchases in the early 1980s. The infant Coastal Conservancy had to get by on a pittance for the first twenty years of its existence. After years of regular expenditures and bond issues for parks, the well had run dry.[16]

The Bay Area resisted the conservative turn. Cities including Berkeley taxed themselves for libraries and schools by popular initiatives that won two-thirds margins required by Proposition 13. Public support continued to be strong for parks and open space. So Bay Area greens picked themselves up and fought back. In 1988 they won a state bond issue for parks, wildlife, and farmland preservation and a major local bond measure for the East Bay Regional Parks District. Then came a hiatus due to the catastrophic statewide recession and budget deficits of the early 1990s. When prosperity returned late in the decade, conservationists took advantage of the upswing to campaign for more funding for parks and resources.

A key player agitating for new funds was Gerald Meral of the Planning and Conservation League (PCL). After the bitter Peripheral Canal battle of 1982 (see chapter 5), Meral left government to head up PCL, where he did yeoman work for park and resources bonds over many years. The secret of success, he discovered, is to convert park bonds into part of pork-barrel propositions in the time-honored way of American politics: using lots of water-resources measures to bring that powerful constituency on board. It helped that the boom economy of the 1990s made people feel temporarily better off and more willing to pay for public goods and that California's growing constituency of people of color was strongly in favor of state and local park bonds.[17]

A real jackpot came with the passage of state Proposition 12 in 2000, for a total of more than $2 billion in resource bonds statewide, including about $350 million for parklands. Environmentalists came back again with Proposition 40 in 2002, garnering $2.6 billion, and Proposition 50 in 2003 for another $850 million. From these, the Bay Area was earmarked for roughly $55 million, $40 million, and $20 million, respectively, to be distributed through the Coastal Conservancy's bay conservancy program. The financial impact was dramatic. More than $100 million in acquisition grants was funneled through the conservancy from 1999 to 2004, leveraging another $100 million from other state funds and $150 million in federal grants.[18]

Remarkably, then, the Bay Area greenbelt was strengthened by a renewed flow of public funds to park and open-space agencies. But the state's fiscal roof fell in again during the recession of 2001–04. The new governor, Arnold Schwarzenegger, refused to raise taxes to cover massive deficits, using the state's borrowing power just to meet short-term obligations. Hence, the ax fell right and left on the operating funds of parks, commissions, and the rest of the state apparatus that business Republicans find repugnant. The horizon looked bleak.

Locally, the wolf was at the door again. The East Bay Regional Parks District failed to secure a two-thirds vote for increased taxes in 2002, losing badly in eastern Contra Costa County. There was another sweeping measure put to Contra Costa voters in 2004, an attempt to get a benefits assessment district such as that won by Santa Clara County's Open Space Authority in 2002. The park district, Save Mount Diablo, Greenbelt Alliance, and the cities put together a grand plan made up of all their park, open space, and recreational demands, so as to build the widest possible coalition of voters. They were opposed by a conservative taxpayers' group and the *Contra Costa Times*, which called it, falsely, an illegal tax—echoes of the lawsuits in Santa Clara County. The weighted—by property value—vote narrowly lost, despite polls showing that a popular vote would have carried by a large majority. The truth is that the whole exercise of seeking a benefits assessment district instead of normal property tax is a perversion of democracy in the name of property and Proposition 13. Such is the dead hand of capital weighing on the hopes of the living.

IN LAND WE TRUST

In the last generation, land trusts have become the new instrument of choice for many conservationists. Former Sierra Club executive director Michael Fischer has called the trusts "the strongest arm of the conservation movement."[19] By 2000, there were more than 1,200 land trusts across the country, ranging in size from small fry with a single land holding to the huge Nature Conservancy. Nationally, small land trusts had protected 6.5 million acres, while The Nature Conservancy had saved 12 million acres in the United States and much more internationally. The second largest, Trust for Public Land, had bought and transferred almost two million acres. Four-fifths of that activity took place in the 1990s alone.[20]

The Logic of Land Trusts

The land trusts have grown for the same reason that park funding and state regulation have been cut back: tax cuts have undermined local and state land conservation agencies and put more money in private pockets. As private wealth soared in the 1980s and 1990s, it left a staggering surplus for philanthropy. Most of this money funneled through charitable foundations such as the Packard Foundation, and a tidy sum passed through land trusts. Among environmental organizations, The Nature Conservancy was far and away the biggest recipient of this largess; it has amassed some $3 billion in assets and has an annual budget ten times that of the Sierra Club's.[21]

Land trusts are mostly private, nonprofit organizations whose purpose is to save land by buying it. Most small land trusts come into being through bequests of the wealthy to protect special parcels; these smaller trusts usually hold onto the land and manage it according to the wishes of the founders. Large land trusts, by contrast, usually pass along land they have purchased to public agencies. The Nature Conservancy is more ambitious and owns around two million acres in 1,400 preserves, the largest private system of land reserves in the world.[22] There are a few government land trusts and conservancies, chartered by states and counties, which buy land to pass along to other government branches.

The basic principle of land trusts is straightforward. In a society based on private property, the best way to secure control over land is to buy it. In this view, government regulation is not enough. But from there, the reasoning diverges for the two kinds of trusts. Local trusts are mostly guardians of a particular heritage and are inclined not to have confidence in government to do a good job in perpetuity. Even such upstanding entities as universities and wildlife services fail regularly to follow the wishes of private donors. Local conservationists have lots of horror stories to back up these fears.[23]

The large-scale trusts, on the other hand, act as partners with government land-management agencies. They engage in "pre-authorization" purchases of property that park departments or open-space districts have their eyes on but haven't yet secured the funding to buy. The logic is that a private land trust is more nimble than a public agency, with its bureaucracy, political balancing, long-range plans, and funding cycles. Trusts can work more closely with landowners without public scrutiny—an important consideration for many rich sellers who value privacy, discretion, and tax deductions. A trust can take an option on a parcel, buy it, and hold it for a period, knowing that eventually the public agency will take the land off its hands. This means a land trust can operate as a revolving fund. Such pass-throughs to public agencies are more popular in the West than on the East Coast.[24]

A variation on outright ownership is the purchase of development rights, or what are called conservation easements. This allows relatively low-impact use, such as grazing, to continue but prevents intensive development. A conservation easement has the advantage of being much cheaper than fee-simple purchase but usually cannot be conveyed to public agencies, which require unencumbered title. Conservation easements have become more popular because of the shift in philosophy away from preservation toward stewardship of land with human activities in place. Property owners like easements because it allows them to continue using the land while financing improvements or gaining tax advantages.[25]

The use of private money to purchase public parks is an idea that was pioneered in New England. The first land trust, established in Massachusetts in 1891, was the work of Charles Elliott, a former partner of Frederick Law Olmsted, son of a Harvard president, member of the Appalachian Mountain Club, and creator of the Boston Metropolitan Parks system. Two other land trusts followed in Connecticut four years later and in New Hampshire a decade later. From there, the model was copied by the English, in their National Historic Trust, but it did not catch on in the United States.

Out in California, the idea seems to have been reinvented independently.[26] The Sempervirens Club used contributions to set up Big Basin State Park, and William Kent donated Muir Woods to the federal government. Save the Redwoods League took the practice further, systematically buying private redwood groves and turning them over to the state. Conservation easements came late to California, with the Scenic Easement Act of 1959 and Conservation Easement Enabling Act in 1975.[27]

The land trust idea was idled by the Great Depression and the decline in great fortunes. It revived after World War II, most notably in the case of The Nature Conservancy, founded in 1946 by a group of activist professors from the ecologists' professional association. The founders gave the organization a strongly ecological bent, but it really took off after 1950 under the tutelage of heiress Katharine Ordway.[28]

Since 1950, the number of land trusts has just about doubled every decade. The most dramatic expansion of funding, however, came in the 1980s, as corporate and foundation money rolled in. The Nature Conservancy ballooned from $58 million to $222 million in annual income over the decade. With hundreds of trusts scattered across the country, there was a need for a trade group, so the Land Trust Exchange was launched in 1981 under the auspices of the Lincoln Land Institute and four local trusts (including the Napa County Land Trust). Now called the Land Trust Alliance, it has published a number of practical monographs advising trusts, donors, and attorneys, and it is beginning to set professional standards for the sprawling world of land trusts.[29]

Bay Area Trusts

The Bay Area has land trusts of all sizes operating on every flank of the greenbelt. They have been responsible for securing more than 250,000 acres—a larger area than in either state or county parks. Somewhat less than half of that has taken the form of easements, more than half as pass-throughs to various public agencies. As a sign of the leading role of trusts recently, about half of *all* new land protection in the region since 1990 has been in conservation easements.[30]

The Nature Conservancy (TNC) has been active in California since 1958 and still has many projects around the state. TNC opened its first West Coast operation in 1963 and made one of its first large-scale interventions in the Forest of Nisene Marks. Since 2000, it has been active around Mount Hamilton (where it has protected 20,000 acres by easement and 60,000 by acquisition), along the Consumnes River in the delta (where it is preserving farms that benefit waterfowl), and in Napa County (a planning project). The Trust for Public Land (TPL) has participated in more than 100 projects of varying size in the Bay Area, adding up to almost 25,000 acres (more in easements). TPL started a major California Coastal Campaign in the 1990s, which acquired the large Coast Dairies property in Santa Cruz, and it has helped in saving Mount San Bruno. It also has several smaller, inner-city projects.[31]

The largest of the local land trusts are in the West Bay: Peninsula Open Space Trust (POST) and the Marin Agricultural Land Trust (MALT). POST favors outright purchases, with around 50,000 acres in San Mateo and Santa Clara counties passed along by 2005 to the Golden Gate National Recreation Area, state parks, San Mateo and Santa Clara county parks, and the Mid-peninsula Regional Open Space District. It also held 50,000 acres in easements. MALT holds only easements, covering 38,000 acres in 2005, roughly a third of the grazing land in northern Marin County.

The North Bay has taken to land trusts in a big way, as well, with the Sonoma Land Trust (1973), Land Trust of Napa County (1976), and Solano Land Trust (1986). In 2005, Sonoma Land Trust had protected roughly 17,000 acres, 5,000 of those in easements. The Land Trust of Napa County held easements on 23,000 acres, had passed along 12,000 to public agencies, and retained 5,000 acres. The Solano Land Trust had around 6,000 acres—mostly in easements. The Sonoma County Agricultural Preservation and Open Space District has acted more like a land trust than a public parks agency, with its 30,000 acres of easements and 40,000 in pass-throughs. Altogether, the North Bay holds more than half the Bay Area greenbelt easements. The East Bay has only a couple small entities—the South Livermore Valley Agricultural Land Trust (now Tri-Valley Conservancy) with 4,000 acres in easements and Muir Heritage Trust—because it relies almost wholly on the East Bay Regional Parks District.[32]

A Civic Trust

The Trust for Public Land was originally a spin-off from The Nature Conservancy. Huey Johnson was hired as TNC regional director in 1964. He moved out west to Mill Valley and was soon steeped in the Marin conservation milieu. During the Marincello fight in the late 1960s, Johnson met attorney Marty Rosen, who hailed from Los Angeles, and the two of them founded TPL in 1972. They bolted from The Nature Conservancy out of frustration with its disinterest in urban green spaces. They were backed by such Marin green stalwarts as Putnam Livermore, Alf Heller, and Martha Gerbode, along with a $10 million line of credit from Bank of America.

Rosen would stay on as executive director of TPL for eighteen years, while Johnson became the state resources secretary under Governors Jerry Brown and Pete Wilson. TPL is headquartered in San Francisco.

TPL's first intervention was in Oakland, where it worked with the Black Panthers on community gardens, and persuaded local finance companies to donate lots in foreclosure; its first land purchase was in Los Angeles for a city park. Its Urban Land Program, run by Peter Stein and Lisa Cashen branched out across the western United States, buying spaces for community gardens, city playgrounds, and tribal reserves. At the same time, it has a Natural Areas Program for conserving large tracts at the urban fringe. TPL keeps no land, passing it all over to public agencies and districts. It is not bound to the ecological model and has always seen the environment as indelibly connected to cities. For this innovation, Johnson received the United Nation's Sasakawa Environmental Prize in 2001.[33]

Another vital function of TPL has been encouraging the spread of local land trusts, a program originally directed by Jennie Gerard. This program was instrumental in getting Bay Area trusts off the ground in Napa, Sonoma, and Marin counties, as well as at Big Sur. In part because of TPL, California has more land trusts (130 plus) than any other state except Massachusetts. Another spinoff is the American Land Conservancy, started by Harriett Burgess. TPL also published handbooks and sponsored conferences before the Land Trust Alliance came into existence. TPL has grown into a national powerhouse, second only to The Nature Conservancy, with an annual budget of around $70 million. Yet "The importance of [TPL's] role in land trust growth doesn't seem widely realized, even within the land trust community."[34]

A Monied Trust

POST has quietly become a powerhouse in Bay Area land conservation and one of the country's largest land trusts. It began as a spin-off from the Mid-peninsula Regional Open Space District, when district directors felt that they had to move faster to save undeveloped lands. So in 1977 the board created an independent nonprofit to accept private donations and negotiate purchases of threatened parcels. Over time POST has grown into the most powerful player in peninsula real estate.[35]

POST quietly buys up properties outside the glare of the public arena. It has been effective in outmaneuvering developers to snap up large parcels, while shielding publicity-shy landowners from disclosure of financial information. Large donors and landowners chary about working with government agencies are willing to work with POST—even if it passes land on to public parks. As with other land trusts, POST uses its capital as a revolving fund and levers additional money from foundations and public sources, such as the state Wildlife Conservation Fund.

POST operates in an overheated real estate market, where it pays an average of $10,000 per acre (much more in last-minute interventions such as Pigeon Point Lighthouse). POST's capacity is clearly linked to the flood of private money pouring out of Silicon Valley. After operating with single-digit budgets for years, it hit stride with a $33 million campaign from 1996 to 1999. That laid the basis for an even more ambitious fund-raising effort launched in 2001 to garner $200 million. The Gordon and Mary Moore Foundation and David and Lucille Packard Foundation each gave $50 million, and the other $100 million was raised by 2006. This kind of money puts POST on an elevated plane, on par with the national conservancies.[36]

Since 2000, POST has focused on protecting San Mateo's magnificent coast. The land trust has been active from Pacifica to Point Año Nuevo. In 2001 it laid out $30 million for Rancho Corral de Tierra, 4,300 acres north of Half Moon Bay on the western flank of Montara Mountain (which will be transferred to the Golden Gate National Recreation Area over several years). POST sometimes sells back to farm operators to allow them to continue farming and to maintain a working landscape along the coast. When it does resell, POST includes deed covenants prohibiting nonfarm development because it recognizes the problems of long-term monitoring of simple easements. Some 20,000 acres are managed by POST on an interim basis, using its Land Stewardship Program to carry out experiments in restoration.[37]

POST's accomplishments are due in large part to its dynamic executive director, Audrey Rust. Rust previously worked as a fund-raiser for the Sierra Club and before that at Stanford University. She runs a tight and effective organization. Nor does it hurt to have a board of directors right out of the peninsula blue book. The status of the organization is manifest in the location of its headquarters on Sand Hill Road—the world center of venture capital and some of the most valuable real estate in the country. Nevertheless, it remains a part of the Bay Area culture zone, in which even the super-rich, including Gordon Moore, may be deeply imbued with green values. As Rust says, "The tradition here is thinking about the public good, and that sets the tone for a lot of political agendas, because the beauty of the environment is linked to our quality of life and the reason people want to live here."[38]

A Working Trust

Marin took a different tack in its use of the land trust mechanism, going after only farmlands. The Marin Agricultural Land Trust (MALT), formed in 1980, was the brainchild of Ellen Straus and Phyllis Faber, veterans of the Marin Conservation League and battles to save West Marin. Faber is a biologist and native plant advocate, and Straus was a dairy farmer who used money from selling the conservation easement on her ranch to start up the first organic dairy west of the Mississippi. The two women were assisted in their plans by TPL and Gary Giacomini, a rancher and county supervisor. MALT has received crucial support along the way from the Marin County Open Space District, the Coastal Conservancy, and the Marin Community Foundation (the Buck Fund). Nevertheless, the majority of its resources have come from state bonds.[39]

MALT is unique because of the remarkable alliance forged between Marin conservationists and the dairy ranchers of the northern third of the county, brokered chiefly by Ellen Straus. In supporting MALT, Marin's greens went beyond the traditional demand for pristine nature, leaving the working landscape of grazing in place. Conversely, the dairy ranchers gave up development rights in order to conserve their way of life. Selling easements provided a monetary boost to the dairy operators, who were feeling the pinch of competition from the San Joaquin Valley. Conservationists have also supported subsidies for dairy farmers and higher milk prices from the state marketing board. The interests of the two factions have been further unified by the shift to organic milk production and consumption, led by Albert Straus, Ellen's son.[40]

This agricultural-environmental alliance has locked in the greens' revolution in Marin County for the foreseeable future. A symbol of what was taking place in Marin was the brilliant *Running Fence* erected by the artist Christo in 1976—white cloth stretched across the open countryside of the northern dairy lands. Not only did it celebrate the landscape, it was a collective construction of ranchers, bohemians, hippie kids, and engineers. Christo's fence was Bay Area cultural politics made manifest.[41]

MALT provided a model for the formation of the American Farmland Trust in 1980. Farmland loss heated up again as an issue in the late 1970s, as farmland values and mortgage debt soared. The American Farmland Trust was founded by Patrick Noonan, former president of The Nature Conservancy, and Doug Wheeler, former executive director of the Sierra Club, and has been headed up since 1986 by Ralph Grossi, a former Marin diary farmer and director of MALT. American Farmland Trust operates nationally and raises about $10 million per year to work on better farm practices and to buy development rights from farmers who want to stay in business. But it has very little presence left in California.[42]

HANGING TOGETHER

While private land trusts have been the news in land conservation, the need for government and regional governance has not gone away. The Coastal Conservancy has been instrumental in keeping government money flowing to land trusts and park agencies, while the Bay Area Open Space Council has provided a united front for the myriad open-space agencies of the region and has proven a worthy advocate for open-space funding at the state level.

Public Trust

The Coastal Conservancy was born in 1976 out of legislative maneuvers to make the California Coastal Commission permanent. The Coastal Conservancy was not an element of the original regulatory scheme but added by legislators who had cottoned to the notion of land trusts. It operates as the "good cop" in relation to the "bad cop" of the state Coastal Commission and avoids the hostility of the business interests that oppose regulation. It stays outside the pale of Sacramento politics by having its office in Oakland.

The conservancy has helped save tens of thousands of acres along the California coast. It holds very little land—about 3,000 acres around the Bay Area. It operates in the same manner as the private land trusts, even though it's publicly funded. It is more nimble than a line agency and can step in to buy key parcels for the state, city, or county park systems. More often, it acts like a public foundation and makes capital grants to private land trusts to aid their acquisitions. It also supports nonprofits doing things such as wet-land restoration, public piers, or trails. The conservancy always has an eye out for ways to leverage its grants with matching funds from foundations, trusts, park agencies, or state programs.[43]

Bay Area open-space advocates saw the virtues of the Coastal Conservancy and wanted some of its magic for themselves, so they went to work on the Legislature. The conservancy was all for it, as was the Bay Conservation and Development Commission, which wanted the advantages the Coastal Commission enjoyed by having a sister agency working behind the scenes. A joint conservancy would also alleviate the competition for state funds among Bay Area open-space agencies. Rather than create another body, the greens pursued the stealth strategy of expanding the purview of the Coastal Conservancy to include the Bay Area. Quietly, a San Francisco Bay Conservancy was set up as a programmatic arm of the parent organization, allowing it to operate in the inland portions of all nine counties around the bay. That program has funded almost 33,000 acres of new land purchases and 89,000 acres of conservation easements.[44]

The Coastal Conservancy is so noncontroversial that it is little known to the general public and its contribution to land conservation little recognized. Yet its model, honed by its first executive director, Joe Petrillo, has been so successful that similar conservancies have been set up to protect the Santa Monica Mountains, Lake Tahoe, the Sierra Nevada, and the San Joaquin River. The public conservancies show that government is still essential to greenbelt protection, even as the purposes of open space change, and that new forms of government action can be invented to meet changing circumstances.

Valued Council

The Bay Area Open Space Council was created as a way of institutionalizing regional cooperation among all the custodians of the greenbelt. Given all the park agencies, open-space districts, and land trusts that had sprouted up, Larry Orman of the Greenbelt Alliance saw the need for making common cause. The idea gestated in informal gatherings of regional open-space managers beginning in 1990. Three years later, Orman got a grant to hire a half-time staffer, John Woodbury—another alumnus of the Berkeley Department of City and Regional Planning—and the council was born.

Today the open space council includes all the major players in Bay Area open space—almost fifty entities (except the cities, which are members of the Association of Bay Area Governments). Woodbury served as executive director until 2005, when Bettina Eng took the helm. The council maintains an inventory of all the open land held in public trust around the region, and maps of these lands are served up in full color on the council's Web site. More important, the council monitors the allocation of millions of dollars to parks, open-space reserves, and conservation easements.[45]

The first challenge for the fledgling Bay Area Open Space Council was to build trust and to convince its members of their common purpose. Woodbury wanted to avoid the pitfalls of regional planning, which had failed so spectacularly. So he cobbled together a "plan" that was nothing more than an assemblage of everyone's wish lists. The council has kept reworking that plan into a more systematic and coordinated guide for overall land acquisition, habitat protection, and conservation management. It has achieved, at long last, something close to the overarching regional open-space planning the greenbelters could never extract from the Legislature.[46]

Times were hard in the recession of the early 1990s, so unity could be forged out of a common financial misery. Woodbury therefore focused on how to gather a regional pot of money that would help everyone and serve as an incentive for working together. Thus, the chief purpose of the open space council is acting as a united front in seeking public funds for parks and open space. Its inaugural proposal in this regard was the San Francisco Bay Conservancy, which Byron Sher of Palo Alto introduced in the Legislature in 1997. The bill passed in a wink because it demanded no new money, no new agency, and no new regulations—but it turned out to be a pot of gold for regional open space. In 1999 the Bay Area Open Space Council pried the first $10 million out of the Legislature for the bay conservancy. Then the council hit the jackpot with the big bond propositions of 2002–04 and began working closely with the bay conservancy to set regional priorities for land acquisition and other projects funded from the $100 million in new money.

Nothing quite like the Bay Area Open Space Council exists in any other metropolitan area in the country.[47] The council's success brought inquiries for assistance from elsewhere in California, and it has helped groups in Los Angeles and on the Central Coast set up similar databases of protected lands. Although the council's purview ends at the nine-county Bay Area, it encourages the formation of other regional councils around the state and cooperates with the Sierra Business Council, Great Valley Center, and others on the theory that the Bay Area has so many resources for open space that its best strategy is to have more allies up and down the state, rather than simply more resources for itself. The Bay Area Open Space Council was instrumental in establishing the California Land Trust Council, headed by one of Woodbury's former staffers, Darla Guenzler.

Open-space districts and land trusts have become a powerful weapon in the arsenal of environmentalists. The immediate results are hard to argue with. In the face of raging growth and rampant land consumption, the Bay Area still has room to breathe in its hills, across its bay, and along its coast-lines. It is relatively compact for an American metropolis—indeed, in the 2000 census it was the second-densest urban area in the country, after Los Angeles. And the regional greensward has been expanding along with the city, despite skyrocketing land prices.

Meanwhile, habitat conservation has allowed a host of wild animals to recolonize the region's open spaces and parks: coyotes, bobcats, tule elk, wood rats, kestrels, eagles, herons, turkeys, boars, gray foxes, and more. The first brown bear in almost a century was sighted on Mount Tamalpais, and mountain lion reports are no longer rare—a couple of the animals have made it all the way to downtown Palo Alto. Fortunately, in the green-minded Bay Area, there is little of the fear of nature biting back that is reported by Mike Davis for L.A.[48]

All of this has been achieved in an environment of greater social conser-vatism and antipathy to taxes, with which green strategists have had to learn to live. To a surprising extent, environmentalists have overcome the barriers of public finance by wise initiatives and with remarkable public support across the Bay Area. At the same time, they have made an end run around government limitations to go directly to landowners through the land-trust option.

Yet there is a troubling undertow in the shift to a dominant role for pri-vate trusts instead of public funding for open-space acquisition and greensward protection. This has its roots in the neoliberal turn against gov-ernment, which swept the country under the Reagan presidency in the 1980s. It is hard to celebrate a shift in the green domain when neoliberalism has been so pernicious in others. Private donors, land trusts, and philanthropic foundations may be valuable allies of green activists, but they are nonetheless a world apart—or a step back—with their staggering wealth, closed-door decisions, and, at best, apolitical stance.[49] They may not, in the end, be the best friends for a militant political culture of conservation and environmentalism of the kind that inspired the Bay Area greenbelt in the first place.[50]

The use of new forms of open-space protection shows the Bay Area at its innovative best but reveals the contradictions within the classes that rule the roost around San Francisco and Silicon Valley. The rich turn over some of their immense wealth to save the lands threatened by the cities where that wealth is generated. The technical-professional elite apply their considerable wits to leading the fight to protect open spaces but never face up to their contribution to the economic powerhouse that overwhelms the land. One thing is certain: while open spaces are places of repose and restoration, the city and its growth machinery never sleep.

8

SOUR GRAPES *The Fight for the Wine Country*

The North Bay—Napa and Sonoma counties—is the least urbanized, most rural portion of the Bay Area. Some definitions of the metropolitan area leave out the northern tier of counties altogether. Many people in the North Bay region still consider themselves a world apart. Nevertheless, the North Bay is an integral part of the Bay Area, part of the countryside within the city. Even beyond the suburbs proper, this is a place actively working to suit the city. The cows give milk for city folk, back lanes are replete with exurban homes, and the grapes tickle the palates of city wine connoisseurs.

In Napa and Sonoma counties, agriculture has not only survived, it has flourished beyond all reckoning. The region has become the Wine Country, one of the premier viticultural districts in the world, generating billions in revenue. It is, moreover, one of the most striking landscapes of the Bay Area, with its vine-draped hillsides, charming farmhouses, and sinuous country roads. It is redolent with scenes of regional prosperity, rural tranquility, and the taste of California. This is, apparently, the best that bourgeois life has to offer.

The North Bay is very much an urban frontier, with some of the fastest growth of the metropolitan region. Development has surged up the U.S. Highway 101 corridor in Sonoma County past Santa Rosa, more than 50 miles north of San Francisco, and suburbs are lapping along the southern edge of Napa County, from American Canyon to the city of Napa. Napa County still holds fast to its agrarian identity, which has given environmentalists a strong lever on land conservation. Napa's seeming rural tranquility is nonetheless being disrupted from within by the commercial success of its wineries and from without by the crushing embrace of its millions of admirers. Sonoma County is more politically fragmented, torn between real estate developers, new urbanites, and rural folk, who are themselves of several minds. Environmental battles have been more eclectic there, from saving the coast, to water wars, to recreational access to open space. But after many years of nibbling around the edges, Sonoma's greens have entered the mainstream, and the county has taken great strides toward greenbelt protection.

DEUS EX VINIFERA

Napa and Sonoma counties are today the center of premium wine production in North America. More than 100,000 acres of vineyards are nestled into the North Bay's valleys and hillsides, their bounty feeding into more than 500 wineries. The North Bay's wine industry brings in an astounding $5 billion per year in revenues, and generates three times that in secondary economic effects.

California wine is bifurcated between premium brands and jug wine for the mass market. Wine grapes thrive in the heat of the Central Valley, but first-rate grapes require the cooling influence of the Pacific. Three-quarters of total production in California comes out of the San Joaquin Valley, where the vines groan with hundreds of pounds of grapes and wineries resemble oil refineries. E & J Gallo, the world's largest winery, spews forth millions of bottles from its Modesto tank farm, with its own glass factory, 30-acre warehouse, and million-gallon concrete holding pit.[1] Fine wine, on the other hand, comes almost entirely from the coast. There are high-end producers scattered from Mendocino to Santa Barbara counties, but the great majority of California's fine wine comes out of the North Bay.

High on the Vine

California wine making took a long time to recover after Prohibition ended in 1933. Still, rebuilding allowed new methods, producers, and places a chance to step forward. The most striking development of the mid-twentieth century was modernization of production using stainless-steel tanks, automatic temperature control, cold fermentation, chemical testing, pristine sanitation, pure yeast, and mechanized sterile bottling. This was done under the tutelage of the oenologists at the University of California, Davis. Gallo was the prime symbol and agent of modernized production, the Ford Motors of mass wine production.

Winemakers had the handicap of a weak domestic market, with per-capita consumption one-tenth that of France and popular tastes veering toward soft drinks. In the immediate postwar era, wines such as white port, made with Thompson seedless grapes, led the field, and the marketers introduced new concoctions including Thunderbird, Ripple, and Bali Hai. The Bay Area lost ground in this era, although a few local wineries—Paul Masson of Santa Clara County and Sebastiani of Sonoma—were in the same league as the Central Valley behemoths.[2]

Nevertheless, a handful of viticultural pioneers in the 1930s and '40s were quietly laying the foundations for Napa Valley's rise to greatness in the world of wine. Louis Martini moved back, John Daniel inherited Inglenook, George de Latour brought André Tchelistcheff from Paris, the Mondavis bought Charles Krug Winery, Hanns Kornell fled Europe, and Lee Stewart founded Souverain. This group got organized as the Napa Valley Grape Growers' Association; later, the Napa Valley Vintners Association was formed. They began the long process of replanting vineyards, matching varieties to soils, learning the vintner's art, bringing in oak cooperage, and the like.

The Napa vintners began to promote their products as something distinctive through tastings, winery visits, labeling, and varietal designations. Consumption of fine wine was painstakingly constructed with the help of taste setters such as Robert Parker of *The Wine Advocate* and Marvin Shanken of *The Wine Spectator*. In valley lore, a 1973 Paris tasting—when a blind sampling by French experts ranked little-known Napa wines above the giants of Bordeaux and Burgundy—has assumed mythic proportions. Sales of premium wine accelerated in the 1970s, then soared beyond anyone's dreams in the 1980s, passing bulk wine in value.[3]

Yet elevated wine consumption had a lot more to do with changing class structure and class culture than international tastings. In the go-go eighties and the NASDAQ nineties, wine became a sign of sophistication for the up-

and-coming bourgeoisie. Today, just 10 percent of people in the United States consume 67 percent of all wine. As that 10 percent got richer, the intake of premium varieties continued to rise even as total consumption leveled off in the chaste age of religious revivalism. Prices shot up for the finest vintages, became outrageous in the 1980s, and bordered on obscene by the end of the boom economy at the end of the twentieth century. As a result, the elite wineries earned excellent profits throughout the 1990s. The wine industry also parlayed its newly acquired international reputation into exports that exceeded half a billion dollars by 1999.[4]

Boutique wineries proliferated. Napa Valley, which had no more than 30 wineries in 1965 and 50 by 1980, exploded to more than 250 by the end of the century. In Sonoma, too, wineries sprouted like weeds, from around 60 in 1970 to more than 250 by 2005. The North Bay has more than half the wineries in the state. Vineyard planting has followed the same upward trend. Acreage skyrocketed in the bubble leading up to 1973, suffered a rude shakeout, then leveled off. The 1980s saw a rapid uptick, which continued right through the Roaring Nineties until overplanting and severe recession resulted in a brief glut in the early 2000s.

As Napa Valley filled up with 40,000 acres in vines, annual grape sales closed in on $400 million. Then, in the 1990s, the forward edge of planting shifted to Sonoma, which hit 60,000 acres by 2000. Grapes became Sonoma County's number-one agricultural product in 1987, passed $100 million in value by 1990, and skyrocketed to more than $400 million in the early 2000s. Vineyards spread up the valleys all the way into Mendocino County, altering the landscape no less than housing tracts have.

North Bay vineyards are, like the wineries, smaller and more specialized than those in the Central Valley. Growers have steadily shifted to high-end varietals. In the first blush of enthusiasm for better varieties, chardonnay and cabernet led the way, then merlot was the rage, followed by zinfandel; in time, more difficult grapes such as pinot noir and grapes for sophisticated blending, including cabernet franc, have increased.

Napa Valley grape prices are the pinnacle in California, with a sharp falloff elsewhere: 28 percent less for Sonoma's grapes, 50 percent less for Central Coast's, 88 percent for San Joaquin Valley's. There is never a glut of Napa Valley grapes. Napa Valley land, which cost $1,000 per acre (planted) in the mid-1960s, hit $40,000 per acre by the mid-1980s and leapt to more than $120,000 per acre in 2000. In 2004 Francis Ford Coppola paid $350,000 an acre for the Cohn vineyard. The North Bay has the most valuable farmland in North America.[5]

The Wine District

The boomlet in boutique wineries was linked to the growing passion among the upper classes for wine making as alternative life's work—or indulgence. John Trefethen was the son of a director of Bank of America; Dan Duckhorn a banker in San Francisco; Reverdy Johnson (Kendall-Jackson) a downtown lawyer; and John Wright (Domain Chandon) a researcher at Arthur D. Little. Three refugees from Los Angeles were Jim Barrett (Chateau Montelena), a corporate lawyer; Mike Robbins (Spring Mountain), a real estate broker; and Donn Chappelet, a vending-machine merchant.[6] This is quite in keeping with the gilt-edged character of the Bay Area economy, its flux of money and talent, and the pursuit of visions of the good life through good work—especially if those visions have a solid payoff.

Making wine is more than an indulgence, however, and the Wine Country is a typical industrial agglomeration—Silicon Valley on the vine. It requires a mix of craft, practical know-how, and research science. It grows through training of skilled workers, on-the-job learning, and inspired creativity. It generates a rich assortment of specialty firms in different wine niches, as well as a panoply of suppliers for everything from oak barrels to vine stocks, machinery to labels. At the downstream end, there is a host of consultants, distributors, and agents moving the product to market. Vineyard managers and winemakers are particularly valued for their expertise, and they often work for several employers at a time. UC Davis and Sonoma State University crank out new talent by the cartload.[7]

Personal networks in the industry are dense. Most of the key practitioners of the vintner's craft cycle through existing wineries, learning the trade, before spinning off on their own: Randy Dunn worked for Caymus, Ric Forman for Mondavi and Sterling, Mike Grgich for Beaulieu and Chateau Montelena. Everyone attests to the spirit of comradeship and friendly competition among the favored circle, sharing equipment and know-how in a freewheeling atmosphere of experimentation and love of the art. André Tchelistcheff left Beaulieu to become a freelance consultant to nearly every start-up in Napa. Some, such as the Portet brothers (Clos du Val), brought their expertise all the way from Bordeaux. Others, including

Richard Peterson, came out of Gallo and Petri in the Central Valley, bringing with them the most advanced techniques and marketing strategies in the business.[8]

The wine industry is knit together through formal organizations, as well. It first organized collectively under the San Francisco–based Wine Institute, which was able, under the sanctions of the State Wine Advisory Board, to put into place some of the strictest production and labeling standards in the world. When Napa vintners became disaffected with the giant wineries, they started their own group, the Premium Wine Producers of California (now the Family Winemakers) in the 1950s. The Sonoma County Wineries Association goes back to 1946, while the Sonoma County Grape Growers Association was put together in 1983. Trade journals, such as *Wines and Vines*, have played an essential part in the exchange of ideas.[9]

Investment capital is crucial to the wine industry. In the early days, you either brought it with you, or you went to the St. Helena branch of Bank of America. Today, loans come from many sources, and specialized financiers in San Francisco track the industry closely. Outside capital began pouring into Napa by the 1960s, when Inglenook and Beaulieu fell into the hands of Heublein distillers. The acquired wineries lost their luster as the parent corporation pumped in capital for mass production and a big sales network. European giants brought a different kind of class when they began investing in the 1970s: Moët et Hennessey led the way (setting up Domain Chandon), followed by Piper-Heidsieck and Mumm; then Philippe Rothschild joined forces with Robert Mondavi. In Sonoma, corporate capital rolled in by the 1980s, as Freixenet set up Gloria Ferrer Champagne, Suntory bought Chateau St. Jean, and Gallo started planting vines for a new, mid-range brand, Turning Leaf. Some of the local companies, such as Jordon, Glen Ellen, and Kendall-Jackson, grew large by grabbing a piece of the mass market, becoming corporate octopi in their own right. Yet small wineries flourished, entering through the premium end of the market, where wines costing over $10 have continued to enjoy double-digit growth in sales in the 2000s.[10]

Wine and Dine

Premium wine fits into a wider pattern of consumer culture in the Bay Area, and local consumption has been pivotal to the winemakers' success. Californians quaff twice as much wine per person as Americans generally, and Bay Area drinkers twice that.[11] San Francisco, which has a record of heavy drinking going back to the Gold Rush, remained notoriously "wet" right through Prohibition. More recently, the gourmet restaurants of the Bay Area food revolution have brought the pleasures of wine to the attention of the consuming classes, as wineries, merchants, and publicists made the sophisticated wine list a natural accompaniment to California cuisine.[12]

Here we enter the domain of symbolic capital and acquisition of status through acts of consumption. What better illustration of the cultivation of class distinction than wine? Witness the ritualized acts of serving and drinking wine: sniffing corks, gazing heavenward, swirling the glass. Observe the long apprenticeship to learn the secrets of the temple: reading holy texts, gathering for tastings, grave conversations with shopkeepers. Marvel at the arcane language of taste, a veritable poetry at the command of a special priesthood. Regard the artful labeling: the bottle as textual archive. Bottled poetry, indeed! And a lot of canned pretense, to go along with it.

Yet consumption is more than symbols, and to regard it only as a projection of hierarchy onto the flat surface of everyday life is to miss the very artfulness of class formation.[13] There is considerable substance in wine appreciation, based on qualities of grapes, soil, climate, cooperage, yeast, tinkering, and time. This material gets worked up deftly by devotees, purveyors, and charlatans into the wispy edifice of the stylized subculture. Because the qualities one admires in wine are not always cheap to create, the cultural arena is easily captured by the wealthy. On the other hand, there is a long history of commonplace wine culture among ordinary people in Europe, which belies any easy association between wine and wealth. Something akin to that everyday culture exists in the Bay Area as well, but it is heavily overlain with deposits of nouveau riche acquisition.

For all its glamour, however, the Wine Country and its magical world of consumption rest on a humble foundation of ordinary labor. Thousands of workers are required to maintain the vineyards, harvest the grapes, do the crush, and prepare the wine. Many are highly skilled in the arts of viticulture and wine making; many more are unskilled, doing heavy labor in fields and sheds. Almost all are poorly paid, by industrial standards—let alone average wages in the Bay Area. These people are almost all Mexican, from states such as Jalisco, Zacatecas, and Michoacan, and a large percentage are undocumented. Many move in circuits of migration between work and home, returning periodically and maintaining homes, families, and farms in Mexico. But many others have settled in the North Bay, marrying and raising families. Their lives are not easy, because of a lack of affordable housing, health-care policies, and social services. And they are virtually invisible in the imagery of Napa and Sonoma's verdant agrarian landscape.[14]

HOLY *TERROIRS*

The transformation of the North bay into the Wine Country would not have been possible without the intervention of Dorothy Erskine, doyenne of the greenbelt. Napa Valley was poised to suffer the same fate as the Santa Clara Valley as the highwaymen made plans to drive two freeways up to Calistoga and link Napa east to Interstate 80 and west to U.S. Highway 101. Subdivisions were springing up in the southern valley by the early 1960s (my college girlfriend's parents lived in one). But Erskine—who had a weekend house in the valley—could see what was coming, and she wanted an agricultural preserve to defend the farmland. In those heady days, what Dorothy wanted, she usually got.

Agriculture Preserved

In 1967, Napa County planners drew a line around the valley, called it an agricultural preserve, and put it in the General Plan. Dorothy Erskine picked up the idea and, with John Horton, put together a conference in early 1968 on agricultural land preservation, invited the county supervisors, and had them convinced by the end of the day. Erskine mobilized Jack Davies of Schramsberg Winery to lead a citizens' committee for the preserve, the Upper Napa Valley Associates, and Volker Eisele to organize the grape growers. They beat back resistance by many of the vintners, whose powers of foresight did not extend from wine markets to the economics of land.[15]

The plan was approved that year by the county supervisors. Nothing else like it existed at the time. It established a minimum parcel size of 20 acres to discourage subdividing and lowered tax assessments to keep farming profitable, in the manner of the Williamson Act of 1965 (see chapter 2). Without the agricultural preserve, there would have been no wine industry in Napa—just as there is no longer a citrus industry in Orange County. It also keeps Napa the least populous county in the Bay Area, with barely 200,000 people.

Napa County hired greenbelt planner Jim Hickey after the Association of Bay Area Governments rejected his grand vision for the Bay Area (see chapter 6). Hickey saw Napa as the last open-space frontier of the metropolitan region and meant to save it. When a majority of slow-growth advocates was elected to the Napa County Board of Supervisors in 1972, the board moved to halt the completion of Highway 29, issued a more protectionist general plan, and established the Land Trust of Napa County (see chapter 7). The greens suffered a setback in 1976 in the midst of the first wine glut. But they recovered two years later to expand the agricultural preserve and double the minimum lot size to 40 acres. A citizens' slow-growth initiative in 1980 tightened limits on housing in unincorporated areas to 1 percent per annum.

When the wine boom lost steam in the late 1980s, slow-growth county supervisors were thrown out of office, along with Hickey. But the day was saved by an agricultural lands preservation initiative, passed by the voters in 1990, which prohibited conversion of vineyards or open space to urban uses up to the year 2020. This time, conservationists had support from the wine industry. These allies also cooperated on a strict hillside-protection ordinance to slow erosion, as well as a right-to-farm ordinance that defended ordinary farming activities against nuisance claims by urban neighbors. The county further established an environmental information center, which serves as a clearinghouse for landowners and environmentalists.[16]

If Napa has been good about protecting its agrarian heartland, it has been remiss in providing recreation open space for its urban populace and working class. It has the fewest acres in parklands among all the Bay Area counties, except Solano. Most of the acreage comes from Robert Louis Stevenson State Park and Knoxville Wildlife Refuge in the far north county. Lake Berryessa, to the east, is a federal irrigation reservoir that flooded out a beautiful agricultural valley in the 1950s; its shoreline is open to public recreation. Napa has a handful of city parks, no county parks, and no county parks department.[17]

Napa County Land Trust's main function is to back up the agricultural preserve by protecting vineyards. It holds about 20,000 acres of vineyard easements, mostly by donation—raising the usual questions about tax breaks for well-to-do landowners for lands with no public access. The trust has passed along roughly 12,000 acres of prime land to the state park and wildlife refuge systems and retains 5,000 acres as habitat management areas. The land trust has become more sophisticated about the problems of long-term easements and insists that donors provide endowments for future monitoring. It has also begun to raise capital more systematically under director John Hoffnagle and to work on wildlife and trail corridors with the Open Space Council.[18]

Pressure has been building on the county to take better care of its non-vineyard lands. The land trust wants a public agency to partner with and has lobbied the Napa County Board of Supervisors and sought aid from Bay Area Open Space Council. An open-space district proposal failed to get off the ground in 1992 but made the ballot in 2000. It failed to get the necessary two-thirds vote for a transit tax to fund a district. So the board of supervisors set up an advisory commission in 2003 to generate public support and hired John Woodbury away from the open space council to do the legwork. The new strategy is to win permission to form a countywide open-space district first and later go back for funding for land acquisition.[19]

A Consuming Place

Wine culture has had a schizophrenic effect on land-use practices and politics in the North Bay. On the one hand, land has come to be considered in terms of *terroir*, the power of soil and sunlight to produce the finest wine in some places and not in others. On the other hand, land has become the terrain of tourism, the pleasuring grounds of weekenders and travelers. The first effect, that of *terroir*, means stricter control of land designation, backed by the law, to assure value in the final product—wine. The second effect, tourism, means greater intrusion of the palaces, palates, and pounding feet of pleasure seekers in the midst of the precious agrarian landscape.

Control over land use in the North Bay has taken an increasingly sophisticated turn within the wine industry itself: the designation of local appellations, or special grape-growing districts. For those who believe in the power of *terroir*, this is a crucial move. In 1983 the Federal Bureau of Alcohol, Tobacco, and Firearms was at last persuaded to sanction exclusive places of origin for wine, in the manner of France's regional labels, or *appellations*. These enhance the cachet and prices of fine wines. Napa growers and vintners were principal backers of the measure, and they quickly declared a Napa Valley appellation. With time, finer gradations, such as Stags Leap and Carneros districts, have been introduced. There are now three dozen appellations in Napa County and a dozen in Sonoma County. Many vintners take the idea even further, denoting the vineyards of origin on their labels.[20]

Concurrently, there was a rearguard action against bottlers using the Napa name on products made from grapes grown in the Central Valley. Napa Ridge, an offshoot of Nestle's Beringer operation, was a prime offender in this regard. The Napa Valley Grape Growers' Association moved to restrict the practice, and in 1990 the county supervisors adopted a winery ordinance that imposed a 75 percent local-origin rule. The matter languished in the courts for fifteen years before this ordinance was finally upheld.

Meanwhile, the Wine Country became the promised land for the wealthy of the Bay Area who were spilling north in search of sylvan landscapes on which to build second homes and weekend hideaways. This is not a new phenomenon, and even Dorothy Erskine had a pied-à-terre and the Livermores a ranch on Mount St. Helena (4,344 feet). But the profusion of wealth in the 1980s and '90s was unprecedented, and monster houses have sprouted all over the hillsides. In the process, St. Helena replaced Ross and Kentfield in Marin County as the chosen place to enjoy one's plenty, to see and be seen among one's class. Events such as the annual Napa Wine Auction and venues such as the Sonoma Mission Inn became staples of the migrations of the beautiful people. In the process, the North Bay became deeply invested with city meanings and city money.[21]

Hard on the heels of the rich came the tourists, lured to the wine shrines for the rituals of tasting amid rural plenitude. The Napa Valley is visited by almost five million people a year; Sonoma, some four million. Arriving by car, they take in vicarious vistas of pastoral wealth, luscious nature enriched by the human hand and the well-stocked bank vault. Bed-and-breakfast inns have sprung up for the romantic overnight stay—some 100 in Napa and 75 in Sonoma. Robert Mondavi Winery has long served as Napa's multicultural center, serving up jazz concerts, art shows, and Great Chef cooking classes; now Mondavi has outdone himself with Copia, a combination museum, amusement hall, and cathedral to the vintner's art, set along the Napa River. As one writer comments, "In the new Napa Valley, image is king and marketing is its handmaiden."[22]

With this spin of the bottle, the quality of memory becomes strained—by the opportunity to make a buck off the wonderful world of wine. Napa is in danger of being turned into an open-air theme park, an upscale version of Marine World–Africa USA in Vallejo. Wineries have been turned into restaurants, gift shops, and general tourist traps. Spring Valley Winery became the fictional site of the television series *Falcon Crest*—J. R. tramping out the vintage. To this has been added hot-air balloons and a Wine Train. Such tourism can easily be its own undoing, as shown by the weekend traffic jams on Highway 29 and arguments over rampant commercialization. The wineries want a free hand to parlay tourism into dollars, while the grape growers have become the main defenders of *terroir*. A measure to limit wineries to wine making, opposed by the vintners, went down to defeat in 1990.[23]

SALVE REGIONA

All is not wine in Sonoma. For more than a century, it has been one of the principal farm districts of California, and it still has a vigorous agrarian economy, more diversified than Napa's. Sonoma is the largest of the bay counties, at one million acres, and its rugged north and west are distant enough to remain sparsely populated to this day. Roughly a third of the county is still rolling cow country, 50 miles of coast remain largely free of building, and redwoods and Douglas fir lie thick upon the western slope. For natural beauty, it is second to none, but the tide of urbanization is sweeping north, making Sonoma County a prime battleground for saving both wild and working countryside close to the city.

Like Marin, Sonoma felt the impact of the Golden Gate Bridge after the Second World War, especially when U.S. Highway 101 was made a freeway in the early 1950s and rammed through the middle of Santa Rosa. By the 1960s, subdivisions began spreading across the valleys, ranchette homes peppered the hillsides, and the new town of Rohnert Park was established, where a bright new California State University campus was located. Population grew by roughly half again in every decade from 1940 to 1980—one of the fastest rates of growth in the region—hitting 200,000 by 1970 and topping 300,000 by 1980. Meanwhile, the weight of settlement began to shift away from the historic centers of Sonoma and Petaluma on the southern edge of the county toward Santa Rosa in the north and along the central highway corridor, where more than half the people in the county lived by the end of the century.

As Marin's stand against growth pushed developers north, Sonoma County did little to protect itself against the onslaught—only Petaluma put up a fight, with its growth-control ordinances (see chapter 6). The county's governing regime welcomed developers and refused all exhortations to join with other Bay Area jurisdictions. The county withdrew from the Association of Bay Area Governments by a 3-to-1 popular vote in 1970, followed two years later by withdrawal from the Bay Area Rapid Transit District. The county board of supervisors refused to engage in serious planning, despite increasingly tough state laws requiring it to. It made little use of Williamson Act agricultural tax-reduction contracts (see chapter 2), and farmers and ranchers evinced little interest in agricultural preservation.

Coastal Access

Sonoma County's greens were few and far between to begin with. The Bodega Head nuclear-reactor fight, which took place on the Sonoma coast in the early 1960s, alerted them for the first time to modern environmental threats. One of the Bodega activists, Karl Kortum, was from an old Sonoma family, and his brother, Bill, was a veterinarian in Sonoma's dairy country. He enlisted Bill to rally the dairy owners against the reactor over fears of radiation contamination of pastures and milk supply (a highly charged issue in the wake of revelations of the time about strontium 90 in milk from atmospheric testing of nuclear bombs).[24]

Bill Kortum was appointed to a statewide citizens committee to look at turning Highway 1 into a scenic road. Although that came to naught, it introduced him to George Collins (leader of the Point Reyes battles) and got him involved in the creation of Salt Point State Park on the northern Sonoma coast in 1966. He rallied support in Petaluma for making it an educational place for schoolkids to learn about tidelands. It was the first new park established on the Sonoma coast since the Second World War.[25]

After all that work, Bill Kortum was shocked to learn of the Sea Ranch, a massive second-home development on a former cattle ranch north of Salt Point. Proposed in 1965, it was quickly rubber-stamped by a progrowth board of supervisors. Ironically, Sea Ranch was a model of Bay Area architectural and landscape naturalism, with small houses of weathered redwood in scattered groupings, designed by Charles Moore and Lawrence Halperin. Despite the aesthetic gloss, Sea Ranch would close off the coastline for 14 miles, blocking access to beaches, tide pools, and abalone gathering that the public had long enjoyed.

In order to fight Sea Ranch and the board of supervisors, Bill Kortum made common cause with skindiver Charles Hinkle and three young professors from Santa Rosa Junior College and Sonoma State College—Chuck Rhinehart, Peter Leveque, and John Crevelli—to form Californians Organized to Acquire Access to State Tidelands (COAAST) in 1968. Dick Day became the group's attorney. They tried to modify county approval of Sea Ranch by a popular initiative the following year, but lost. Undaunted, they looked to the state for redress and found an ally in Napa's state senator, John Dunlap. Bill Kortum helped to found the California Coastal Coalition (later, California Coastal Alliance), which rallied the public to the cause of coastal protection. After three years to trying to get a bill through the Legislature, the alliance opted for the California Coastal Initiative of 1972 (see chapter 5).[26]

COAAST was the first environmental organization in Sonoma County, and it remained active for more than twenty-five years before folding up with the retirement of its stalwarts, Rhinehart and Crevelli. During that time, they bird-dogged county officials and state agencies up and down the North Coast. In 1970 they allied with the Jenner Coalition, led by Virginia Hechtman, to halt Utah Construction's gravel mining at the mouth of the Russian River. Once the new state Coastal Commission was in place, COAAST was able to secure public access at Sea Ranch and reduce the number of houses there by half. COAAST took on another large development, called Bodega Harbor, and had it downsized too. Along the way, they secured Bodega Head and Willow Creek campground for the Sonoma Coast Beach Park, a 25-mile-long noodle of a state park running from the Russian River to Tomales Bay. As a result, today three-fifths of the Sonoma coast is protected in parks.[27]

Bill Kortum has been the leading green warrior of Sonoma County and a consistent generator of new ideas and organizations for forty years. The son of a chicken farmer, he got a dose of politics early from his father's involvement in a fight to reroute the freeway around the best chicken-farming land near Cotati. He came to conservation out of a love of the rural landscape and the coast, not prior ideological conviction, and engaged the world with a classic American sense of republican virtue, democratic intelligence, and personal surety. Kortum has never budged from Cotati, but he has had an indelible impact on the whole state. His coastal work brought him to the attention of Bill Evers of the Planning and Conservation League, and he joined the PCL board in 1969. His efforts to have sewage used to grow forage for cows alerted Dorothy Erskine to his work, and he was invited onto the People for Open Space board in 1971. When the California Coastal Alliance morphed into the California League of Conservation Voters in 1973, he moved over to that board.

In 1983, when Bill Kortum thought the California Coastal Commission was getting too soft on development, he initiated a new campaign, called Coastwalk, to bring public attention to the coast. Under the leadership of Richard and Brenda Nichols of Sebastopol, Coastwalk became a statewide organization getting people out to the seaside and succeeded in winning legislative approval of a 1,000-mile Coastal Trail in 2003. Sonoma County is the pilot project for the trail.[28]

Unsupervised Growth

When the green forces moved inland to take on the board of supervisors directly, the going got rougher. Day and Hinkle ran for county supervisor in 1972, and Hinkle won; Kortum followed in 1974. They immediately began to press for a sound general plan, in accordance with the new state comprehensive planning mandate. But they ran into a buzz saw. The developers, county farm bureau, Santa Rosa Chamber of Commerce, and Sonoma County Taxpayers' Association organized a recall election that evicted Hinkle and Kortum from the board in June 1976, then cleansed the county planning commission of green sympathizers. A new plan was hastily issued that summer that gutted land protections.

Things looked bleak. But the good old boys who had run the county for so long overplayed their hand. There was resentment about the way they were running roughshod over legitimate land-use concerns. In the fall election of 1976, Helen Rudee, a moderate Santa Rosa Republican, was elected, becoming the first woman on the board, and was joined by Eric Koenigshofer, a student from Sonoma State. As a result, the board could muster a moderate green majority on a regular basis. With the addition of Helen Putnam, former mayor of Petaluma, in 1978, and Ernie Carpenter, who replaced Koenigshofer in 1980, that majority held until 1988. In 1978, a new general plan with real substance was approved that included community separators (space between towns), controls on lot splitting (rural subdivision), and limits on the density of septic tanks.[29]

Petaluma's innovative growth controls suffered a setback when the slow-growth city council was voted out in 1978. But a fight soon broke out over a major development approval for Frates Ranch on the east side of town, and the progrowth council had its decision overturned in 1982. This dispute propelled liberal Democrat Lynn Woolsey onto the City Council and then to Congress.

Water supply was controversial in Sonoma County, just as it had been in Marin and San Mateo counties (see chapter 4). Developers, the county, and the city of Santa Rosa were enthusiastic promoters of damming the Russian River, as well as tapping into the headwaters of the Eel, and the Army Corps of Engineers was happy to accommodate them. Plans were drawn up in the 1960s for the Warm Springs Dam, and the Sonoma County Water Agency was created by the supervisors to act as water wholesaler. All through the 1970s, the nascent environmental movement in Sonoma campaigned against the Warm Springs project. But efforts to stop it went down to defeat at the polls in 1972 and even more dramatically in 1979. The dam was completed in 1983, creating a large flatwater reservoir and recreation area. Although the dam was justified as a flood-control measure, the Russian River still floods regularly downstream.[30]

While planning improved in the 1980s, there was no slowing the pace of growth in Sonoma County. Development was still welcomed with open arms from Rohnert Park to Cloverdale, and especially in burgeoning Santa Rosa. Meanwhile, thousands of exurban houses were popping up on subdivided farm and woodlands: on the west side around Sebastopol, on Sonoma Mountain to the east, and along the Russian River to the north. The county's economic base was changing, too, from agrarian to urban, marked by the appearance of industrial parks all along U.S. Highway 101. There was even talk of Sonoma becoming a new electronics center, "Telecom Valley." The social makeup was changing as well, with more well-educated refugees from the central Bay Area, more professional and technical workers, and more Latino immigrants drawn in as a low-wage labor force in malls, restaurants, and warehouses. Sonoma County was on the verge of a new era that would prove friendlier to environmentalists, in which it awoke to find itself an urban county at last.[31]

Agricultural protection has not been as pivotal in Sonoma County as it has in Napa. Farm interests are too splintered. The wine industry has never dominated the county and was slower to develop its own cachet. It jostles with dairying, specialty farming, and horticulture. Organics, in particular, have grown to serve a more health- and taste-conscious Bay Area, with epicurean delights such as preserves, duck, foie gras, cheese, and olive oil. A substantial nursery sector thrives, as well, by supplying rhododendrons, roses, and fruit trees to city gardeners.

Agricultural-land protection was first bruited during the general-plan debates of the 1970s. After his ouster as a county supervisor, Bill Kortum organized the Sonoma County Farmlands Group in 1977, along with Jim Sullivan, and promoted a farmlands protection initiative in 1980. They lost because farmers and ranchers were still unconcerned about urban pressures. After another decade of intense urbanization, agrarian views began to change. The farmlands group got an audience when it issued a report showing that agriculture was no longer the leading sector in the county. In 1988, the county supervisors passed a right-to-farm ordinance and inserted farmlands protections in the general plan.[32]

A unique project to strengthen the specialty-farm sector, created in 1989, is the Sonoma County Agricultural Marketing Program. SCAMP borrowed the vintners' idea of appellation with its "Sonoma Select" label, in order to enhance the image of local products. This effort is complemented by a private program called Sonoma County Farm Trails that links together small farms wishing to plug into the flow of tourists by offering a taste of the country for those with the leisure to stop and chat awhile. Sonoma's programs have been copied around the country.[33]

The boom in vineyards has had a dramatic impact on the Sonoma landscape and a paradoxical effect on county politics. In place of the unity found in Napa, the expanding wine industry has driven a further wedge between agriculture interests and environmentalists. During the 1980s and 1990s, planting proceeded furiously. As acreage doubled and redoubled, vineyards rushed up steep hillsides, displaced old apple orchards, and replaced woodlands. Townsfolk and exurbanites began to take offense at the slash-and-burn approach of giants such as Gallo and Kendall-Jackson, as well as dot-com millionaires wanting a bit of vineyard chic around their country estates. A notorious case from the mid-1990s is Twin Valley Ranch, where Gallo hastily cleared 500 acres of Douglas fir and redwood. The devastation so outraged the public that Gallo was forced to the negotiating table with environmentalists of the Watershed Protection Alliance to hammer out a system of regulation. This produced a vineyard-erosion ordinance in 1996 and a stronger, voter-approved hillside ordinance three years later.[34]

Another bone of contention has been the freewheeling application of pesticides to vineyards. Such practices were highlighted by a feared invasion of the glassy-winged sharpshooter, bearer of Pierce's disease—deadly to grapevines. The sharpshooter, an insect recently introduced to California, has been moving steadily northward. Around 2000, Sonoma growers went on a war footing and were prepared for an assault on residential neighborhoods, where the sharpshooter can live on ornamental plantings. This did not set well with environmentalists, organic farmers, or fearful homeowners, who strung together a formidable no-spray network. On the other hand, a radical effort to ban genetically modified crops in Sonoma County (as had previously been done in Mendocino County) was decisively beaten by the Farm Bureau and its allies in 2005.[35]

GREEN SONOMA ASCENDENT

By the end of the 1980s, Sonoma County had become the flash point for the encounter of city and country in the North Bay. By then, the perception was widespread that the county was in danger of losing its rural character. Ranching was in decline and suburban and exurban dwellers had grown more numerous; both were increasingly uneasy with further growth. Sonoma had reached the turning point hit by Marin County and the San Francisco Peninsula a generation earlier.

In Taxes We Trust

Through the 1980s, Sonoma County had done little to establish its own parks or open spaces. It relied on state parks and recreation areas, but had few county parks of its own. A Sonoma Land Trust was established in 1973 by Otto Keller, a wealthy philanthropist (with help from Trust for Public Land). It took off under the leadership of Joan Vilms in the 1980s and has since built up a diverse collection of holdings for agricultural preserves,

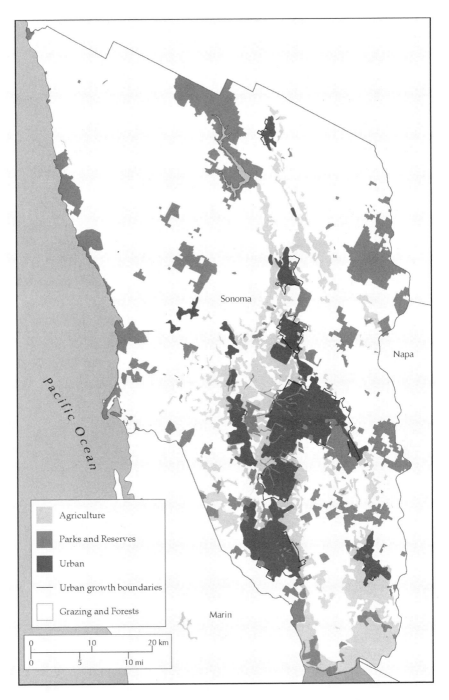

Pacific Ocean

Sonoma

Napa

Agriculture

Parks and Reserves

Urban

—— Urban growth boundaries

Grazing and Forests

Marin

0   10   20 km

0   5   10 mi

MAP 12. Sonoma County Land Use, 2005. Drawn by Darin Jensen. Courtesy of Greenbelt Alliance and Sonoma County Farm Bureau, San Francisco, 2004; Dyett & Bhatia, *Smart Growth Strategy/Regional Livability Footprint Project*.

urban separators, forest management, and generic open space, with about 6,000 acres in conservation easements and 6,000 in fee simple. Another 5,000 acres have been transfered to public agencies. In recent years, the Land Trust has moved to more professional staffing, a permanent steward-ship program, and a membership strategy to tap local wealth.[36]

A major step forward was the establishment of the Sonoma Agricultural Preservation and Open Space District in 1990. This initiative was the work of a coalition of farmers, environmentalists, and local businesses, with a push from Joan Vilms, businessman Charles Cook, and supervisor Jim Harbison. While legally a special district, it is overseen by the Sonoma County Board of Supervisors, with a citizens advisory committee drawn from various interest groups. It was the first open-space district in the United States to be funded by a sales tax. Although sales taxes are regressive, it was a brilliant tactic to avoid the two-thirds majority rule on property taxes and to blunt opposition from property-rich, cash-poor ranchers. Revenues have expanded steadily as the county has grown, giving the district upward of $100 million over ten years to work with, a remarkable sum.

The open-space district uses its considerable revenues on four categories of land use: agricultural, recreational, community separators, and natural resources. It has protected 67,000 acres (as of 2005), passing along more than half that to state parks, wildlife refuges, and county parks. It retains some 27,000 acres under easement and fee simple, a number it wants to double by 2010. Its general manager since 1996 is Andrea Mackenzie, formerly with the East Bay Regional Parks District.[37]

The main thrust of the open-space district in its first years was agricultural preservation, followed by habitat and scenic protection, and the chief tool was conservation easements. Among its purchases is the single largest easement in the Bay Area, covering half the 19,000-acre Cooley Ranch on the Sonoma–Mendocino county border. By the mid-1990s, however, discord was heard over the public benefit of private easements and tax breaks to vineyards; many county residents thought they had voted for parks and recreation, not just idle land. The district's focus began to shift away from vineyards toward dairy and ranch lands and toward acquiring more land for public recreation, especially county parks (it is, effectively, the land agent for the Sonoma County Regional Parks Department). Recreational lands are mostly acquired in fee simple.

Three women—Caryl Hart, Dee Swanhuyser, and Sandra Learned

Perry—started a new nonprofit, LandPaths, in 1996 in order to manage open-space district recreational sites. LandPaths has since expanded its sphere to organizing public access on private lands with conservation easements and acting as interim manager of parks being acquired by public agencies (it is the only private entity in the state doing public parks management). LandPaths has also developed outreach programs to bring schoolkids, minorities, elderly people, and the disabled out to enjoy public open spaces. Open-space policy in Sonoma has undergone an important change of heart.[38]

A County in Transit

An important step for the green forces was the formation of Sonoma County Conservation Action in 1991 by Bill Kortum, Joan Vilms, and Dick Day. They hired Mark Green, an energetic young activist from the Planning and Conservation League, to create a network of grassroots canvassers. Green successfully mobilized the public behind green lines (urban-growth boundaries), which were adopted by eight of the nine cities in the county during the 1990s (only Cloverdale refused). Sonoma County had suddenly leapt to the head of the pack in growth controls in the Bay Area. When Green left in 2001 and Day died a year later, Sonoma County Conservation Action suffered, though it has regrouped since.[39]

Attention turned in the 1990s to salvaging the Russian River from the gravel miners and city sewage from Santa Rosa. The river has suffered massive erosion and bank collapse, as well as foul water and disappearing wildlife, even as recreational use has gone up. Protests to the board of supervisors by Marty Griffin and others in the 1970s went nowhere. But legal suits against gravel miners in the 1990s triggered the first serious hydrologic study of the basin. Friends of the Russian River was founded in 1993, in keeping with the urban creeks restoration movement around the Bay Area (see chapter 2). The California Coastal Conservancy was poised to undertake an ambitious river-management plan. When the conservancy backed out, Sonoma County renewed permits for gravel mining in 1994. Two years later, threatened with another suit by the friends of the Russian River, the California Environmental Protection Agency began a statewide Watershed Management Initiative, with the Russian River as its principal model. In 1999 and 2000, the federal EPA declared the salmon and steelhead runs endangered and designated parts of the river as critical habitat under the Endangered Species Act. Watershed management initiatives have since become a valuable tactic for green activists around the Bay Area in promoting better land-use practices.[40]

Transportation heated up as a public-policy concern in the 1990s, recognized for its role as a driver of suburbanization. When a prohighway tax measure was put to county voters in 1990, the greens mobilized to defeat it, forming a Transportation and Land Use Coalition among COAAST and the local branches of the Sierra Club, Audubon Society, and League of Women Voters. They even got the county Chamber of Commerce to come on board. The coalition got CalTrans to fund a comprehensive transportation plan for the county, which was undertaken by Peter Calthorpe, famed advocate of Smart Growth, in 1995. It called for rail transit up the central corridor and housing clustered in transit villages. A measure to implement the plan was put on the ballot in 1998 and passed by the voters, but a companion measure to approve parcel tax financing didn't get the needed two-thirds majority. When business pulled out of the coalition to back a highways-only tax measure in 2000, the greens took their revenge; it went down to defeat, too.

In 2004 a mixed-mode transportation funding measure was approved that included widening of U.S. Highway 101 along with money for buses, bike trails, and a SMART train. The new bi-county Sonoma–Marin Rapid Transit District (SMART) turned into a force for coherent planning on major development projects. The towns of Petaluma, Cotati, and Windsor began planning for transit hubs, affordable housing, and greater density, while Santa Rosa, the center of developer power, has had to be pushed in that direction, kicking and screaming. An "accountable development coalition" was formed in 2005 by Sonoma County Conservation Action, Sierra Club, the North Bay Labor Council, Greenbelt Alliance, and the Living Wage Coalition to promote Smart Growth with equity, to gain a community-benefits agreement for the Santa Rosa SMART train hub development, and to win a sales tax increase to fund the SMART train. This coalition went far beyond anything assembled to back the 1998 transportation corridor initiative.[41]

Despite widening support, Sonoma environmentalists have not yet gained the upper hand in county politics. Caught between the Scylla of the builders and the Charybdis of the ranchers and growers, the greens have never entirely broken the hold of the county power structure and progrowth slant. In addition, the greens are tossed by the stormy politics of the rural-exurban fringe, where the foundational faith is property rights, low-density living, and personal independence. Sonoma's greens lack the easy upper-class access of the Marin Conservation League or the professional connections of the peninsula's Committee for Green Foothills. They have been unable to make common cause with the grape growers, as in Napa. Nor have they forged a sustained connection to the working class, as has the East Bay Regional Parks District—though great progress has been made on that front.[42]

Noreen Evans, who went from the Santa Rosa city council to the state assembly in 2004, has been agitating for unity between greens and labor, with considerable success. For a long time, the county's greens failed to see that workers in rural areas appreciate outdoor life, too, and have common grievances against the property-owning classes. Marty Bennett, college professor and union activist, has worked through New Economy–Working Solutions (NEWS) and Sonoma County Conservation Action to win support for Smart Growth from workers who don't want their kids moving to Lake County and for affordable housing, better wages, and community-benefits agreements from environmentalists, who need working-class votes for their programs.[43] Sonoma County politics are on the verge of a major transformation.

9

TOXIC LANDSCAPES *Beyond Open Space*

L̲arge cities foul their own nest as they grow, putting the health of their citizens in jeopardy. The density of people, industry, and transportation puts immense pressure on the capacity of air, water, and land to absorb waste. By the end of the nineteenth century, American cities were beginning to grapple with the massive works necessary to collect wastewater, displace garbage, and dissipate fumes, and a public health movement had arisen in an effort to cope with the human cost of industrial filth. But little was done to regulate discharges, and the principal civic policy was collection of refuse and sewage so that it could be shunted a little farther away. Almost nothing was done to solve air pollution.[1]

In the tidal wave of industrial and urban expansion that swept over America after World War II, the discharge of wastes was choking rivers, washing up on beaches, and blotting out views. The waste stream also changed form. Automobile exhaust challenged industrial smoke as the major air pollutant, and photochemical smog hung over the cities. Wastewater oozed out of septic tanks in suburbs, while aging city sewers overflowed. Laundry detergents put a froth on the mess and fed downstream algal blooms. Meanwhile, garbage became an unsorted hash of organic refuse and indestructible trash, the bulk of which was cardboard and plastic containers. All this ended up in voluminous dumps covered with dirt and called, ironically, sanitary landfills. The flood of waste was soon apparent to all.[2]

Out of the growing burden of postwar waste came a revival of a second kind of environmentalism: pollution control. Nationally, the struggle to cope with pollution followed a different course from the open-space movement, and land conservation groups such as The Nature Conservancy and Wilderness Society were not in the forefront. A new generation of organizations, such as Environmental Action and the Natural Resources Defense Council would take up the fight (although older ones including the Sierra Club would soon join the fray, as well).

Steps Ahead

Air pollution first caught the public's eye with the killer fogs of London and Donora, Pennsylvania, and air alerts in New York and Los Angeles. Suburban homeowners were frequently discomfited by sewage backing up in their basements and backyards, and trips to the beach were often marred by fouled waters and drifting refuse. Popular attention was captured by antilitter and highway beautification campaigns, since roadside trash was a visible and ubiquitous part of the waste stream. But much of the agitation for federal intervention came from local governments—quite the opposite of their resistance to state and federal land-use controls—because they were overwhelmed by their responsibility for disposing of the growing mass of waste.

The first federal air-pollution act came in 1955, and the original Clean Air Act was enacted in 1963; a Motor Vehicle Pollution Control Act followed in 1965 and an Air Quality Act in 1967. The first national aid program for sewers and landfills was the federal Water Pollution Control Act in 1948, and over the next twenty years Congress kept passing amendments to ratchet up the level of assistance to localities. The first attempt to regulate water pollution came in the Water Quality Act of 1965.[3]

But these half measures, lacking sufficient funding or regulatory teeth, were not enough to stanch the flow of muck into the air, water, and landfills. Local governments were getting more desperate and air and water quality was growing worse. Furthermore, over the course of the 1960s, public awareness of pollution was on the upswing, environmental militancy was rising sharply, and political will had taken a quantum leap ahead. The lead in mobilizing for stronger regulations was taken at the end of the decade by Ralph Nader's team of young activists, Nader's Raiders, who put out the report *Vanishing Air* and helped organize the Clean Air Coalition. Earth Day activists from Environmental Action joined the coalition, along with local clean-air movements in Pittsburgh and Chicago. In Washington, D.C., Democratic Senators Edmund Muskie of Maine and Gaylord Nelson of Wisconsin shepherded new legislation through Congress. [4]

The result was a remarkable set of laws passed in 1970 and 1972. The keystone was the Clean Air Act (Amendments) of 1970. The same model was followed in the Clean Water Act (federal Water Pollution Control Act Amendments) of 1972. These two laws took a radical stand: pollutants were named, sources were targeted, and cleanup had a timeline. The water and air were to be made breathable, swimmable, and drinkable so as not to harm people and other living things. The new Environmental Protection Agency (EPA), created by executive order of President Nixon in the wake of the 1969 National Environmental Policy Act (NEPA), was given responsibility to enforce the laws. One can argue with the liberal, regulatory model, but these pushed the envelope to the limit. Industry would soon push back.[5]

While less dramatic, action was also taken to deal with solid waste. The first hesitant step was the Solid Waste Act of 1965 (as an amendment to the Clean Air Act), followed by a national survey of solid-waste practices. This led to stronger legislation in 1970: the Resource Recovery Act, which prodded the states into taking responsibility for local-government waste management and resource recovery. Oregon passed the first bottle deposit bill, in 1972, and several states followed. The feds were little help, so most recycling campaigns bubbled up from below. New Jersey enacted the first recycling law, followed by New York and several other states. The 1972 Marine Protection, Research, and Sanctuaries Act was popularly known as the Ocean Dumping Act because its main thrust was to stop sewage and garbage from going into coastal waters.[6]

The New Environmentalism

The banner years of federal pollution laws—the early 1970s—marked the dawn of a new era of environmental politics. Indeed, the term "environmentalism" only came into common use at this time as a blanket term covering rather disparate conservation and antipollution campaigns.

Along with the new laws came a host of new organizations whose main business was lobbying Congress, hounding EPA to enforce the law, and taking polluters to court. The most important of these were the Environmental Defense Fund (1967), Natural Resources Defense Council (1970), and Sierra Club Legal Defense Fund (1970). They grew out of the first environmental lawsuits over Storm King on New York's Hudson River, Mineral King in the Sierra, and interpretation of NEPA. All three groups were created with help from the Ford Foundation. They would be immensely successful over the next decade in making the promise of the pollution acts come true.[7]

The green lawyers brought a new level of professional talent and scientific expertise to the table. But their success depended on the legal provisions of the Clean Air and Clean Water acts allowing citizens to sue for compliance if industry balked and the government did not do its job. And sue they did, in scores of successful actions. Their legalistic approach made perfect sense in the wake of the Warren Court's dramatic support for civil rights and other progressive causes. And the move among conservation groups including the Sierra Club and Audubon Society to set up Washington, D.C., offices made equally good sense in light of the expansion of federal powers under the Great Society. These were not simply elitist moves by professionals who did not care about the grassroots.[8]

Nevertheless, the national organizations faced real problems of outgrowing their popular base and becoming too oriented to the beltway. The Sierra Club quadrupled its membership in the 1970s and hit 600,000 members in 1990, while the Wilderness Society had grown to 350,000 and Environmental Defense Fund to 150,000. The old-line outdoor recreation groups, such as Audubon Society, Izaak Walton League, and the National Wildlife Federation, were even larger, the latter hitting a million associates and more than five million subscribers by the end of the 1980s. All became more bureaucratic as they hired more managers, professionals, and lawyers. Moreover, their staff were from the upper classes, mostly white and male, and increasingly financed by corporations and foundations.[9]

Not surprisingly, a schism broke out in the environmental movement, beginning with Dave Forman's exit from the Wilderness Society to found Earth First! in 1978. The big green groups came in for more and more criticism from below.[10] Nonetheless, national organizations were not all of a piece. The legal eagles were much more involved with pollution control, while the older groups were still primarily associations for land conservation and outdoor recreation. Nor did the 1970s mean the end of grassroots efforts around the country. Greenpeace is the case most often cited, since it grew so extraordinarily in the 1980s, becoming the world's largest environmental organization, but it began humbly enough in 1971 as a local uprising in Vancouver, British Columbia.

In the Bay Area, the sense of a local movement remained strong despite the activity in Washington, D.C. Environmental Defense Fund and National Resources Defense Council almost immediately established West Coast offices in Berkeley and Palo Alto (now in Oakland and San Francisco), with staffs who acted quite independently of the national leadership. These included key figures in California air and water policy, such as Tom Graff, Mary Nichols, and David Roe. For example, Roe was the author of California's hazardous-materials warning law, Proposition 65, which passed in 1986. On the other hand, these people often cycled into government and corporate jobs in a way that staff at grassroots environmental organizations did not.

The Sierra Club had a split personality, thanks to its deeply ingrained culture of local chapters, which could pursue their favorite issues with minimal control from the central staff. The club has always been much more grassroots and less beltway-oriented than any of the other national organizations. Its national headquarters remained in San Francisco. David Brower's breakaway group, Friends of the Earth (FOE), was also headquartered in San Francisco and deeply involved in California politics during the 1970s. FOE became increasingly schizophrenic, however, between the beltway professionals led by Rafe Pomerance and local militants, and it split apart in 1984. Brower was ousted again and went on to establish the Earth Island Institute in San Francisco, while FOE left for Washington, D.C.

AIRING GRIEVANCES

Though Los Angeles is notorious for its bad air, smog had grown severe in the Bay Area by the 1960s, particularly in the burgeoning South Bay. Given the state's acute air-quality problems, California began to act on air-pollution control earlier than any other state. In 1955 the state established an Air Resources Board (ARB) and regional air-pollution control districts. The ARB led the way in framing new car-emissions standards, a feature taken up in the federal Clean Air Act of 1970. California remains the leader in air-pollution control to this day, with many states following in step.[11]

Regional air-quality and emissions standards are set by the Bay Area Air Quality Management District (BAAQMD), whose board of directors is made up of local government officials. The board issues permits to refineries, power plants, and other point sources of air emissions. It can issue regulations and bans, such as vapor capture at filling stations, ending backyard burning, and restricting oil-based paints. It must draw up (and update) regional plans on how it can meet the ambient air standards set by the state Air Resources Board and the federal Environmental Protection Agency.

The People's Lawyers

Air-pollution regulation in the Bay Area was poor before 1970.[12] When the Clean Air Act passed, local environmentalists leapt into the breach to get its provisions enforced in the bay region. Their principal tactic has been to challenge BAAQMD permits and plans, but they have also used the California Environmental Quality Act (CEQA), which requires an environmental impact statement on major projects, and the community right-to-know provisions of state and federal laws. The Sierra Club's Bay Chapter has been deeply involved, as have the Sierra Club Legal Defense Fund, Friends of the Earth, National Resources Defense Council, and environmental justice groups. But the key watchdog around the bay for nearly thirty years has been Citizens for a Better Environment (now Communities for a Better Environment).

CBE began in Chicago in 1971 as a grassroots organizing group put together by a Saul Alinsky disciple, Mark Anderson, who invented door-to-door canvassing for environmental causes. CBE set up a San Francisco office in 1977 to join the fight against the Diablo Canyon nuclear plant and soon started an air program under Nick Arguimbau. Arguimbau became CBE's attorney, after which the air program was led by Jeff Gabe in the 1980s and Julia May in the 1990s. CBE opened its first office in Los Angeles in 1982 and then separated from the Midwest group by mutual consent. Its Bay Area office was much larger than the Southern California one until the late 1990s.[13]

CBE has set many precedents in air-, water-, and toxic-pollution control over the years. It has had a string of brilliant attorneys, including Arguimbau, Alan Ramo, Richard Toshiyuki Drury, Adrienne Bloch, and Suma Peesapati, who have won 90 percent of their lawsuits and maintained a legal department out of penalty fees extracted from polluters (as well as directing penalty money to community benefits). CBE has had a strong and committed staff of scientists, such as Julia May and Greg Karras, going head to head with the experts for regulatory boards and industry. It has also been committed to grassroots canvassing and organizing, from Mike Belliveau in the 1970s through Denny Larsen in the 1990s to Carla Perez in the 2000s. Says one former staffer:

> CBE has been a great secret. It's unusual to have a grassroots group going down in the trenches, combined with scientists doing research to show what's possible and lawyers doing hard-hitting enforcement of environmental law. It's hard to combine those three fields in one organization—everyone has different approaches and there have been tensions—but it is extremely effective.[14]

Fuels Rush In

The most dramatic early struggle over air pollution came against Dow Corporation, which proposed a petrochemical complex on the north shore of the Sacramento River, across from Pittsburg, in 1975. The legal fight was spearheaded by Nick Arguimbau of CBE and Lauren Silver of Sierra Club Legal Defense Fund, backed by People for Open Space and Friends of the Earth. The key question was whether Dow would violate regional air-quality targets for hydrocarbons and ozone. As things heated up, Governor Jerry Brown and the state Air Resources Board were drawn into the fray. Opposition was so intense that the ARB forced BAAQMD to deny Dow a permit, and the company withdrew in anger, denouncing the Brown administration for "bureaucratic red tape" (which, of course, had nothing to do with the facts).[15]

A prime target for air-pollution control was the string of five oil refineries in Contra Costa and Solano counties, which make the Bay Area one of the largest refining districts in the country. Contra Costa County has more toxic chemicals in its air than anyplace else in California, except Los Angeles County, and the highest asthma and cancer rates in the Bay Area. The refineries were big, dirty, and unscrupulous—and backed by some of the largest corporations in the world: Chevron, Shell, Exxon, Unocal. The Bay

Area Air Pollution Control District was a classic captive agency, under the influence of corporate lawyers and with a staff often drawn from former refinery employees.[16]

Yet CBE began to make headway. It forced BAAQMD to lower sulfur-dioxide emissions, got vapor recovery on tankers loading petroleum products, won better regulations on thousands of refinery valves, and got tighter controls on flaring (burning off excess gases). Although these may seem like technicalities, in fact they have had a big impact, and such controls have spread to other air pollution control districts around the country. Julia May calls this "trickle-up" regulation, and it has given CBE influence far beyond its limited resources—which would never allow a direct assault on national legislation or regulations.[17]

CBE continued to hammer away. They nailed industry-leader Chevron in 1994, forcing a 30 percent reduction in total emissions and extracting a multimillion-dollar penalty (which built a community health clinic in Richmond). But the staff at BAAQMD became more recalcitrant in the Governor Pete Wilson era, so CBE started to go directly to the refiners to negotiate good-neighbor agreements. A breakthrough deal came with Unocal (now Conoco-Phillips), which had an egregious record of bad maintenance. After a sixteen-day leak of fumes left more than a thousand people seeking medical aid and Unocal under criminal indictment, the company agreed to emission reductions, monitoring at their property line, and other improvements. BAAQMD staff opposed the agreement, but after they were caught shredding public documents, the board fired the top managers.[18]

Power and Transit

Another big target industry is electric power. Power plants are among the worst sources of air pollution, and the Bay Area's worst was long the aged PG&E station at Hunter's Point in San Francisco. This is a neighborhood with two to three times the average rates of cancer, asthma, and emphysema for the state. Activists, including CBE, Greenaction, and local community groups, got the city to strike a deal with PG&E to close the plant as soon as an alternative source of electricity could be found (a new feeder line up the peninsula). Environmentalists also pushed a city bond initiative in 2001 to spend $100 million on solar power. The Hunter's Point plant finally shut down in 2006.[19]

The same alliance was able to stop the expansion of a Potrero Point peak-power plant owned by Mirant Corporation. Mirant wanted to add capacity and promised that by burning natural gas, it could reduce emissions. But CBE's staff showed that the new plant would put more pollution into the air because of its size and time of operation. Furthermore, they argued that there was no need for the energy. California's air has been put in greater jeopardy by energy deregulation, enacted in 1996. That legislation gives the governor the power to waive air-pollution regulations for peak power. Republican Governor Arnold Schwarzenegger, elected in 2004, immediately sought greater authority to allow one-stop permit shopping to speed the way for new power plants. Says one CBE attorney, "The governor is letting the oil companies run wild in California."[20]

Highways and transportation have been key targets of air-pollution activists because of the role of car and truck traffic in ambient air quality, especially ozone and particulates. Because the state Air Resources Board sets new-car standards, the main tool locally has been to target transportation policy through the Clean Air Act's regional planning procedures.

The Sierra Club's Bay Chapter was a pioneer in air-quality battles, led by Becky Evans, as was the California League of Conservation Voters under Carl Pope (later executive director of the Sierra Club). The club's first foray was to join with the National Association for the Advancement of Colored People to stop Interstate 880 from going through West Oakland (it was delayed but finally built). In the late 1970s, the club sued CalTrans over the widening of Interstate 580 through the East Bay hills and got the agency to put in the Bay Area's first carpool lanes. Although CalTrans got Congress to change the rules for a time, carpool lanes have since become a normal part of the freeway system.[21]

During the first decade of the Clean Air Act, the Bay Area made considerable progress in improving air quality. But then regulators began to run out of easy solutions. Population and auto usage were still growing (the number of cars in the region went from one million in 1950 to almost five million by 2000, a doubling of per-capita auto ownership). After ozone levels bottomed out in the late 1980s, they began to rise again. Under the Reagan administration, the federal EPA was trying to evade the 1987 deadline for meeting the ozone standard for ambient air, so CBE's Mark Abramowitz sued them—when no other environmental organization would take on the case—and won. The case was of national importance. Three years later, a compliant Congress watered down the Clean Air Act by eliminating deadlines and putting in emissions trading.[22]

Nonetheless, CBE, Sierra Club, and the Sierra Club Legal Defense Fund went after BAAQMD in 1989 over ozone attainment and transportation planning. They proved that the Bay Area Air Quality Management District's plan was inadequate and had to be rewritten—shutting off new highway construction for six months. That got everyone's attention. But it didn't do much to wean Californians from their automobiles. Ozone levels have been on the edge of noncompliance ever since, depending on the business cycle and the weather.[23]

Environmentalists therefore turned to pressuring the Metropolitan Transportation Commission to fund more mass transit. MTC, like Cal-Trans, resists the notion that it should be held accountable for the effects of its policies on land use, environmental quality, or social justice (the defeat of regional government still haunts the Bay Area; see chapter 6). In the Intermodal Surface Transportation Efficiency Act of 1991, Congress finally imposed a regional transportation-planning mandate on agencies such as MTC (amazingly, in twenty years MTC had never come up with a financially realistic plan for disbursing billions in federal and state funds). This gave an opening to the greens.

In 1992 John Holtzclaw of the Sierra Club, John Woodbury of the Alameda–Contra Costa Transit District (later of the Open Space Council), and Matt Williams (later on the AC Transit board) put together the Regional Alliance for Transit to offer alternatives to MTC's regional transportation plans. These include more rail transit, bus transit, parking rebates, and cluster development—the antisprawl, smart-growth package. To continue bird-dogging MTC, the alliance set up a permanent organization, the Transportation and Land Use Coalition (TALC), based in Oakland. TALC tries to offer a unified position for groups such as Greenbelt Alliance, Urban Ecology, Sierra Club, and Urban Habitat, such as the Great Communities Initiative for smart growth and affordable housing, with local community input.[24]

In 2001 public health became a point of leverage on transportation policy. A new coalition, called the Bay Area Clean Air Task Force, brought together twenty environmental and social-justice groups under the leadership of Linda Weiner of the American Lung Association's San Francisco office. Earthjustice attorneys working for the coalition sued MTC over its failures to raise transit ridership levels to those promised twenty years earlier; the plaintiffs won in federal district court, but lost on appeal. A new suit in 2002–03 by CBE and Sierra Club was successful in holding MTC and BAAQMD jointly responsible for the failure to meet ozone standards. This gave environmentalists and health activists the right to participate in the MTC's transportation planning process.[25]

One of the most exciting grassroots mobilizations in the Bay Area in years is resurgent enthusiasm for city cycling. Bicyclists are in the vanguard challenging the American mania for cars, promoting bikes as the clean-air, energy-conscious alternative. Local groups, such as the the East Bay, Peninsula, and Berkeley Bicycle Coalitions, lobby local governments for bike lanes, car-free days in Golden Gate Park, and alternative vehicles. Critical Mass, which began its mass rides in San Francisco in 1992, has spread around the world. They also engage in militant actions to interfere with auto traffic. The bicycle groups are the moral equivalent of nineteenth-century hiking clubs.[26]

CLEANSING WATERS

The most telling aspect of the waste plume of postwar urbanization was the fouling of San Francisco Bay. Unlike the rest of the country, where septic tanks were suburbia's Achilles heel, in the Bay Area housing was sewered, but the sewage was largely dumped untreated into the bay. Only Palo Alto had secondary treatment; the rest of the cities had primary treatment or none at all. San Francisco, which mingled its storm and septic sewers, disgorged vast plumes of untreated waste with every winter rain. San Jose had purposefully built a gigantic collector sewer to serve its canning industry; as the city grew, its modern *cloaca maximus* became the largest stream flowing into the South Bay. Adding to the load were outfalls from dozens of industrial plants.[27]

The Cost of Clean Water

Once again, California moved to the forefront of environmental regulation. Just as Bay Conservation and Development Commission (BCDC) was being made permanent in 1969, the Porter-Cologne Act put in place a strong regulatory apparatus for water pollution. It established a set of regional water-quality control boards along the lines of air pollution control districts and revamped the state Water Resources Control Board (formerly concerned only with water allocation). The system of waste-discharge permits established by the Bay Area Regional Water Quality Control Board became the model for the federal regulatory system embodied in the Clean Water Act of 1972. The board was also the first in the country to use bioassays to test effluent quality.[28]

Nonetheless, water cleanup efforts were moving at a crawl until the passage of the 1972 federal Clean Water Act. Its main contribution to water quality was $30 billion in grants to cities for sewer construction and sewage treatment—the biggest federal pork-barrel program of the day. Around the Bay Area, three dozen new municipal sewage works were built. By the end, every collector sewer had at least secondary treatment (a level still not reached in Southern California). San Jose upgraded. Palo Alto and Sunnyvale claimed tertiary treatment levels. East Bay Municipal Utility District put in a huge plant at the foot of the Bay Bridge. San Francisco, in one of the two biggest projects in the country, spent $1 billion to separate its sewers from storm drains and build two great treatment plants.

This massive investment in clean water paid off, as bay water quality improved in San Francisco Bay and fetid mudflats ceased to emit sulfurous vapors. The bay took a step back from the grave. Yet nearly 200 municipal and industrial outfalls still emptied into it as of 1980. Activists at Citizens for a Better Environment identified three keys to cleaning up the bay: upgrading ambient water standards, cutting back on refinery discharges, and regulating industrial wastes dumped into municipal sewers.

Crude Discharges

In order to get better water-quality standards, CBE put together a coalition to encourage the Bay Area Regional Water Quality Control Board to engage in regional planning, as called for in the Clean Water Act. Dozens of groups joined the Basin-Plan Coalition, each according to its specialty. CBE focused on the toxic threat, Save San Francisco Bay on wetlands, National Resources Defense Council and Environmental Defense Fund on water supply, Greenpeace on wildlife, and so forth. As CBE's water specialist, Greg Karras, discovered when he moved to the Bay Area in 1984, there was already a large and cooperative assemblage of bay protection groups in the region, in sharp contrast to his experience in Los Angeles. The coalition succeeded in pressuring the regional water board to adopt the first standards in the country for toxics in ambient waters.[29]

The coalition also persuaded the water board to introduce the first permitting system for nondomestic discharges into municipal sewer systems and to upgrade its bioassays of effluent quality from using sticklebacks (a notoriously indestructible fish) to trout and invertebrates over longer periods to check for reproductive hazards. Then the attorneys and scientists at CBE went after industrial discharge permits issued by the regional water board. Chevron's refinery in Richmond was the most visible political target and the industry leader. Moreover, it was the biggest bay polluter, still dumping huge quantities of undiluted refining wastes into the bay at Castro Cove.[30]

CBE sued. Then, in 1988, Karras reached a historic settlement with Chevron in which the company agreed to an audit of its production complex to see what could be done to reduce total effluent and residue of toxic and other pollutants. As a result, Chevron's refinery was able to cut discharges of heavy metals such as chromium by 90 percent—confirming the view that the best way to keep such hazardous pollutants out of the environment is to change technology. Karras has since done a couple hundred such hazard-reduction agreements with industrial polluters. These agreements take the radical—and necessary—turn of going around emissions standards to address production directly.[31]

But the refiners threw everyone a curve by shifting to heavier crude as California's oil supplies declined. That this meant more selenium in the effluent was discovered only by happenstance. A biologist tracking selenium runoff from the San Joaquin River in the wake of the Kesterson disaster (wherein water birds were dying from selenium in agricultural drainage water) found that the bay showed higher selenium concentrations than the river. No one knew why until CBE's Karras spotted something suspicious on a permit application by Chevron. The industry denied everything, of course, and the regional water board backed down. So CBE, along with the Bay Institute, San Francisco BayKeeper, and Audubon's Bay Chapter, sued Unocal and Exxon. After ten years, they won in 1996, gaining the biggest clean-water settlement ever around San Francisco Bay: $10 million–$20 million in cleanup and $4.8 million in penalties, which went to the San Francisco Foundation as a fund for bay restoration grants. Selenium discharges went down dramatically.

Chip Solutions

The next big offensive took place in Silicon Valley. The southern bay suffered from two problems: too much freshwater from sewers (since water was imported to the basin) and too many heavy metals (copper, nickel, gold, and silver) from electronics production. Neither municipal treatment nor industrial pretreatment programs were dealing with the problems, so CBE joined with Silicon Valley Toxics Coalition and the South Bay Audubon chapter to form the Coalition for Effluent Action Now (CLEAN) South Bay in the late 1980s.

CLEAN South Bay undertook a couple of innovative strategies. They created a database of all industrial discharges going into the sewers and discovered that actual releases were thirty times what companies had reported. That caught the public's attention. Then they used a feature of the Clean Water Act amendments of 1987 (passed over President Reagan's veto) to petition the water board to implement a toxics-control strategy and to put a flow cap on San Jose's outfall. After being rebuffed by the regional board, they won on appeal to the state Water Resources Control Board.

Its back to the wall, San Jose began to negotiate. The answer, as with the refiners, was to work with electronics assemblers that used metal-plating processes in order to introduce new technology and reduce water use. That cut heavy-metal discharges by 80 percent, while saving money and improving product quality (although most disk-drive and circuit-board production would flee Silicon Valley for Asia in the 1990s).[32]

San Francisco's sewage discharges have also come in for criticism. The Hunters' Point sewage plant and outfall put a heavy burden on the city's poorest neighborhood, Bayview–Hunters' Point, from odors and overflows during storms. A coalition of twenty-one environmental and neighborhood groups convinced the city in 2001 to work on a decentralized wastewater treatment plan instead of expanding the Hunters' Point plant.[33]

CHOOSE YOUR POISON

Toxic metals and chemicals present an indelible threat to people that far exceeds the discomfiture of other kinds of environmental degradation. This threat has been vastly multiplied by industrial progress, both in the quantity of hazardous materials released and in the quality of man-made chemicals, especially the synthetic organics. Radiation wastes became a major hazard, as well, thanks to the mid-twentieth-century breakthroughs in atomic fission. The danger posed by chemical dumps, stack fumes, and contaminated drinking water is so immediate that few would knowingly accept the risk.[34]

While Rachel Carson's *Silent Spring* raised national consciousness in the 1960s through its revelations about far-reaching pesticide poisoning, the sad truth is that things changed very little over the next decade. Americans (including mainstream greens) continued to see the chief environmental issues as land conservation and generic air and water pollution—until the lid blew off the toxic cesspool in the late 1970s. The epoch-making clean air, clean water, and solid waste laws barely touched the problem of toxics. The crisis would come to a boil in the 1980s.[35]

Pesticide contamination was bruited by Congress in the wake of Carson's revelations but only brought under government regulation for the first time in 1971 with the passage of the Federal Insecticide, Fungicide, and Rodenticide Act, a piece of legislation as limp as its acronym, FIFRA. Congress enacted a more general Toxics Substance Control Act five years later that allowed EPA to review chemical products as they came to market. But the agency was overwhelmed by the thousands of substances needing study, and few restrictions followed.[36]

The 1970 Resource Recovery Act raised the problem of disposing of hazardous wastes, but little was done. The 1976 Resource Conservation and Recovery Act (RCRA) finally brought the federal government into the business of finding adequate dumpsites and called for tracking the disposal of hazardous wastes. No one fully appreciated what this would mean since there were still no comprehensive data on the overall waste stream. In fact, RCRA led to the closure of two-thirds of the landfills in the country, opened up a few superdumps full of toxics, and drove waste managers back to an old solution: incineration. Dumps and incinerators would become the flash points of antitoxics struggles.[37]

Workers took the brunt of the toxic tide, so it was left to unions to raise the first protests against the effects of hazardous substances on humans. The United Mineworkers of America was furious about black-lung disease and silicosis due to poor mine ventilation and won the Federal Coal Mine

Health and Safety Act of 1969. Thereafter, the unions led the charge for a comprehensive Occupational Safety and Health Act, passed by Congress in 1970. The instrumental figure in that victory was Tony Mazzocchi of the Oil, Chemical, and Atomic Workers, who went on to inspire a generation of doctors and health workers to organize local committees on occupational safety and health around the country.[38]

Worker health was still seen as something apart from mainstream environmentalism until the Love Canal disaster broke in 1978, hitting a complacent American political order in the solar plexus. California discovered its own Love Canal a year later at the Stringfellow Acid Pits in Riverside County. It soon became apparent that the country was saturated with hazardous wastes, dumped here, there, and everywhere for decades. Toxic dumps and their victims had an element of human drama that the media leapt on. A national antitoxics movement was born, and it would sweep the nation over the next decade—offering one of the few popular revolts against the tightening rule of neoliberalism in the Reagan era. Neighborhood groups formed by the hundreds to protest dumps, incinerators, and poisoned industrial sites. Congress was forced to respond with a cleanup program launched under the Superfund Act of 1980. Congress reauthorized the program in the Superfund Amendment and Reauthorization Act (SARA) of 1986 and strengthened RCRA the same year—the last major pieces of federal pollution legislation to this day.[39]

The popular struggle to control toxics was led by Lois Gibbs of Love Canal and her National Clearinghouse on Hazardous Wastes (now called the Center for Health, Environment and Justice). They were joined by the Boston-based National Toxics Coalition, which led an aggressive grassroots campaign, the Superdrive for Superfund, to win SARA. These two organizations wove together a national network of outraged citizens and did yeoman work to organize and empower a largely working-class base. Another arm of the antitoxics uprising, the environmental health movement, would emerge later around health professionals and scientists. It is represented by such national organizations as the Environmental Working Group and the Collaborative on Health and Environment.[40]

In the Bay Area, the antitoxics movement developed around four axes: pesticides, toxics in the bay, groundwater contamination in Silicon Valley, and general environmental health. Each of these posed distinct problems and engaged different groups of activists.

Toxic Farming

The toxics movement took off in the Bay Area somewhat earlier than in most places. Hazardous dumps were not its focus because they were moved out to the San Joaquin Valley at Corcoran and Buttonwillow (the largest in the United States), where population and political protest were lower. Most incinerators, such as the one built near Modesto to burn the world's biggest pile of used car tires, were also moved out to the valley.[41] Antitoxics struggles in the Bay Area began, instead, over pesticide exposure of farmworkers—not surprising, since agribusiness was drenching California in pesticides at a volume higher than in any other place in the world.

The United Farm Workers made pesticides one of its frontline issues, and it was the farmworkers—not the environmentalists—who succeeded in having DDT banned by the federal EPA in 1970. The decisive lawsuit was brought by Ralph Abascal and Ralph Lightfoot of California Rural Legal Assistance (CRLA) in San Francisco. Farmworker agitation reverberated with the work of scholarly critics of agricultural pest control at University of California, Berkeley, such as Robert van den Bosch,[42] and echoed off the 1970s occupational health movement, represented in the Bay Area by such activists as Amanda Hawes of Santa Clara Committee on Occupational Health (SCCOSH), Robin Baker of the Labor Occupational Health Project at UC Berkeley, and Ellen Widess at CALOSHA. Things came to a head with the scandalous poisoning of workers by the nematicide DBCP at a plant in Lathrop, near the delta. Government action cooled, though, once Governor Deukmejian clamped down on the regulators after his election in 1982.

But the contamination wouldn't go away, and neither would the activists. One branch went international with the Pesticide Action Network (PAN), a project spun off from Food First Institute by Monica Moore and Greta Goldenman, with the help of David Chatfield of Friends of the Earth and journalists David Weir and Marc Shapiro.[43] PAN launched the Dirty Dozen campaign against the most-common pesticides worldwide. A second branch went after pesticide residues on food. Laurie Mott of National Resources Defense Council's San Francisco office shook the apple industry to its roots by forcing a suspension of Alar by EPA in 1989. A third branch stayed local, winning spraying bans in cities around the Bay Area. Standouts in this effort were women such as Donna Sheehan of Mow our Weeds in

Marin, Nancy Skinner of Berkeley Citizens Action, and Sheila Darr of the Bio-Integral Resources Center.[44]

In 1996 a statewide coalition was formed to revive and strengthen the campaign for pesticide control in California. Called Californians for Pesticide Reform, it has come to embrace some 170 groups, including CRLA, PAN, and NRDC. Executive director David Chatfield is a veteran of Friends of the Earth, Greenpeace, and Clean Water Action. The coalition puts out reports, maps, and action plans for pesticide activists up and down the state. The hot-button topic for the Bay Area in the 2000s is the chemical burden in the bodies of everyone—even those seemingly most immune by class, locale, and habits.[45]

The Toxic Bay

The antipesticide campaign dovetailed with the concerns over water quality in San Francisco Bay. One of the biggest cumulative hazards is pesticide runoff from Central Valley agribusiness. In addition, the bay suffers a toxic assault from industrial outfalls, street refuse, automobile exhaust, mining wastes, and old dumps. Bay waters are laced with oil, mercury, copper, dioxin, and PCBs, which slowly eat away at wildlife populations by harming reproduction, weakening fry, and increasing morbidity. As toxics concentrate in the food chain, fish and shellfish are rendered unfit for human consumption.[46]

The danger had been minimized until Citizens for a Better Environment's report *Toxics in the Bay* blew the lid off in 1981. The study was done by Mike Belliveau, a CBE volunteer who stayed on to head its toxics program. Using a seed grant from Save San Francisco Bay Association, Belliveau cobbled together publicly available data to show the cumulative danger to wildlife and public health. The impact was dramatic, because the public and media were primed by years of save-the-bay agitation to hear of any threat to the bay.[47]

Belliveau followed with a 1987 report, *Toxic Hot Spots*. This relied on research from the U.S. Geologic Survey and plugged into the Superfund idea. The study again made a media splash, and the map of toxic hot spots was widely reprinted. The short-term result was a Bay Protection and Toxic Cleanup Act out of the Legislature in 1989, but it had a sunset provision and was killed by newly seated Governor Pete Wilson in the early 1990s. As toxics became the focus of CBE's work, Belliveau was elevated to executive director in 1990. A longer-term outcome was CBE's work with subsistence fishermen to put up warning signs about the high levels of toxics in their catch.[48]

Richmond, at the head of the refinery belt, became a major site of popular organization against toxics. In the early 1980s, local activists such as Ernie Witt, Amadea Thomas, Jean Siri, and Henry Clark were working on community mobilization against industrial pollution when Belliveau of CBE became involved in the town. The cause at hand was an incinerator near an elementary school in North Richmond, where asthma rates were startlingly high among the schoolkids. The California Waste Management Board had joined the incineration bandwagon and industry jumped on board as a way of disposing of hazardous waste. Rebuffed by the regulators, the activists from CBE and Contra Costa County formed the West County Toxics Coalition in 1984. Henry Clark became the executive director, a position he still holds.[49]

CBE put out a report, *Richmond at Risk*, which provided a snapshot of chemical hazards and their impact on the impoverished black residents of the city. As a result, Richmond would figure in the famous environmental justice report issued by the United Church of Christ in 1987 (see chapter 10). The West County Toxics Coalition and CBE have continued to do battle with Richmond's dozens of toxic polluters, such as General Chemical, Ortho Corporation, and Myers Drum—which still incinerates thousands of drums of toxic waste a year.[50]

A prime target for toxics activists has been the Contra Costa refinery belt. All five of its refineries are among the twenty-five most toxic industrial polluters in the state, and Chevron is California's top emitter of toxics, at a million pounds per year. The refineries present a host of problems, including episodic leaks, flares, and explosions, as well as normal emissions. Citizens in all the working-class refinery towns have organized to cope with toxic releases and have gotten the county to create a Hazardous Materials Commission, a Hazardous Materials Ombudsman, and an early-warning system for dangerous releases, but have not been able to stop the refineries' misdeeds. Denny Larsen of CBE help start up bucket brigades of citizen air monitors and hatched the idea of a national Refinery Reform campaign.[51]

A further reverberation of the fight to clean up the refineries has been the controversy over MTBE, a chemical additive to gasoline that improves engine performance. When leaded gas was banned in California in the

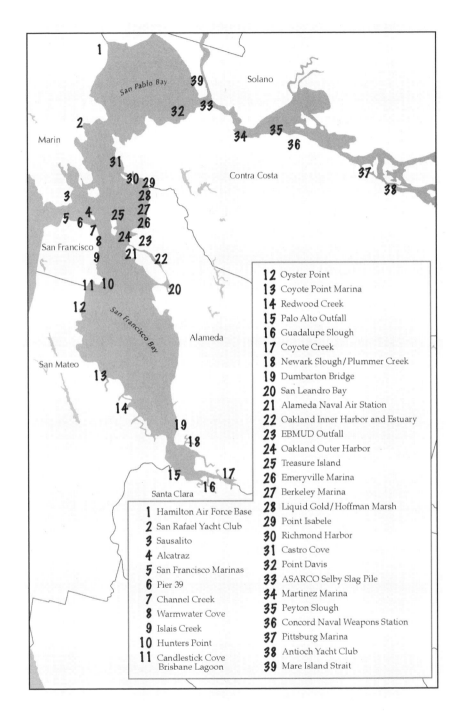

1 San Pablo Bay
39
32 33
Solano
34 35
36
2
Marin
31
30 29
28
27
26
Contra Costa
37
38
3
4 25
5 6
7
24 23
8
9 21 22
San Francisco
11 10
20
12
San Francisco Bay
Alameda
San Mateo
13
14
19
18
15 17
16
Santa Clara

12 Oyster Point
13 Coyote Point Marina
14 Redwood Creek
15 Palo Alto Outfall
16 Guadalupe Slough
17 Coyote Creek
18 Newark Slough/Plummer Creek
19 Dumbarton Bridge
20 San Leandro Bay
21 Alameda Naval Air Station
22 Oakland Inner Harbor and Estuary
23 EBMUD Outfall
24 Oakland Outer Harbor
25 Treasure Island
26 Emeryville Marina
27 Berkeley Marina
28 Liquid Gold/Hoffman Marsh
29 Point Isabele
30 Richmond Harbor
31 Castro Cove
32 Point Davis
33 ASARCO Selby Slag Pile
34 Martinez Marina
35 Peyton Slough
36 Concord Naval Weapons Station
37 Pittsburg Marina
38 Antioch Yacht Club
39 Mare Island Strait

1 Hamilton Air Force Base
2 San Rafael Yacht Club
3 Sausalito
4 Alcatraz
5 San Francisco Marinas
6 Pier 39
7 Channel Creek
8 Warmwater Cove
9 Islais Creek
10 Hunters Point
11 Candlestick Cove
   Brisbane Lagoon

MAP 13. Toxic Hot Spots of San Francisco Bay, 1986. Drawn by Darin Jensen.
Courtesy of Citizens (Communities) for a Better Environment, Oakland.

1970s, the oil companies turned to the chemical MMT. When that turned out to be a widely diffused hazard, it too was banished. So they came up with a substitute, MTBE—a refinery waste product the engineers wanted to make use of. CBE staff could see another catastrophe in the making, because MTBE eats through seals, leaks out of storage tanks, and diffuses rapidly in groundwater—and is highly poisonous and carcinogenic. A campaign against it began in 1994 and caught fire in 1997, with labor unions and environmental justice groups joining the fight. By the time they made headway, however, MTBE was everywhere in California's land and water. The Bush administration, true to form, later tried to shield companies from lawsuits over its hazards.

The Toxic Valley

The third branch of the toxics movement in the Bay Area came out of the South Bay. The chief hazard came from the solvents used to clean silicon chips and disk drives of dust and contaminants on the microcircuits. Leaking solvents from a Fairchild Semiconductor plant and IBM disk-drive factory in San Jose's Los Paseos neighborhood had gotten into well water, where they were discovered in 1982 and quickly linked to serious health problems. A local mothers' group, led by Lorraine Ross, raised their concerns, and a *San Jose Mercury* reporter picked up on it. Her 1982 story caught the eye of Ted Smith, a Stanford-trained lawyer working with Chicano and union activists, whose wife, Mandy Hawes, was an occupational-health lawyer. Smith decided to get involved, starting up the Silicon Valley Toxics Coalition (SVTC), as a project of SCCOSH.

The fledgling toxics coalition mobilized valley residents as the truth came out that there were scores of leaking solvent tanks at electronic production sites all through Silicon Valley and that they were contaminating the groundwater. SVTC won a county ordinance in 1983 that forced companies to adopt double-lined tanks. This groundbreaking measure included one of the first-ever community right-to-know provisions for hazardous materials (an idea drawn from the occupational-health movement). That provision was incorporated into California state law a year later, then into Proposition 65, the California Community Right-to-Know Act, and into SARA, the national Superfund reauthorization act, in 1986.[52]

The SVTC pressured the federal EPA to list many of the solvent leaks as

Superfund sites, making Santa Clara officially the most-contaminated county in the nation. This stripped away the "clean industry" mask behind which high-tech companies had long operated. Another path-breaking local ordinance followed, regulating the use of toxic gases for doping microcircuits.

In 1986 SVTC joined the fight for Proposition 65 and worked with the National Toxics Coalition in support of SARA. Then, using data from the Toxic Release Inventory under the right-to-know provisions of SARA, SVTC put out a damning report on the quantity of toxic materials used by local industry. This led to SVTC's leading role in the CLEAN South Bay coalition, the group that dramatically reduced the influx of toxic metals going into San Francisco Bay.

Smith and SVTC worked with other antitoxics groups over the next decade, partnering with National Toxics Coalition in its Save the Ozone campaign for Earth Day 1989, along with CBE, Greenpeace, Clean Water Action, and the statewide Toxics Coordinating Project. Silicon Valley became a special target of the Save the Ozone movement at Earth Day 1990 when it turned out that IBM and other Silicon Valley factories were among the world's largest users of ozone-killing chlorofluorocarbons (CFCS). In a classic pattern of corporate evasion, IBM first denied the problem, then said there was no alternative, then that alternatives would be too expensive, and finally begrudgingly agreed to a four-year phaseout. In the end, an engineer working in his kitchen at home discovered that a perfectly good solution to cleaning off silicon wafers was, shockingly, soap and water.[53]

Toxic Bodies

The fourth branch of the antitoxics movement in the Bay Area concerns environmental health. It takes a militant stance on the environmental sources of endemic diseases coursing their way through the American populace, in contrast to the genetic explanations favored by so much of the medical establishment. It is a natural extension of the occupational health and anti-toxics movements into realms otherwise considered isolated medical disorders. Environmental health activists have benefited from the local presence of two very rich foundations doling out money in the health field: the California Endowment and the Wellness Foundation.

One environmental health threat of long concern is lead, which disproportionately affects children exposed to leaded paint and pipe solder in older buildings. Leaded gasoline was banned in California by the ARB in the 1970s, but this had not eliminated the danger facing city-dwellers. In 1994, Lead Safe California was formed, with Ellen Widess as executive director, to bring together a broad coalition to develop lead-control policies. It succeeded in drafting legislation satisfactory to everyone from mortgage lenders to tenant groups, but the bills died in the state senate in the mid-1990s.[54]

Asthma is another health scourge that has reached epidemic levels around the world. So many community organizations had arisen around asthma that the California Endowment funded a statewide Community Coalition to Fight Asthma in 2002, including a Regional Asthma Management and Prevention Initiative for the Bay Area, led by Joel Ervice. The Center for Environmental Health, founded in Oakland in 1996 by Michael Green, addresses a wide range of toxic health hazards around the state and brings lawsuits against corporations violating Proposition 65, emitting asthma-causing chemicals, or selling consumer products containing lead, among other things. Settlements go into a fund to support community environmental justice groups.

Breast Cancer Action is another grassroots advocacy group, formed by Elenore Pred, Susan Claymon, and Belle Shayer in 1990. The Breast Cancer Fund was founded two years later, by Andrea Martin. These two organizations function as education, support, and advocacy groups, as in Breast Cancer Action's push for the California Breast Cancer Act in 1993 or the Breast Cancer Fund's campaign for safe cosmetics. Both groups are based in San Francisco but have a national profile. Their force grows out of the extremely high incidence of breast cancer in the Bay Area (for reasons never fully determined) and the long-term instantiation of women's-rights activism here.[54]

In short, Bay Area antipollution environmentalists have been very active since the mid-1970s. Citizens for a Better Environment took up the cudgels against industrial corruption of air and water. West County Toxics Coalition has been busy harassing Chevron and other Richmond polluters. Silicon Valley Toxics Coalition blew the whistle on the electronics industry. Some of these actions have had national reverberations, such as SVTC's pioneering community right-to-know regulations, CBE's hazard reduction agreements, and CRLA's suit to ban DDT. California got the first universal toxic-labeling law, and the Bay Area has led the way on targeting breast cancer as an environmental problem.

All in all, it is an admirable record, but not without a downside. The bay remains a toxic mess, thanks chiefly to uncontrollable, low-level discharges from a million and one nonpoint sources. San Jose still has one of the worst levels of air particulates in the nation. Garbage still piles up at Altamont Pass, Pittsburg, Suisun Bay, North Richmond, and Fremont. The load of toxics in human bodies continues to grow. And the social injustice of the disproportionate impact of toxicity on people of color in the inner cities has not been satisfactorily addressed.

10

GREEN JUSTICE *Reclaiming the Inner City*

The biggest environmental news of the 1990s was the rise of the environmental justice movement, which has its roots in popular uprisings over toxic hazards that endanger the lives and health of working-class communities. But it went even farther by joining together concerns about toxic wastes and public health with those over racism and unequal exposure to hazards. A new generation of environmental advocates of color stepped forward in the 1990s to lead the way toward cleaning up the urban heartland.

Environmental justice advocates put the green mainstream on notice that its achievements were not enough. Protecting the land and waters, forests and wetlands, did not exhaust the meaning of "environment," nor did it protect the people most exposed to the dangers of pollution: people of color, working people, poor people. The urban environment is a place of intense human habitation and the greatest concentrations of industry, commerce, and transport. It is a brown environment as much as a green one, but it is no less a living, pulsing mix of nature and society for all that. If environmentalism is really a city-based battle cry against the ravages of capital, then it is only logical that it addresses the most heavily impacted urban spaces.

Moreover, such places are occupied by those with the least ability to protect themselves from harm, escape the depredations of modern life, or enjoy the virtues of open space at the urban fringe. These people have a very different view of what constitutes "the environment" from traditional conservationist organizations, which are overwhelmingly white and upper class. Proponents of environmental justice have, therefore, an alternative perspective on what constitute the greatest threats to human well-being and what ought to be done to improve their lives. It is not quite true that we all drink the same water, breath the same air, or enjoy the same bay views.

BROWN ENVIRONMENTALISM

As toxics became major news in the 1980s, the disproportionate exposure of African Americans, Mexican Americans, Native Americans, and Asian Americans became a rallying cry for organizing against racism in America and in defense of the health and safety of communities of color. Brown environmentalism was born. One foundation of environmental justice is the civil rights movement, and most of the leaders of the environmental justice movement are steeped in the ethics and politics of racial justice. These activists view the toxic threat not simply as a technical issue of disposal, but as another dimension of the inequities of the social order. Nonetheless, the toxic threat is what prompted a new movement, as people of color, along with everyone else, discovered hazardous wastes.[1]

Environmental justice has its origin stories, just as green conservation does.[2] One is Linda McKeever Bullard's civil-rights suit on behalf of Houston's Northwood Manor housing project against a neighboring toxic waste project in 1979; her husband, Bob Bullard, went on to be a chief intellectual force in the movement. A second one is Chicano organizers in Albuquerque who, in coping with emissions from a particleboard factory, formed the Southwest Organizing Committee in the early 1980s, led by Richard Moore. A third story begins about the same time in Louisiana, where Pat Bryant and the African American Gulf Coast Tenant Project started to take on the vast refining and petrochemical strip along the lower Mississippi River. A fourth beginning occurred in North Carolina in 1982 with protests against a PCB-laced landfill in an African American community; this drew in Benjamin Chavis from the United Church of Christ's Committee on Racial Justice. Chavis solicited the path-breaking report by Charles Lee on systematic racial bias in the siting of hazardous-waste dumps and incinerators.[3]

A self-conscious environmental justice movement was consolidated in

1991 at the first National People of Color Environmental Justice Summit, held in Washington, D.C. Six hundred activists gathered to compare experiences and draw up an action agenda. The summit took its cue from three precedents. The first was making connections among local movements, as the Southern Organizing Project had done in the 1980s in pulling together black protests against toxic dumps. The second was bringing the cause to the attention of the white world, as Pat Bryant and Richard Moore had done just before the summit when they sent letters, signed by one hundred community leaders, to the top ten national environmental organizations that denounced their lack of people of color in leadership positions and their obliviousness to environmental racism; the letters created a national media sensation. The third was engaging the federal government directly, as had been done by a group that included Bob Bullard and Charles Lee when they sent a declaration on environmental racism to William Reilly, director of the Environmental Protection Agency in 1990; they ended up meeting with Reilly regularly.

Bringing together so many like-minded activists generated a new level of energy and launched a host of organizations. The Southwest Organizing Committee created the Southwest Network for Environmental and Economic Justice, Native American activists created the Indigenous Environmental Network, and Asian American militants organized the Asian Pacific Islander Environmental Network (APEN). The impact of the new movement was apparent with the establishment of an Office of Environmental Equity at EPA, rechristened the Office of Environmental Justice by the incoming Clinton administration. Moreover, President Clinton signed an Executive Order in 1994 that required every federal agency to consider its impact on environmental racism and instigate policies to correct such racial discrimination in siting and other ways. Environmental justice had come of age.

BAY AREA ENVIRONMENTAL JUSTICE

The Bay Area gets scant attention in histories of environmental justice,[4] yet the region has played a significant part in the evolution of the national movement. Places such as Richmond and Silicon Valley have been featured in national discussions of environmental racism, and local groups such as Urban Habitat and Asian Pacific Islander Environmental Network have figured prominently in nationwide environmental justice programs and projects. It should also be remembered that long before environmental justice became a catchword, the campaign to protect farmworkers from pesticides was playing out here, the Trust for Public Land had been formed expressly to work in inner-city Oakland and San Francisco, and Judi Bari and Phil Burton were promoting a brand of environmental class justice in the redwoods. The political base for mixing the politics of environmental quality and social justice would appear to be broader here than almost anywhere in the country. In a moment of enthusiasm, attorney Luke Cole, founder of the Center on Race, Poverty, and the Environment in San Francisco, makes the claim that "the San Francisco Bay Area has the highest density, per capita, of environmental justice activists in the United States." While that may be true, it has not always translated into the most effective or unified environmental justice movement on the ground. Brown environmentalism has had to face a more daunting set of problems than its green equivalent and has less to show for its efforts so far.[5]

The Bay Area branch of the environmental justice movement is best known for the work of Carl Anthony and Urban Habitat. Anthony, an African American, is an architect, civil rights activist, and former professor at Berkeley who joined the board of David Brower's Earth Island Institute in 1988 at the behest of Karl Linn, a longtime leader of the national community gardening movement who had retired to Berkeley. Two years later, Anthony launched Urban Habitat, the first program in the country to be aimed at the environment of inner cities. Urban Habitat grew out of the first Eco-Cities Conference organized in Berkeley by Richard Register, a longtime advocate of radical overhaul of cities.[6]

Anthony wedded his long involvement in community aid projects in the ghettoes of New York, Philadelphia, and Oakland to a new realization of the importance of environmental quality for all people. He admired how Judi Bari had been able to wed social justice and forest protection in Redwood Summer. Funded by the San Francisco Foundation, Urban Habitat quickly became the second-largest project at Earth Island Institute. A new journal, *Race, Poverty, and the Environment*, coedited by Luke Cole, filled a critical niche for the widest discussion of environmental justice issues and has served as the principal forum for the national justice movement ever since. Urban Habitat brought 45 of the 600 delegates to the 1991 summit.[7]

Asian Americans had not been visible in toxics and environmental justice work before the summit, and virtually their only representatives came from the Bay Area delegation. Out of that experience was born the Asian Pacific Islander Environmental Network, based in Oakland. APEN was founded in 1993 by Pam Tau Lee, Peggy Saika, Jack Chin, Marta Matsuoka, and Francis Calpatura—all veterans of Third World College strikes, labor organizing, immigrant rights, and other Asian American ferment. Saika became the first executive director; Vivien Chang is the current director. Pam Tau Lee was appointed to the federal EPA's Environmental Justice Advisory Council in 1994, along with Carl Anthony.

APEN began organizing among Laotian refugees in Richmond over exposure to toxics through fishing, gardening, and industrial air pollution. It has since opened a second front in Oakland's mixed Asian communities and has broadened its work to include housing, the home environment, and workplace health and safety. For example, it organized workers at the American Xtal Technology (AXT) semiconductor plant in Fremont to call in CALOSHA over arsenic poisoning. APEN remains focused on grassroots organizing and community needs, but is open to alliances with progressive groups around the Asian communities of the Bay Area. It leaves regional environmental work to other organizations. APEN remains the sole Asian American environmental group nationally.[8]

Another grassroots environmental justice group is the West County Toxics Coalition (WCTC), which grew out of organizing in Richmond and is deeply committed to the African-American community there (see chapter 9). Not only has it confronted Chevron over air quality, it has had to address severe and immediate threats such as the sulfuric acid–plant explosion in Richmond that sent 20,000 people to the hospital in 1993. In that case, WCTC, in alliance with Citizens for a Better Environment and California Rural Legal Assistance, won a class action suit against General Chemical and got a toxic release early warning system put in place in Richmond. Another long-term battle was to rescind the license that the Bay Area Air Quality Management Board had kept renewing for the incinerator at the Ortho Corporation (a subsidiary of Chevron) pesticide factory, which was finally shuttered in 1997.[9]

Several organizations have sprung up in Oakland as local mobilizations by communities of color against toxic hazards. East Oakland's People United for a Better Oakland (PUEBLO) grew out of a lead-poisoning screening project of the Center for Third World Organizing in the early 1990s, which created a model county program for lead-poisoning education and assessment. PUEBLO moved on to successful campaigns for mandatory measles vaccination, a good-neighbor agreement with Owens Brockway glass factory in Fruitvale to eliminate its dioxin emissions, and a brownfield-site cleanup of an old lead-battery plant. It also started a police watch project and a campaign for just-cause eviction, showing its multi-issue, multi-ethic approach. At the end of the decade, it led a fight to clean up the Integrated Environmental Systems incinerator in East Oakland, the last medical-waste incinerator in the state. A broad coalition formed between local and regioinal groups, but PUEBLO pulled back when it became apparent that the goal was to close IES, with attendant loss of jobs. In 2004, an accounting review found evidence of financial mismanagement, and outside funding evaporated. The board was forced to terminate all staff, and many went on to other projects around the city. PUEBLO's board and staff were reconstituted the following year.[10]

West Oakland has been a focus of activists, due to its combination of location, pollution, and populace. Long a center of African American habitation and organizing, the area holds a special place in Bay Area racial politics. At the same time, West Oakland is rife with toxics hazards from past and present industrial sites, encircling freeways, and the vast Port of Oakland complex. The people of the area are exposed to a witch's brew of toxics and have a number of ailments, such as asthma, at much higher rates than elsewhere. Several organizations have raised the environmental justice banner in West Oakland. Closest to the grass roots are the Chester Street Block Club Association, headed by Renee Morrison; Citizens for West Oakland Revitalization, led by Monsa Nitoto; and the Seventh Street Initiative, a community planning network. These groups grew up around struggles to improve neighborhoods afflicted by disinvestment, city neglect, freeway reconstruction, and other inner-city ills.

A model victory was won in the case of the AMCO Chemical site, in the South Prescott neighborhood of West Oakland. In 1995, utility workers struck an underground plume of vinyl chloride left by AMCO. After detecting vapors in nearby homes and the local playground, EPA set up an incinerator to burn off gas extracted from the soil. The Chester Street Block Club, backed by Greenaction, pressured EPA for a better solution and won Superfund listing in 2003, which meant soil removal and replacement, with residents relocated during the cleanup.[11]

West Oakland's ills have also drawn the attention of professional groups such as the Urban Strategies Council and PolicyLink in downtown Oakland—both started by Angela Blackwell—liberal think tanks that have taken an interest in environmental justice and are backed by Hewlett Foundation grants. A significant intervention has been by the Pacific Institute, an Oakland nonprofit and another Hewlett grant recipient, whose main focus is the global environment and sustainable development. Pacific Institute launched an environmental indicators project in 2000, a community-guided research project on problems of the neighborhood, covering everything from air quality to property ownership. Key toxics issues are brownfield sites, factory emissions, and diesel fumes. Institute staff such as Meena Palaniappan and Azibuke Akono partnered with West Oakland activist groups and sought local volunteers to take air samples, monitor hazardous areas, and report emissions to the Bay Area Air Quality Management District. The indicators project helped the locals win a long battle to close the Red Star Yeast factory over its emissions of carcinogenic acetaldehyde.[12]

Attention turned next to diesel fumes, which are particularly bad because of heavy truck traffic in and out of the Port of Oakland. Volunteer monitoring in West Oakland revealed fine particulate levels ninety times the state average. Rallying to the cause, regional activists from TALC, Sierra Club, and Urban Habitat, among others, formed the Transportation Justice Coalition to mobilize support for the West Oaklanders' campaign to stop trucks from idling endlessly outside the port while waiting to load and unload containers.[13]

In San Francisco, the highly polluted southern waterfront and the Bayview–Hunters Point neighborhood (mostly African American and poor) have been the center of environmental justice mobilization since the mid-1990s. The worst offender is the decommissioned Naval Shipyard at Hunters Point, a Superfund site riddled with poisonous waste, including lead, PCBs, and radioactivity; other sore spots are the two bayside power plants, the city's sewage outfall at Yosemite Slough, and diesel fumes. The Southeast Alliance for Environmental Justice, cobbled together from community groups in 1995, blocked a proposed third power plant before disbanding. The fight against PG&E was carried on by Mothers' Environmental Health and Justice Committee, led by Marie Harrison (now with Greenaction), and Bayview–Hunters Point Community Advocates, led by Olin Webb and Karen Pierce, which has gone on to installing alternative energy systems in homes and businesses. Literacy for Environmental Justice, a youth group headed by Dana Lanza, does air monitoring and oversees Heron's Head Park (a restored piece of the bayfront). Alliance for a Clean Waterfront and Community First Coalition unite local and citywide organizations pushing for cleanup of Yosemite slough and the shipyard. Crucial support has come from regional organizations such as CBE, Golden Gate Audubon, Greenpeace/Greenaction, Earthjustice, and the Environmental Law and Justice Clinic at Golden Gate University.[14]

The most active group in the Mission district has been People Organized to Defend Economic and Environmental Rights (PODER), led by Antonio Diaz, which operates on several fronts, including toxic brownfield sites and housing displacement. Latino Issues Forum has also frequently joined in environmental justice battles, such as the Potrero power station expansion. The Chinese Progressive Association, headed by Warren Mar, is a long-standing force for social justice in the city, and it too has gotten involved in community toxics hazards and community organizing in Chinatown and other Asian neighborhoods.

Two San Francisco-based environmental justice organizations, spun off from Greenpeace, have been energetic in support of community groups at Hunters Point, Oakland, Alameda, and East Palo Alto, as well as around the West and abroad. Greenaction was established by local activist Bradley Angel in 1998, after Greenpeace dropped its U.S. environmental justice program. ArcEcology was founded in 1983 by Saul Bloom, also formerly with Greenpeace, to focus on arms control and the environmental damage caused by the military. Major achievements of the city-wide movement have been a $13 environmental justice grant program and the passage of the country's first "Precautionary Principle" legislation in 2003, committing the city to exposing residents to as little toxic harm as possible.[15]

A unique institution is the community gardening group, the San Francisco League of Urban Gardeners. SLUG began as a largely white community-gardening coalition in the northern part of the city. But in the 1980s, it morphed into an environmental justice organization centered in the southern edge of the city, under the guidance of executive director Jeff Miller and board member Karl Linn. Its functions broadened along with its constituency, from gardening to community mobilization, job creation, and brownfield remediation—including reclaiming toxic tidelands for community parks. Under the direction of Mohammed Nuru, SLUG grew tremendously, fostered by grants from the city's environmental justice fund.[16]

A MOVEMENT TRANSFORMED

Environmental injustice had a profound impact on the antitoxics movement. Things started out well enough, as the Citizens Clearinghouse and National Toxics Coalition worked with local groups of all colors and were deeply sympathetic to the demand for racial justice. For example, Ted Smith and Richard Moore first met at a workshop called by Lois Gibbs and went on to collaborate on projects including Sacred Waters. The NTC sent its organizers around the country, including Richmond, and was involved in the founding of West County Toxics Coalition. After the 1991 environmental justice summit, NTC's staff insisted on more people of color on the board, and Pam Lee, Richard Moore, and Peter Cervantes-Gautschi were recruited. But then things turned sour. White board members, other than Kate Kyker and Ted Smith, were hostile to the move, and the internal pressure blew the organization apart in 1993.[17]

Similar upheavals occurred in the Bay Area. A dramatic rupture took place when Urban Habitat broke away from Earth Island Institute. Carl Anthony could not get white colleagues who were agitating for forest protection in New Mexico to take notice of Hispanic anger over the loss of historic rights to the woods. He recounts:

> After you think about the struggle for 150 years of people who have some relationship to the land to have these environmentalists show up—not that they were wrong in their prognosis, but they were incredibly wrong in their rudeness and their lack of capacity to understand that people had a relationship to this land that had to be dealt with. . . . There was an arrogance there that was really unbelievable to me. It was very painful to me. I expected a certain level of racism, but I didn't expect this kind of really deep arrogance.[18]

Even though Anthony was president of Earth Island Institute, he could not move the zero-cut fanatics, and David Brower straddled the fence. In disgust, Anthony took Urban Habitat independent in 1997. Later, the office moved to Oakland.

Citizens for a Better Environment went through a radical makeover from within to become an environmental justice organization. Impressed by what he had learned in Richmond and by the critique of the big green organizations, Mike Belliveau was determined to make CBE's staff look more like the people of California and to do more work with communities of color. He brought on board Carlos Poras, Wendell Chin, and Richard Tokyushi Drury, and by the time Belliveau departed, half the staff were of Latin, African, or Asian descent; that proportion subsequently rose to two-thirds.

Poras, a former labor organizer, reoriented CBE's Southern California operation by targeting poor communities and organizing around their demands, bringing scientific and legal expertise to bear on local struggles as needed. It was a very successful strategy, the foundations liked it, and CBE came to be widely seen as a leading force for environmental justice in Los Angeles. The southern office soon outgrew the northern one. The Bay Area office remained whiter and more top-down in its strategies but was also integrating community organizing into its tactics.

The headquarters of CBE eventually moved to L.A. and its name was changed to *Communities* for a Better Environment because the word "citizen" was felt to be distasteful to immigrants. The financial, legal, and scientific staff are about evenly split between the north and south offices.

CBE's transition was not without drama, however. When Stephanie Pincetl became chair of the board of directors, she was determined to push the process ahead faster—a reflection, no doubt, of the successful developments in L.A., where she lives. The board was made over and Belliveau ousted in 1998. A series of interim executive directors cycled through and many new staff were brought on board, so programs suffered from inexperienced staff. A serious drop in funding hit during the sharp recession of 2001–03. Poras was made executive director but replaced not long thereafter.

If CBE did not implode as the National Toxics Coalition had, it did limp a bit. Nonetheless, the work continued and people's commitment to the cause was undiminished. And although interpretations of events differ, everyone associated with CBE believes the organization is back as strong as ever and is better for the makeover. Indeed, they are proud that CBE is one of the only groups in the country to make the environmental justice transition successfully. Northern California director A. J. Napolis is trying to follow much the same model of community organizing as in the south, but differences persist.[19]

Meanwhile, Silicon Valley Toxics Coalition, which has long partnered with environmental justice groups, brought on board more staff and directors of color. It has a strong health and environmental justice program, which organizes an annual People's Earth Day. Yet Ted Smith's lack of color—despite his long association with civil rights and labor militants—has led some people to define svtc out of the environmental justice movement. In 2005, Smith stepped down and Sheila Davis, a woman of color, took the helm. This has allowed Smith to move away from local organizing to focus on global high tech through the Campaign for Responsible Technology and Clean Computer campaign. While the electronic campaigns have given the organization a firmer claim to expertise and allowed it to make a quantum jump in foundation funding and staff size, they also took it farther afield from grassroots organizing.[20]

ONE NOTION, DIVIDED

Environmental justice has expanded the scope of the environmental movement, as communities of color discovered that green concerns are not just an indulgence of well-favored and as white environmentalists came to see that they were missing an essential dimension of the pollution problem: its disproportionate impact on the poor and people of color. Carl Anthony is a good example of someone who underwent a change of heart as he came to grips with the ideas of environmentalists such as David Brower and Wendell Berry; as he puts it, "[A] central question to me is, what is my relationship to the land as an urban person? It's a very fruitful question. I'm able to see things now because I have traveled this path that I couldn't see at the beginning." Conversely, cbe's Greg Karras lost his blinders by talking to fishermen about the toxics threat and hearing their anger about the necessity of fishing to supplement their meager diets, knowing that the fish were poisoning their families.[21]

Nonetheless, environmentalism is not one big tent over a joyous revival meeting, everyone locked arm in arm. The demand for environmental justice has just as often divided the wider movement into green and brown parties, regional and local groups, professionals and grassroots organizers, and a throng of competitors scrambling after the same paltry grant funds. One despairs at times of the fragmentation and falling out that has too often impeded the environmental justice movement.

Racial Divides

There is no getting around the reality of a deeply divided society in America. An essential dimension of this is the segregation of urban space, as shown by every study of the residential topography of American cities. This is not simply an unintended consequence of income inequality, but a systematic product of the actions of the privileged. The richer and whiter people are, the better able they are to insulate themselves from factories, pollution, and toxics. At the same time, they flee from the poor and the dark-skinned into white and wealthy enclaves such as Marin County, Blackhawk, or Palo Alto, and they erect subtle but effective barriers to entry for those less privileged. As a consequence, the confinement of people of color and working-class folk to less-desirable neighborhoods and their greater exposure to toxics is not an accident of fortune.[22]

When people have widely different circumstances, their views about the environment are going to be at odds. Some people are more likely to see green foothills as their backyard; other people, abandoned industrial sites— hence, the greater interest among the upper classes in open-space conservation than in brownfield remediation. The schism is rendered worse by the sense of entitlement of the upper classes, who have a much more expansive sense of their "backyard," imagining as their rightful domain the whole city and even far beyond to distant natural wonders. At the same time, European Americans are notoriously blind to their own white-skin privilege, as upper-class people are to the struggles of working-class survival. For most white environmentalists, the key problem is too much growth, not too much inequality, and the central threat to their way of life is too little breathing space, not too little justice. Worse yet, there are not a few who feel that it is other people—especially dark-skinned immigrants—who are the greatest threat to their golden lands.[23]

People of color and workers are, by contrast, much more constrained in their movement and experience, more laden with everyday burdens, and more concerned about a decent job and basic safety. Their notion of the environment is likely to be narrower in geographic scope and broader in definition, including housing, jobs, and schools. Hence, brown environmentalists believe that a better environment requires social amelioration more than wildlife habitat, community resources more than scenic hillsides. Nevertheless, inner-city folk are quite appreciative of the pleasures of natural beauty, parks, and wildlife when they have the chance to enjoy them.

The social chasm between green and brown environmentalists has, as one would expect, led to frequent clashes, mutual incomprehension, and wounded pride. Some environmental justice militants simply refuse to work with their white counterparts. In Henry Clark's sober view:

> It's probably a long ways before there's any real working together or cooperation. . . . You can put up a lot of pretense that we're working together and that we all are sitting in a room together, but when it comes down to it, you know, the people of a certain class or a certain color or a certain wealth, they're going to be running things and making the decisions, and the same old racism and classism and divisions among people in the broader society is going to be right there. It's going to work against us working together.[24]

Ted Smith is blunt in his assessment of how hard it has been to get support for urban toxics-control campaigns from traditional conservationists such as the Committee for Green Foothills. Worse, one still hears horror stories of people of color invited onto mainstream boards or staffs and running into good old-time racism from some pillar of the ruling class.[25]

In the political circumstances of the Bay Area, it is rare to find green activists who are unsympathetic to the idea of social justice. Marty Rosen of Trust for Public Land, for example, sees himself blending the philosophies of Martin Luther King Jr., Aldo Leopold, and Frederick Law Olmsted. As we've seen, key toxics activists, such as Ted Smith, Greg Karras, and Mike Belliveau, wholeheartedly embraced the lessons of civil wrongs and environmental injustice in America. Bradley Angel made the leap from Greenpeace to Green Action. Bay Area foundations, such as Gerbode, Heller, Vanguard, San Francisco, Rosenberg, and East Bay Community, have been supportive of environmental justice. The fact of racial divides in urban space and in mental frames of reference remains daunting, however, even here.[26]

Class Divides

Tension can run high *within* the environmental justice movement as well. This is often put in terms of locals against outsiders. Local organizers see themselves as rooted in the local community and committed for the long term, while outsiders flit in to capture a high-profile moment and depart. This can lead to resentment, such as that felt by PUEBLO when environmental health and anti-incinerator groups descended on its turf in East Oakland. Henry Clark, of West County Toxics Coalition, grumbles:

> When outside groups come into a community, they need to work with the organizations that are working on those issues already. They need to work under the leadership of groups that are already there. Not in the arrogant sense, but no one can replace the WCTC or upstage us because we pioneered the way.[27]

This warning can apply just as well to a big-picture group such as Urban Habitat, even if it is run by people of color. Nevertheless, it is not true that *only* neighborhood groups have roots among the people. Some regional organizations such as CBE and Sierra Club have been embedded in the environmental community of the Bay Area for a long time. All community is not at the micro scale. So what makes someone an outsider?

A lot of it has to do with internal class differences, which gives some people more standing and greater access to the corridors of power. Community groups normally lack the kind of scientific credentials, legal expertise, and knowledge of government protocols that larger organizations enjoy. The latter get those skills by hiring college-educated staff, usually from outside local communities. Nevertheless, it is not possible for community groups to carry the fight into the courts, to regional boards, or on to Sacramento by themselves. WCTC, for example, has had to rely repeatedly on legal help from CBE and others. Organizations in West Oakland need the technical aid and resources of Pacific Institute to advance their causes against Red Star Yeast and the Port of Oakland. Working together is necessary, everyone acknowledges, but it easily engenders resentment on the part of those who feel dependent or a sense of being unappreciated on the side of those who help.[28]

There is, furthermore, the question of who gets to speak for the movement and who gets to set priorities. The directors of large organizations such as Urban Habitat and CBE are better placed to be sought out by the press and are more at home in the world of professional, business, and political affairs. Carl Anthony and Juliet Ellis, for example, are eloquent advocates of environmental justice who have made Urban Habitat a model of enlightened leadership, capacity for compromise, and ability to work with the widest possible range of allies. Yet Urban Habitat is far from a community-organizing group in the classic mold and not representative of the Bay Area environmental justice movement as a whole. Indeed, Anthony and Ellis are not without critics among movement activists for being too far removed from the grassroots. In 2005 there was a mass departure of staff from Urban Habitat over just such disagreements. On the other hand, young organizers are born rebels who often fail to understand the importance of the kind of firm leadership and a consistent direction that has made Urban Habitat so successful for years.[29]

Money Divides

Funding is perhaps the biggest source of tension both within the brown environmental movement and between it and the mainstream greens. Memberships and donations don't amount to much in poor neighborhoods, so community groups are forced to look outside for financial support from foundations, state programs, or city funds. Even regional organizations such as CBE have had to abandon the idea of supporting themselves through canvassing, turning instead to foundations, just as the mainstream greens have (see chapter 7). But the money available for environmental justice grants is modest compared to that given to the green mainstream, and it became even scarcer when foundation funds shrank in the economic slump of the early 2000s. Competition for those funds is fierce and entails a good deal of internecine discord, particularly from community groups that find it very difficult to win grants. Not only are small groups less likely to have direct channels to foundations, it is a heavy burden to allocate precious staff time cultivating foundation connections, writing grants, and monitoring budgets rather than doing community organizing.[30]

An irony for the environmental justice movement is that the more successful of its organizations have become part of the same nonprofit economy as the big green organizations the movement has long criticized. Urban Habitat, in particular, is a creature of the foundations, thanks to connections cultivated by Carl Anthony while at Earth Island Institute. When Anthony left the Bay Area in 2001 to run an international fund at the Ford Foundation, his successor at Urban Habitat (and onetime assistant), Juliet Ellis, had just spent three years at San Francisco Foundation. The success of Urban Habitat is undoubtedly tied to its ability to keep the money flowing from foundations to projects, including those, such as *Race, Poverty, and the Environment*, that benefit the wider environmental justice movement.

CBE is another case of success through good funding, including cost recovery on its legal victories. Paradoxically, CBE's much greater reliance on foundations today than twenty years ago makes it less financially grassroots-based at the same time that it has gone over to being a fully environmental justice organization run by people of color. Is it not ironic that Juliet Ellis praises A. J. Napolis for making CBE "less confrontational" at the same time that CBE has become less white? And is it not a mixed blessing to be featured by the Ford Foundation as a model environmental justice organization?[31]

No doubt compromises have been made, and have to be made, to please mainstream funders, and no doubt those compromises grate on the sensibilities of the younger, more radical among community organizers. Nevertheless, the unpleasant reality for the environmental justice movement, as for any popular mobilization based in poor barrios and ghettoes, is that there are too few resources, too many intractable social problems, and too few tangible successes. Under the circumstances, it is no easy task to hold organizations together and keep the flame of hope alive. No wonder the movement is fraught with internal fractures and prone to resentments. One can well admire the tack taken by Van Jones, a young and charismatic environmental justice advocate and founder of the Ella Baker Institute in Oakland, that one should try to speak well of everyone in the environmental movement and to avoid the kind of schisms that have always plagued the militant left, but the conditions for such behavior lie more in adverse circumstances than in ideological failings.[32]

THE URBAN HABITAT

The more aware one becomes of the problem of environmental injustice, the more it seems that hazards to community health and well-being are to be found at every turn: toxic brownfield sites, diesel-truck exhaust, contaminated military bases, leaking gasoline storage tanks, and power plant plumes. These hazards are close to where people of color live and work. But they are not the only hazards the lower classes face. The poor have to deal with neighborhoods rife with crime, dangerous traffic, run-down schools, lack of hospitals, and unemployment. To the city dweller, the term "environment" implies all these things, and one's exposure to hazards depends on the whole arrangement of urban space and place. This forces environmental justice advocates to think like geographers and city planners.

Carl Anthony offers a particularly capacious view of the urban environment, thanks to his background as an architect, planner, and teacher. Anthony's comprehensive perspective is at the heart of Urban Habitat's regional vision. At the 1991 summit, he noticed that most environmental justice activists were based in rural areas and had little grasp of urban issues. He started thinking about the inner city in environmental terms, focusing on problems such as the reuse of contaminated industrial sites and old military bases. In keeping with this, Urban Habitat put out two major reports, one on redeveloping brownfields and the other a critique of transportation policy. From these followed the Brownfields Leadership and Community Revitalization projects and the Transportation and Environmental Justice Project (now the Transportation Justice Working Group). Anthony also took on the task of chairing a public planning effort for the conversion of the decommissioned naval air base in Alameda to civilian uses.[33]

Anthony's thinking goes even further, however, to a consideration of the whole of urban growth and decline. As he puts it, "[A] big challenge is to be thinking about our metropolitan regions holistically. . . ." He is critical of unchecked urban sprawl and argues that "This pattern of fragmented land-use decision making and squandering of land has incredibly negative environmental consequences, but it also has important social justice consequences." He therefore sees attempts to control sprawl through smart-growth policies as an opportunity, "a potentially important, strategic alignment," and is critical of his own constituency for ignoring it. "The environmentalists and middle-class people are pretty well onto [Smart Growth], but communities of color are not. They generally operate under the assumption that whatever those people are doing in the suburbs, it has nothing to do with us."[34]

Anthony is not quite correct in his assessment of what other activists of color are thinking, however. For one thing, they have sometimes supported

Smart Growth in the form of inner-city densification and transit villages, as Monsa Nitoto has in West Oakland. At the same time, they have frequently been critics of urban-limit lines, a favorite tool of open-space advocates by the 1990s. The Greenlining Institute, in particular, has taken up the cause against zoning and planning restrictions that adversely affect the poor and people of color. The reasoning against green lines is the common fear that they limit the supply of land, thereby driving up housing costs within the cities, and higher housing costs are a matter of enormous account to working people throughout the Bay Area.

There are others thinking holistically about the urban habitat, as well. A coalition of environmental justice groups put their heads together to issue a manifesto, *Building Healthy Communities from the Ground Up*, in 2003.[35] The coalition consisted of APEN, CBE, Environmental Health Coalition, PODER, and SVTC. The manifesto lays out a sweeping set of policies for achieving environmental justice: reduce exposure to harmful pollution, ensure decent housing, protect workers' safety, promote community-based land-use planning, oversee transportation development, and provide good schools. And it calls for greater state capacity, increased foundation support, and more solidarity with social and economic justice movements. Ambitious, to be sure, but it follows directly from the logic of environmental quality and equality in the city.

A practical example of what these urbane activists are after is Richmond Vision 2000, which joined environmental justice, labor, and church groups into a broad alliance for clean air, a living wage, and just-cause evictions. Another approach, pursued in San Jose, Emeryville, and Sonoma County, is to secure community benefits agreements, which extract concessions from developers on wages, jobs, and community aid in return for political support for major urban in-fill projects.[36]

For its part, Urban Habitat under Juliet Ellis continues to promote regional thinking. Its biggest program is called the Social Equity Caucus, which brings together a host of environmental justice and related groups to create joint campaigns on everything from the Racial Privacy Initiative of 2003 to regional transportation policy. Included are Citizens for West Oakland Revitalization, CBE, Richmond Vision 2000, Contra Costa Central Labor Council, and TALC, among others. It is one of the broadest coalitions of its kind in the country and one that other regions are looking to imitate.[37]

Ellis has even bigger ambitions for Urban Habitat. She believes that the environmental justice movement is "stuck in a hole because it's too small" and that its "evolution has to be multisector and multi-issue."[38] She is enthusiastic about a grand coalition on regional policy—the Alliance for Sustainable Communities—put together by Urban Habitat, Sierra Club, ABAG, and the Bay Area Council. This alliance has raised an impressive $160 million from banks, corporations, and foundations for a Community Capital Investment Initiative, which includes a Smart Growth Fund and a Brownfield Fund. That kind of money is forthcoming because the Bay Area Council is the leading big-business forum for the metropolis.

The Alliance for Sustainable Communities was also made possible because the Bay Area Council had a sympathetic director from 1997 to 2003 in Sunne McPeak, a liberal politician with a green streak and a sense of racial justice. Such bridges across the gulfs of class and race seem more possible here than most anywhere else in America. As Ellis observes, "The Bay Area is unique in what you can do: if you can't make it work here, then it's not going to work anywhere."[39]

This kind of regional coalition with big business harkens back to the schemes of People for Open Space and the Vision 2020 Commission a generation earlier. Not surprisingly, the pursuit of environmental justice in the inner city has led back to the same big questions about urbanism that were once posed by the Modernist planners. The programs of those seeking a better urban habitat have once again converged, as we saw when greens made common cause with social justice advocates and organized labor over affordable housing, Smart Growth, and transportation policy in Silicon Valley in the 1980s or in Sonoma County in the 2000s (see chapters 6 and 8).

But such alliances have a cost. At best, working with mainstream greens and capitalists can be irksome for environmental justice advocates, even for someone as open-minded as Juliet Ellis. Not surprisingly, Ellis has come in for criticism from some environmental justice militants for cozying up to big business. They have questioned the wisdom of allying with the Bay Area Council instead of focusing on the grassroots, given the kind of compromises that are needed to secure capitalist backing for projects in communities of color. Maybe, contrary to Ellis, cross-class and cross-race alliances cannot work here, as long as the country remains as divided and unequal as it is.

In a similar fashion, West Coast activists rebelled against tendencies in the national environmental justice movement after the second National

People of Color Environmental Justice Summit, put together by Bob Bullard and Beverly White in 2002. The California delegations were not in accord with the easterners on two counts: a black versus white view of race, and a policy-oriented and legalist strategy. This was, in fact, the trigger for the formation of the California Alliance for Environmental Justice and the writing of *Building Healthy Communities*. Local militants wanted more, not less, grassroots community organizing.[40]

Nonetheless, a danger lies in the path to community development, social justice, and urban political economy. It may be biting off more than the groups can chew. If so, they could well fall victim to normal politics in America, which has swallowed so many leftists before them. Their deep critique of capitalism and racism is correct, but (short of a revolution) there is still the matter of the best strategy for effecting change. There is no easy answer. Environmental justice militants often criticize white environmentalists for their narrow vision, but the fact is that American politics rarely rises above single issues, short-term crises, and campaign personalities. Diluting one's message by trying to do too much can lead to incapacitation. Militants would, therefore, do well not to forget the *environment* on the road to environmental justice.

CONCLUSION *City and Country Reconciled?*

Looking over the Bay Area, one sees a remarkable amount of open space—green, blue, and golden—on every side. Environmental protection is well integrated into the fabric of urban development through the institutionalization of land planning and preservation by governments, nonprofit organizations, and the law. It is, moreover, a part of the life of Bay Area residents, who make heavy use of beaches, trails, bay fronts, and swimming holes. And, while this is an urbane countryside in most ways, some reaches are still remarkably wild.

Green ideology and politics have been strong in the Bay Area for more than a century. Conservation began as—and to some extent remains—a bourgeois conceit, to be sure. One cannot deny the upper-class sources of the open-space ethic, nor the upward tilt of the Bay Area's class structure, with its abundant sources of wealth, large slice of technical and professional labor, ample opportunities to pass through the entrepreneurial eye of the needle into the capitalist class, and thick strata of well-paid workers. A green city is a sort of bourgeois utopia.[1] No wonder there is such an immense reservoir of green sentiment among the highest ranks of the bourgeoisie.

On the other hand, Arcadian dreams are not the exclusive province of the well-to-do. Here in the Bay Area, such ideas have captivated a wide cross section of the citizenry, who made them their own and learned to love the urban countryside with a passion. Here more than anywhere in the United States, environmentalism developed a mass popular following that caused conservation organizations to burst their upper-class seams and propelled politicians to put in place laws that are still the bane of the Rambo capitalists and neoliberals. This entrenched ecotopian outlook is worthy of respect.

Bay Area environmental militancy demonstrates that radical politics are not limited to the liminal, as some postmodernists would have it, nor solely the property of the proletariat, as many of the Old Left once believed. Radical politics can show up, unexpectedly, among elements of the most favored folk. A crucial aspect of the whole story is that you don't save redwoods, wetlands, or clean air by being nice and playing the game by the rules. Even upper-class conservationists often went head to head with the forces of capital and growth and beat them back, changing the rules of engagement along the way.

The Bay Area planning movement was not a typical exercise in Modernist fantasy and arrogance. The growth-control rebellions were not a "Watts Riot of the Middle Class." The bay and coast were not saved through liberal reform, but by the strictest regulation of private property in the public interest. The greens have not ignored the inner city but have been advocates of urban in-fill, affordable housing, and public transit. And the call for environmental justice has been answered from many quarters, including some from the world of white privilege.

There is nothing magical in this kind of commitment. It does not arise from the fresh ocean breezes or congeal among the mists of the forest primeval. It is a material product of a long history, it is cumulative, and it is a collective project. Bay Area environmentalism remains vibrant because of the lessons of past green politics and the sense of possibility bred by repeated success. It is also lived every day in the views of the open bay, birds on the mudflats, and unspoiled hillsides, and it is enjoyed regularly on walks on park trails, swims in recreational lakes, or even flights into the airports.

But that taken-for-granted faith in the greensward can be a danger. It makes us lax about the rampant urbanization of the Central Valley, the way supposedly safe areas are chipped away as at Mount San Bruno, and the unacceptable air quality in West Oakland. And forgetting what has been achieved, a loss of historical memory, is a sure route to political defeat in the future. Look at American politics forty years after the Gulf of Tonkin Resolution and the Civil Rights Act.

The Bay Area experience argues against the view of some critics that the environmental movement has been downhill since the sixties. That argument is quite ironic, since it is the greens who are usually accused of "declensionist narratives"—saying the world is going to hell in a handbasket—and now there is a declensionist narrative about the environmental movement itself. This complaint was made first by white radical environmentalists of the 1980s and then embroidered in the early 1990s by environmental justice advocates. It has been picked up by Bob Gottlieb and, more stridently, Richard White, among environmental historians. It got a wide play yet again after the 2004 election in an essay by two young, brash East Bay green provocateurs, Ted Nordham and Michael Shellenberger, called "The Death of Environmentalism."[2]

The argument is that the big green organizations moved to Washington, D.C., cut their connection to a mass base, and lost their radical edge. Trying to be players inside the Washington beltway, they acted as unelected ventriloquists for nature and the people while being bought and paid for by Ford Foundation and other rich donors. As they narrowed the possibilities of environmentalism, they fell prey to the snares of scientism, compromise, and legalism, which limited their effectiveness. And, by playing the liberal game, they were condemned to fighting a long retreat against the corporations and their armies of lawyers, lobbyists, and house experts.

No doubt, the beltway is corrupting, part of the corruptions of power that are rife in Washington, D.C. Nonetheless, the critique misses several essential things. First, it fails to recognize that the move to Washington made sense in its time, following legislative success in Congress and legal victories in the U.S. Supreme Court. Circumstances have changed dramatically in national politics, rendering former tactics obsolete. Second, the greens have recognized this and altered their tactics over the last decade to more electoral and grassroots work, as well as forming coalitions with labor and social-justice forces.[3]

Third, the critics define the big green machine far too narrowly. By ignoring the multitude of grassroots groups outside the big ten and putting environmental justice outside the pale of environmentalism altogether, they do violence to the richness and breadth of an enormous movement. There has been more cross-fertilization between the land conservation, pollution control, and social justice movements than this categorization allows for. Fourth, the declensionist critique overlooks the vigorous tradition of internal dissent that has always prevailed at the green grassroots. The Sierra Club has been split apart several times, over big dams, nuclear power, and, in the Bay Area, the Peripheral Canal.

But, most of all, circumstances have been altered by the historic defeat of New Deal liberalism and the sixties Left. They have been done in by vast, sustained offensives against government, workers, the poor, secularism, and civil rights, attacks that go under such names such as Reaganism, the New Right, and the Christian Right, but most accurately they should be seen as a marriage of economic neoliberalism, social neoconservatism, and political dedemocratization. We cannot, in good faith, criticize the greens in isolation from this overwhelming political fact of our time. What is amazing, in fact, is how well the greens have resisted the frontal assaults of the Reagan–Bush years. But if things don't seem to be going as well as they once did, the blame lies at the feet of the Heritage Foundation, Ronald Reagan, and their ilk more than the shortcomings of the big ten conservation organizations.

Globally, environmentalism has spread like wildfire since the 1970s, both out of the rich countries and through local uprisings from Bhopal to the Danube. Bay Area environmentalism has been a player in this expansion to the wider world, promoting the green gospel internationally as it once did across the United States.

Friends of the Earth spawned chapters in seventy countries. Friends of the River morphed into the International Rivers Network, a leading combatant against damming and disruption of rivers worldwide. Rainforest Action Network took on the ravaging of the world's tropical forests. The International Marine Mammal Project spun off from local protection efforts. The Environmental Project on Central America demonstrated the deadly effects of the U.S. counterinsurgency war of the 1980s. Earth Island Institute incubated all four, as well as the cause of Lake Baikal as David Brower's last pet project. Amazon Watch is headquartered here, as is the broad antimultinational group CorpWatch. The Goldman Prize has become the Nobel Prize of international environmental activism. The Gordon and Mary Moore Foundation devotes millions to global problems through Conservation International.

Pesticide Action Network is a global force for limiting toxic exposures. Silicon Valley Toxics Coalition has become a leading voice internationally warning of the immense toxic load in the buildup of electronic trash. Pacific Institute has tackled international water supply problems. Blue Water Network began by challenging boat pollution on the bay and Lake Tahoe and now works globally as part of Friends of the Earth. Carl Anthony is now running an international program in sustainable development for the Ford Foundation. The Center on Race, Poverty, and the Environment is bringing lawsuits all the way to Alaska. APEN participates in the World Social Forum. Berkeley has an office of the International Council for Local Environmental Initiatives—and so on. This part of the tale, too, needs its narrator.

All is not well with environmentalism in the United States, however. Green organizations face tremendous pressure to roll back regulations and dismantle legislation such as the Clean Air Act and Endangered Species Act. By virtue of their success in setting up a new regulatory regime from 1964 through 1980, green organizations became a large, visible, and hated target of the corporations and the political right. The George W. Bush administration has systematically undermined as many environmental laws as it could, including such matters of importance to California as roadless areas, logging, fish protection, and oil drilling.[4]

Back on the home front in the Bay Area, things are not so peaceful, either. With the latest waves of urban expansion, the Bay Area has exploded in size, reaching far beyond the old redoubts such as Point Reyes. Growth has slipped its traces again, running amok for 100 miles along every transportation artery. The metropolis is cascading into the Central Valley and up and down the coast, just as happened a generation earlier in Southern California. Indeed, the rate at which American cities consume land has gone up, not down. The future of open space, agriculture, and waterways in the Central Valley is very much in doubt, and that region already vies with Los Angeles for the worst air quality in the nation.[5]

Politically, the surrounding counties are more conservative than the Bay Area, and environmentalists are at a premium, without an organized constituency for greenbelts, clean air, or land-use planning. The developers are correspondingly powerful, leading classic local growth coalitions to pave the valley with suburban sprawl. Notably, Richard Pombo (R-Tracy), from the fast-growing edge of the Bay Area, became one of the leading voices in Congress against environmental regulation and for private property rights.

Meanwhile, there is a complacency in the old core now that Marin is well fortified, Napa rich beyond anyone's dreams, and the peninsula's foothills and coastline secure. Too few are listening to green city visionaries such as Richard Register and Peter Berg about the need to radically overhaul cities to end sprawl, reduce auto usage, and end oil dependence.[6]

Environmentalists understand the importance of land and its protection but are less astute about the dominance of capital—ever fungible, ever nimble, and ever hungry for growth. Only the experience of battles for environmental protection against the bulldozers can reveal how rapacious capital and American society can be and how hard the fight to hold back the onslaught must be if anything is to be salvaged, however naively dreamed or innocently enjoyed in the greensward all around. As long as capital accumulation calls the tune, the juggernaut of urban growth will continue to roll on, and no city will ever be reconciled with the countryside.

APPENDIX

TABLE 1. California State Parks in the Bay Area

County	Name	Date	Acres
Alameda	Bethany Reservoir (State Recreation Area)	1974	608
	Carnegie (State Vehicular Recreation Area)	1979	3,810
	Eastshore State Park (State Seashore) (including Emeryville State Marine Reserve)	1985	442
	Lake Del Valle (State Recreation Area)	1967	3,732
	Robert W. Crown Memorial (State Beach)	1961	132
	Total for Alameda County		8,726
Contra Costa	Albany State Marine Reserve	1995	14
	Franks Tract (State Recreation Area)	1959	3,523
	John Marsh Home (Park Property)	1981	3,664
	Mount Diablo (State Park)	1931	20,103
	Total for Contra Costa County		27,304
Marin	Angel Island (State Park)	1955	747
	China Camp (State Park)	1976	1,514
	Marconi Conference Center (State Historic Park)	1989	62
	Mount Tamalpais (State Park)	1928	6,243
	Olompali (State Historic Park)	1977	700

TABLE 1 *continued*

County	Name	Date	Acres
	Samuel P. Taylor (State Park)	1946	2,707
	Tomales Bay (State Park)	1950	2,431
	Total for Marin County		14,404
Napa	Bale Grist Mill (State Historic Park)	1974	1
	Bothe-Napa Valley (State Park)	1960	1,780
	Robert Louis Stevenson (State Park)	1949	4,058
	Sugarloaf Ridge (State Park)	1931	139
	Total for Napa County		5,978
San Francisco	Angel Island (State Park)	1963	9
	Candlestick Point (State Recreation Area)	1972	205
	Total for San Francisco County		214
San Mateo	Ano Nuevo (State Reserve)	1958	1,319
	Ano Nuevo (State Park)	1985	2,896
	Bean Hollow (State Beach)	1958	44
	Big Basin Redwoods (State Park)	1915	904
	Burleigh H. Murray Ranch (Park Property)	1979	1,325
	Butano (State Park)	1956	4,548
	Castle Rock (State Park)	1994	25
	Gray Whale Cove (State Beach)	1966	3
	Half Moon Bay (State Beach)	1956	181
	Montara (State Beach)	1959	780
	Pacifica (State Beach)	1979	21
	Pescadero (State Beach)	1958	700
	Pigeon Point Light Station (State Historic Park)	1981	76
	Point Montara Light Station (Park Property)	1982	6
	Pomponio (State Beach)	1960	421
	Portola Redwoods (State Park)	1945	2,528
	San Bruno Mountain (State Park)	1980	298
	San Gregorio (State Beach)	1958	172
	Thornton (State Beach)	1955	58
	Total for San Mateo County		16,205

County	Name	Date	Acres
Santa Clara	Castle Rock (State Park)	1968	104
	Henry W. Coe (State Park)	1959	61,133
	Martial Cottle Park	2004	137
	Pacheco (State Park)	1995	640
	Total for Santa Clara County		62,014
Solano	Benicia (State Recreation Area)	1957	447
	Benicia Capitol (State Historic Park)	1951	1
	Total for Solano County		446
Sonoma	Annadel (State Park)	1971	5,093
	Armstrong Redwoods (State Reserve)	1934	752
	Austin Creek (State Recreation Area)	1964	5,927
	Fort Ross (State Historic Park)	1909	210
	Jack London (State Historic Park)	1959	3,393
	Kruse Rhododendron (State Reserve)	1934	317
	Petaluma Adobe (State Historic Park)	1951	41
	Robert Louis Stevenson (State Park)	1965	1,544
	Salt Point (State Park)	1968	5,685
	Sonoma (State Historic Park)	1909	64
	Sonoma Coast (State Beach)	1934	9,711
	Sugarloaf Ridge (State Park)	1920	3,645
	Total for Sonoma County		36,382
Santa Cruz	Big Basin Redwoods (State Park)	1906	17,129
	Castle Rock (State Park)	1968	5,113
	Henry Cowell Redwoods (State Park)	1953	4,316
	Lighthouse Field (State Beach)	1978	38
	Manresa (State Beach)	1948	138
	Natural Bridges (State Beach)	1933	59
	New Brighton (State Beach)	1933	157
	Rancho San Andreas (Castro Adobe)	2002	1
	Santa Cruz Mission (State Historic Park)	1959	2
	Seacliff (State Beach)	1931	87
	Sunset (State Beach)	1931	302

TABLE 1 *continued*

County	Name	Date	Acres
	The Forest of Nisene Marks (State Park)	1963	10,222
	Twin Lakes (State Beach)	1955	95
	Wilder Ranch (State Park)	1974	7,000
	Total for Santa Cruz County		44,659
	Total for 9-county area (59 units)		171,673
	Total for 10-county area (73 units)		216,332

NOTE: 96 percent of park acreage is owned outright by the state. Date is first acquisition; most parks have been expanded over time.

SOURCE: State Department of Parks & Recreation. Data as of 7/27/2005

TABLE 2. County Parks in the Bay Area

County	Park	Acreage	First Acquired
Contra Costa County	Alamo/Danville border	48	1987
	Discovery Bay	10	1980
	County Acreage	58	
Marin	Deer Park	53	
	McNears Beach	54	1969
	John McInnis Park	441	1972
	Paradise Beach	19	1967
	Stafford Lake	64	1968
	Tiburon Uplands	24	1958
	White House Pool	25	1972
	Homestead Valley	48	1974
	Creekside Park	26	1975
	Walker Creek Ranch	1,740	1985
	County Acreage	2,494	
Napa	No parks		
San Mateo	Coyote Point Recreational Area	727	1941
	Edgewood Park	467	1980
	Fitzgerald Marine Reserve	56	1967
	Flood Park	21	1937
	Heritage Grove	38	1976
	Huddart Park	974	1944
	Junipero Serra Park	108	1956
	Memorial Park	499	1924
	Midcoast Beaches	110	1971
	Pescadero Creek Park	5,973	1967
	Sam McDonald Park	1,002	1958
	San Bruno Mountain Park	2,326	1978
	San Pedro Valley Park	1,150	1972
	Sanchez Adobe Historical Site	5	c.2000
	Sawyer Camp Trail	32	1892

TABLE 2 *continued*

County	Park	Acreage	First Acquired
San Mateo	Southcoast Beaches	14	1977
	Wunderlich Park	942	1974
	County Acreage	14,444	
Santa Clara	Almaden Reservoir	112	
	Almaden Quicksilver Park	4,147	1973
	Alviso Marina	17	1964
	Anderson Reservoir	4,771	unknown
	Calero Reservoir	3,912	1977
	Chesbro Reservoir	216	c. 2000
	Coyote Creek Bicycle Trail	1,974	1963
	Coyote Reservoir/Harvey Bear	1,520	1985
	Ed Levin Park	1,562	1964
	Field Sports Park	93	1976
	Guadalupe Reservoir	107	
	Hellyer Park	205	unknown
	Joseph D. Grant Park	9,553	1975
	Lexington Reservoir	1,072	1990
	Los Gatos Creek	80	1966
	Mt. Madonna Park	3,688	1927
	Motorcycle Park	459	unknown
	Penitencia Creek Park	134	1977
	Rancho San Antonio	165	1977
	Sanborn - Skyline Park	3,688	1962
	Santa Teresa Park	1,627	1960
	Skyline Recreation Route	27	>2000
	Lower Stevens Creek Park	1,077	1924
	Upper Stevens Creek Park	1,276	unknown
	Sunnyvale Baylands	217	>2000
	Uvas Canyon Park	1,133	1961
	Uvas Creek Preserve	95	unknown
	Vasona Lake Park	152	1961
	Villa Montalvo Arboretum	132	unknown
	County Acreage	43,320	

County	Park	Acreage	First Acquired
Solano	Belden's Landing	10	1980
	Lake Solano Regional Park	177	1968
	Sandy Beach Regional Park	40	1980
	County Acreage	227	
Sonoma	Alder Park	5	1968
	Arnold Field	10	1978
	Bouverie Wildflower Preserve	22	1974
	Cloverdale River Park	40	1996
	Crane Creek Regional Park	129	1974
	Doran Park	102	1940
	Ernie Smith Park	12	1987
	Foothill Regional Park	208	1990
	Forestville River Access	13	1999
	Gualala Regional Park	195	1970
	Healdsburg Beach	12	1950
	Helen Putnam Regional Park	216	1978
	Hood Mountain Regional Park	1,430	1973
	Hudeman Slough	5	1962
	Hunter Creek Trail	18	2005
	Joe Rodota Trail	68	1988
	Keiser Park	20	1977
	Larson Park	7	1988
	Maddux Ranch Park	10	1987
	Maxwell Farms Regional Park	85	1981
	Pinnacle Gulch Trail & Beach	25	1977
	Ragle Ranch Regional Park	156	1976
	Riverfront Park	305	2005
	Shiloh Ranch Regional Park	860	1988
	Soda Springs Reserve	49	1994
	Sonoma Valley Regional Park	168	1973
	Southwest Regional Park	20	1980
	Spring Lake Park	306	1970
	Steelhead Beach	47	1996
	Stillwater Cove	226	1972

TABLE 2 *continued*)

County	Park	Acreage	First Acquired
Sonoma	Westside Park	25	1967
	Wohler Bridge Fishing Access	33	1988
	County Acreage	4,827	
San Francisco	Alamo Square	13	n/a
(city and county)	Alta Plaza	12	n/a
	Balboa Park	27	n/a
	Bay View Park	36	n/a
	Bernal Heights Park	24	n/a
	Buena Vista Park	36	n/a
	Candlestick Park	84	n/a
	Corona Heights	13	n/a
	Crocker Amazon Playground	56	n/a
	Glen Canyon Park	58	n/a
	Golden Gate Park	1,018	1868
	India Basin Shoreline Park	10	n/a
	Interior Green Belt	21	n/a
	John McClaren Park	333	n/a
	Julius Kahn Playground	12	n/a
	LaFayette Park	11	n/a
	Lake Merced Park	701	n/a
	Lincoln Park	113	n/a
	Lower Great Highway	22	n/a
	Marina Green and Yacht Harbor	162	n/a
	Midtown Terrace Playground	12	n/a
	Mission Dolores Park	14	n/a
	Moscone Recreation Center	12	n/a
	Mountain Lake Park	14	n/a
	Mt. Davidson Park	39	n/a
	Ocean View Playground	10	n/a
	Palace of Fine Arts	17	n/a
	Park Presidio Blvd	16	n/a
	Potrero Hill Recreation Center	10	n/a
	Pine Lake Park	30	n/a

County	Park	Acreage	First Acquired
	San Francisco Zoo	134	n/a
	Sigmund Stern Recreation Grove	34	n/a
	St Mary's Recreation Center	14	n/a
	Twin Peaks	32	n/a
	West Sunset Playground	17	n/a
	89 small parks and playgrounds, 1–10 acres	300	
	80 pocket parks and playgrounds less than 1 acre	60	
	County Acreage	3,528	
	Total for 8 counties	68,898	

NOTES: Table omits minor holdings, most less than 1 acre. Parks with lakes and creeks include water-surface acreage in total. Year refers to date of first acquisition. Most parks have been expanded through additional purchases; those dates are not provided here. Some parks shrink from time to time because of land swaps and boundary adjustments. Some minor acreage is leased from other agencies.

SOURCE: GreenInfo Network and County Parks and Recreation Agencies and their Web sites, as of 2005. Data not complete. County agencies claim more than totals of major parks; this may be due to minor holdings or discrepancies in data.

TABLE 3. East Bay Regional Parks

Park	Acreage	Date of Acquisition
Antioch-Oakley Regional Shoreline	9	2005
Ardenwood Historic Farm	208	1981
Bay Point Regional Shoreline	150	2003
Big Break/Delta Shore	1,656	1987
Bishop Ranch Regional Preserve	444	1978
Black Diamond Mines Regional Preserve	6,286	1973
Briones Regional Park	6,117	1966
Brooks Island Regional Shoreline	373	1968
Brown's Island Regional Shoreline	595	1978
Brushy Peak Regional Preserve	1,833	1998
Carquinez Straits Regional Shoreline	1,415	1981
Anthony Chabot Regional Park	5,065	1949
Claremont Canyon Regional Park	208	1979
Contra Loma Regional Park	779	1971
Coyote Hills Regional Park	978	1967
Crockett Hills Regional Park	1,720	1988
Crown Beach Regional Shoreline	383	1966
Cull Canyon Regional Recreation Area	360	1961
Del Valle Regional Park	4,315	1974
Diablo Foothills Regional Park	1,060	1976
Don Castro Regional Recreation Area	101	1961
Dry Creek Pioneer Regional Park	1,626	1978
Dublin Hills Regional Park	572	1996
Garin Regional Park	3,142	1965
Hayward Regional Shoreline	1,697	1977
Huckleberry Botanic Regional Preserve	240	1973
Kennedy Grove Regional Recreation Area	221	1965
Lake Chabot Regional Park	4,927	c. 1951
Las Trampas Regional Wilderness	5,092	1971
Leona Heights Regional Open Space Preserve	289	1986
Little Hills Ranch Regional Park	100	1976
Martin Luther King, Jr. Shoreline	741	1976

Park	Acreage	Date of Acquisition
Martinez Regional Shoreline	342	1976
Middle Harbor Shoreline Park	38	2002
Miller/Knox Regional Shoreline	306	1970
Mission Peak Regional Park	2,999	1976
Morgan Territory Regional Preserve	4,547	1975
Ohlone Regional Wilderness	9,736	1976
Oyster Bay Regional Shoreline	195	1981
Pleasanton Ridge Regional Park	4,848	1984
Pt. Isabel Regional Shoreline	23	1975
Pt. Pinole Regional Shoreline	2,315	1971
Quarry Lakes	538	1975
Redwood Regional Park	1,831	1940
Roberts Recreation Area	82	1951
Round Valley Regional Preserve	1,910	1988
San Pablo Bay Regional Shoreline	322	1988
Shadow Cliffs Lakes	266	1972
Sibley Volcanic Regional Preserve	678	1941
Sobrante Ridge Botanic Preserve	277	1985
Sunol Regional Wilderness	6,858	1962
Sycamore Valley Regional Preserve	695	1989
Tassajara Creek Regional Recreation Area	27	1971
Temescal Lake	49	1967
Tilden Regional Park	2,079	1940
Vargas Plateau Land Bank	1,030	1993
Vasco Caves Regional Preserve	1,339	1997
Waterbird Regional Preserve	198	1992
Wildcat Canyon Regional Park	2,430	1967
All Parks		59
Total Acreage		98,659

NOTE: Date refers to initial purchase at that site; most parks have been enlarged through further acquisitions over time. Eastshore State Park is also operated by EBRPD. Regional trails are not included.

SOURCE: East Bay Regional Parks District (prepared by Leddie Renwick)

TABLE 4. Midpeninsula Regional Open Space District Preserves

Open Space Preserve	Acreage	Date
Bear Creek Redwoods	1,343	1999
Coal Creek	493	1980
El Corte de Madera Creek	2,821	1984
El Sereno	1,412	1975
Felton Station	43	2005
Foothills	211	1974
Fremont Older	739	1975
La Honda Creek	2,078	1984
Long Ridge	1,985	1978
Los Trancos	274	1976
Mills Creek	n/a	2000
Monte Bello	2,943	1974
Picchetti Ranch	308	1979
Pulgas Ridge	366	1983
Purisima Creek Redwoods	3,120	1982
Rancho San Antonio	3,800	1975
Ravenswood	373	1980
Russian Ridge	1,822	1978
St. Joseph's Hill	268	1984
Saratoga Gap	1,291	1974
Sierra Azul	16,879	1979
Skyline Ridge	2,143	1982
Steven's Creek Shoreline	55	1980
Teague Hill	626	1988
Thornewood	163	1978
Windy Hill	1,306	1981
Total	46,819	

NOTES: Date is initial acquisition; most preserves have been expanded by further purchases. The district bought the Phleger estate and turned it over to the Golden Gate National Recreation Area in 1993.

SOURCE: MROSD

NOTES

INTRODUCTION *Saving Graces*

1. The nine counties are San Francisco, San Mateo, Marin, Alameda, Contra Costa, Santa Clara, Sonoma, Napa, and Solano; Santa Cruz County is not usually included in the Bay Area, though I refer to it here and there throughout this book.

2. Greenbelt Alliance, *At Risk* (San Francisco: Greenbelt Alliance, 1991, 1994, 2000, 2006). Open space is growing at an average rate of 10,000–15,000 acres per year.

3. Greenbelt Alliance, "Wall of Sprawl Threatens the Greenbelt," *Greenbelt Action: The Newsletter of the Greenbelt Alliance*, special ed. 2000. Adam Rome, *The Bulldozer in the Countryside: Suburban Sprawl and the Rise of the American Environmental Movement* (New York: Cambridge University Press, 2001). Edward Soja, "Inside Exopolis: Scenes from Orange County," in *Variations on a Theme Park: The New American City and the End of Public Space,* ed. Michael Sorkin (New York: Hill and Wang, Noonday Press, 1991).

4. George Henderson, *California and the Fictions of Capital* (New York: Oxford University Press, 1998), 18–27. See also James Vance, *The Merchant's World* (Englewood Cliffs, N.J.: Prentice-Hall, 1970); and Gunter Barth, *Instant Cities: Urbanization and the Rise of San Francisco and Denver* (New York: Oxford University Press, 1975).

5. Raymond Williams, *The Country and the City* (London: Chatto and Windus, 1973). The "pastoral" has been present in urban thought, at least back to ancient Rome. See Paul Alpers, *What Is Pastoral?* (Chicago: University of Chicago Press, 1996).

6. Roderick Nash, *Wilderness and the American Mind* (New Haven, Conn.: Yale

University Press, 1967); Leo Marx, *The Machine in the Garden: Technology and the Pastoral Ideal in America* (New York: Oxford University Press, 1964); Henry Nash Smith, *Virgin Land: The American West as Symbol and Myth* (Cambridge, Mass.: Harvard University Press, 1950); Kenneth Jackson, *The Crabgrass Frontier: The Suburbanization of the United States* (New York: Oxford University Press, 1985); and Peter Schmitt, *Back to Nature: The Arcadian Myth in Urban America* (New York: Oxford University Press, 1969). Recent efforts to bridge the gap include Robert Gottlieb, *Forcing the Spring: The Transformation of the American Environmental Movement* (Washington, D.C.: Island Press, 1993); Rome, *Bulldozer,* 2001; and Matthew Gandy, *Concrete and Clay: Reworking Nature in New York City* (Cambridge, Mass.: MIT Press, 2002).

7. Dorothy Ward Erskine, "Environmental Quality and Planning: Continuity of Volunteer Leadership," in *Bay Area Foundation History Series,* vol. 3 (Berkeley: Regional Oral History Office, Bancroft Library, University of California, Berkeley, 1976), 141; T. J. Kent Jr., *Open Space for the San Francisco Bay Area: Organizing to Guide Metropolitan Growth* (Berkeley: University of California, Berkeley, Institute of Governmental Studies, 1970), 5. On the physical geography, see Doris Sloan, *Geology of the San Francisco Bay Region* (Berkeley: University of California Press, 2006). On the region's beauty, see Galen Rowell, *Bay Area Wild: A Celebration of the Natural Heritage of the San Francisco Bay Area* (San Francisco: Sierra Club Books, Mountain Light Press, 1997).

8. Mike Davis, *Ecology of Fear: Los Angeles and the Imagination of Disaster* (New York: Henry Holt, Metropolitan, 1998); Greg Hise and William Deverell, editors, *Land of Sunshine: An Environmental History of Metropolitan Los Angeles* (Pittsburgh: University of Pittsburgh Press, 2005); and Stephanie Pincetl, "The Politics of Growth Control: Struggles in Pasadena, California," *Urban Geography* 13, no. 5 (1992).

9. Gottlieb, *Forcing the Spring,* 1993, ch. 2; Glenda Riley, *Women and Nature: Saving the Wild West* (Lincoln: University of Nebraska Press, 1999); Susan Schrepfer, *Nature's Altars: Mountains, Gender, and American Environmentalism* (Lawrence: University of Kansas Press, 2005). On women in Third World environmentalism see Carolyn Merchant, *Earthcare: Women and the Environment* (New York: Routledge, 1996); and Mary Joy Breton, *Women Pioneers for the Environment* (Boston: Northeastern University Press, 1998).

10. On garden clubs and conservation, see Buckner Hollingsworth, *Her Garden Was Her Delight* (New York: Macmillan, 1962); Shana Cohen, "American Garden Clubs and the Fight for Nature Preservation, 1890–1980" (PhD dissertation, Department of Environmental Science, Policy, and Management, University of California, Berkeley, 2005); cf. Anne Scott, *Natural Allies: Women's Associations in American History* (Urbana: University of Illinois Press, 1991). On women, parks, and playgrounds, see Suzanne Spencer-Wood, "Turn-of-the-Century Women's Organizations, Urban Design, and the Origin of the American Playground Movement," *Landscape Journal*

13, no. 2 (1994); and Louise Mozingo, "A Century of Neighborhood Parks" (manuscript, Department of Landscape Architecture, University of California, Berkeley, n.d.). On women and California progressivism, see William Deverell and Tom Sitton, editors, *California Progressivism Revisited* (Berkeley: University of California Press, 1994).

11. Helen Rudee (Sonoma County Board of Supervisors, 1976–88), interview with author, July 16, 2004. For confirmation of the number of women conservation activists, see John Hart and Nancy Kittle, *Legends: Portraits of 50 Bay Area Environmental Elders* (San Francisco: Sierra Club Books, 2006).

12. On women's domestic responsibilities, see Arlie Hochschild, *The Second Shift: Inside the Two-Job Marriage* (New York: Viking, 1989); and Marjorie DeVault, *Feeding the Family: The Social Organization of Caring as Gendered Work* (Chicago: University of Chicago Press, 1991). On women moving from the domestic into the public sphere, see Maureen Flanagan, "The City Profitable, the City Livable: Environmental Policy, Gender, and Power in Chicago in the 1910s," *Journal of Urban History* 22, no. 2 (1996), 163–90; Delores Hayden, *The Grand Domestic Revolution: A History of Feminist Designs for American Homes, Neighborhoods, and Cities* (Cambridge, Mass.: Harvard University Press, 1985); and Lisa McGirr, *Suburban Warriors: Origins of the New American Right* (Princeton, N.J.: Princeton University Press, 2001).

13. On the development of political consciousness from everyday experience, see John Gaventa, *Power and Powerlessness: Quiescence and Rebellion in an Appalachian Valley* (Urbana: University of Illinois Press, 1980); Doug McAdam, *Political Process and the Development of Black Insurgency, 1930–1970* (Chicago: University of Chicago Press, 1982); Ann Bookman and Sandra Morgan, editors, *Women and the Politics of Empowerment* (Philadelphia: Temple University Press, 1988); and Saba Mahmood, "Rehearsed Spontaneity and the Conventionality of Ritual: Disciplines of Salat," *American Ethnologist* 28, no. 4 (2001).

14. The different gender responses to mountains and wild nature are well documented in Schrepfer, *Nature's Altars*, 2005. Cf. John McPhee, *Encounters with the Archdruid* (New York: Farrar, Straus and Giroux, 1971); and Stephen Fox, *The American Conservation Movement: John Muir and His Legacy* (Madison: University of Wisconsin Press, 1981).

15. Raymond Williams, *Country and the City*, 1973; Denis Cosgrove, *Social Formation and Symbolic Landscape* (London: Croom Helm, 1984); Robert Fishman, *Bourgeois Utopias: The Rise and Fall of Suburbia* (New York: Basic Books, 1987); Michael Heiman, *The Quiet Evolution: Power, Planning, and Profits in New York State* (New York: Praeger, 1988); Heiman, "Production Confronts Consumption: Landscape Perception and Social Conflicts in the Hudson Valley," *Society and Space* 7, no. 2 (1989); and Laura Pulido, "Rethinking Environmental Racism: White Privilege and Urban Development in Southern California," *Annals of the Association of American Geographers* 90, no. 1 (2000).

16. Flanagan, "City Profitable, City Livable," 1996.

17. James Duncan and Nancy Duncan, *Landscapes of Privilege: The Politics of the Aesthetic in an American Suburb* (New York: Routledge, 2004). On the importance of public space and the right of access—indeed, the right to flourish—in the city, see Don Mitchell, *The Right to the City: Social Justice and the Fight for Public Space* (New York: Guilford Press, 2003). In short, the Bay Area greenbelt is in many ways a monumental class project, much like Haussmann's Paris or Robert Moses's New York, but it is one with a remarkably public face. I disagree with Mitchell's view (2003, 143) that open space is more regulated and less politicized than city streets.

18. Richard White, "Are You an Environmentalist, or Do You Work for a Living? Work and Nature," in *Uncommon Ground: Toward Reinventing Nature,* ed. William Cronon (New York: W. W. Norton, 1995); and Andrew Hurley, *Environmental Inequalities: Class, Race, and Industrial Pollution in Gary, Indiana, 1945–1980* (Chapel Hill: University of North Carolina Press, 1995).

19. Cf. Gottlieb, *Forcing the Spring,* 1993, 15.

20. On Williams's politics, see David Harvey, *Spaces of Hope* (Berkeley: University of California Press, 2001); on locally rooted movements, see Allan Pred, *Place, Practice, and Structure* (Cambridge, UK: Polity Press, 1986); John Agnew, *Place and Politics: The Geographical Mediation of State and Society* (Boston: Allen and Gavin, 1987); and Stephen Pile, *Geographics of Resistance* (London: Routledge, 1998).

21. The term "political culture" goes back to liberals celebrating American democracy in the postwar era and was revived by liberal urban theorists such as Clarence Stone in *Regime Politics: Governing Atlanta, 1946–1988* (Lawrence: University of Kansas Press, 1989). For a review, see Philip Ethington, "Mapping the Local State," *Journal of Urban History* 27, no. 5 (2001), 686–702. An alternative, radical tradition of thinking about politics and culture goes back to Antonio Gramsci in *Prison Notebooks* (New York: International Publishers, 1971). It reappears among current urban historians such as Philip Ethington, *The Public City: The Political Construction of Urban Life in San Francisco, 1850–1900* (New York: Cambridge University Press, 1994); Mary Ryan, *Civic Wars: Democracy and Public Life in the American City During the Nineteenth Century* (Berkeley: University of California Press, 1997); and Sven Beckert, *The Monied Metropolis* (New York: Cambridge University Press, 2001).

22. For sustained arguments on the character of places, see Pred, *Place, Practice, and Structure,* 1986; Mike Davis, *City of Quartz: Excavating the Future in Los Angeles* (London: Verso Press, 1990); Michael Johns, *The City of Mexico in the Age of Diaz* (Austin: University of Texas Press, 1997); and Harvey Molotch, William Freudenberg, and Krista Paulsen, "History Repeats Itself, But How? City Character, Urban Tradition, and the Accomplishment of Place," *American Sociological Review* 65 (2000). For my views on the character of the Bay Area and California, see Richard Walker and Bay Area Study Group, "The Playground of U.S. Capitalism? The Political Economy of the San Francisco Bay Area in the 1980s," in *Fire in the Hearth: The Radical Politics of Place in America,* ed. Mike Davis Hiatt, Steve Davis, Michael Kennedy, Susan Ruddick,

and Mike Sprinker (London: Verso Press, Haymarket, 1990); Richard Walker, "California Rages Against the Dying of the Light," *New Left Review* 209 (1995a); Richard Walker, "Another Round of Globalization in San Francisco," *Urban Geography* 17, no. 1 (1996); and Richard Walker, "California's Golden Road to Riches: Natural Resources and Regional Capitalism, 1848–1940," *Annals of the Association of American Geographers* 91, no. 1 (2001).

23. On the importance of class political organization in changing circumstances, see Melani Cammett, "Fat Cats and Self-made Men: Globalization and the Paradoxes of Collective Action," *Comparative Politics* 37, no. 4 (2005), 379–400.

24. On the language of political cultures (variously called "discourses," "dispositions," "frames," "narratives," etc.), see also Michael Rogin, *Ronald Reagan, The Movie, and Other Episodes in Political Demonology* (Berkeley: University of California Press, 1987); and Benedict Andersen, *Imagined Communities: Reflections on the Origins and Spread of Nationalism* (London: Verso, 1983).

25. See Mahmood, "Rehearsed Spontaneity," 2001; Andrew Sayer, *The Moral Significance of Class* (Cambridge, UK: Cambridge University Press, 2005); and Alisdair MacIntyre, *After Virtue* (Bloomington: Indiana University Press, 1984). This is a more supple notion of culture than used by Pierre Bourdieu, *Distinction*, trans. Richard Nice (Cambridge, Mass.: Harvard University Press, 1984).

26. Kenneth Olwig, *Landscape, Nature, and the Body Politic: From Britain's Renaissance to America's New World* (Madison: University of Wisconsin Press, 2002). See also Yi-Fu Tuan, *Space and Place: The Perspective of Experience* (Minneapolis: University of Minnesota Press, 1977), and Patricia Price, *Dry Place: Lanscapes of Belonging and Exclusion* (Minneapolis: University of Minnesota Press, 2004). Taken together, this sense of place and moral order, this political culture and institutional fabric is the flesh and blood of history draped on the bare bones of class structure, individual action, and economic interest—in the sense of E. P. Thompson's *The Making of the English Working Class* (London: Penguin, 1964). It is, of course, complicated by gender and race in ways that Thompson did not confront adequately.

27. On the local and global, see Richard Walker, "Another Round of Globalization," 1996; Jane Wills and Roger Lee, editors, *Geographies of Economies* (London: Edward Arnold, 1997); and Kevin Cox, editor, *Spaces of Globalization: Reasserting the Power of the Local* (New York: Guilford Press, 1997).

28. On the Nazi connection, see Gray Brechin, "Conserving the Race: Natural Aristocracies, Eugenics, and the U.S. Conservation Movement," *Antipode* 28, no. 3 (1996). On struggles within the Sierra Club, see Carl Pope, "The Virus of Hate: The Sierra Club and the Immigration Debate," *Sierra* 89 (May–June 2004), 14. On Orange County politics, see McGirr, *Suburban Warriors*, 2001. On San Diego, see Mike Davis, Kelly Mayhew, and Jim Miller, *Under the Perfect Sun: The San Diego Tourists Never See* (New York: The New Press, 2003).

29. It is simply not true that urban greens today are the naive wilderness wor-

shippers subject to so much criticism by scholars in recent years. For example, see Neil Smith, *Uneven Development* (Oxford: Basil Blackwell, 1984); David Harvey, *Justice, Nature, and the Geography of Difference* (Oxford: Basil Blackwell, 1997); and William Cronon, "The Trouble with Wilderness: or, Getting Back to the Wrong Nature," in *Uncommon Ground: Toward Reinventing Nature*, ed. William Cronon (New York: W. W. Norton, 1995).

30. Hal Rothman, *The Greening of a Nation? Environmentalism in the United States Since 1945* (Fort Worth, Texas: Harcourt Brace College Publishers, 1998b), xii; Rome, *Bulldozer*, 2001, 13.

31. Robert Bullard, *Dumping in Dixie: Race, Class, and Environmental Quality* (Boulder, Colo.: Westview Press, 1990); Laura Pulido, "A Critical Review of the Methodology of Environmental Racism Research," *Antipode* 28, no. 2 (1996a).

32. Mike Davis, *Prisoners of the American Dream* (London: Verso Press, 1986); Kim Moody, *An Injury to All: The Decline of American Unionism* (London: Verso Press, 1989).

33. Jon Halper, editor, *Gary Snyder: Dimensions of a Life* (San Francisco: Sierra Club Books, 1991); Linda Hamalian, *A Life of Kenneth Rexroth* (New York: W. W. Norton, 1991). On the Left Coast, see Richard DeLeon, *Left Coast City: Progressive Politics in San Francisco, 1975–1991* (Lawrence: University of Kansas Press, 1992); Walker et al., "The Playground of U.S. Capitalism?" 1990.

34. On the Western sublime in America, see Nash, *Wilderness and the American Mind*, 1967; Kate Ogden, "Sublime Vistas and Scenic Backdrops: Nineteenth-century Painters and Photographers at Yosemite," *California History* 69, no. 2 (1990); Patricia Limerick, "Disorientation and Reorientation: The American Landscape Discovered from the West," *Journal of American History* 79, no. 3 (1992); Rebecca Solnit, *Savage Dreams: A Journey into the Hidden Wars of the American West* (San Francisco: Sierra Club Books, 1994); and John Sears, *Sacred Places: American Tourist Attractions in the Nineteenth Century* (Amherst: University of Massachussetts Press, 1998). On California domination, see Earl Pomeroy, *The Pacific Slope* (New York: Albert Knopf, 1965); Gray Brechin, *Imperial San Francisco: Urban Power, Earthly Ruin* (Berkeley: University of California Press, 1999); and Richard Walker, "Golden Road to Riches," 2001.

35. Gottlieb, *Forcing the Spring*, 1993, 102, 105–13. Similarly, Rome, *Bulldozer*, 2001, never recognizes how many of the critics he cites, such as Jack Kent and Raymond Dasmann, were from the Bay Area. Schrepfer, *Nature's Altars*, 2005, 31, is also too quick to reduce the Sierra Club to just another hiking group.

1 / OUT OF THE WOODS *Stirrings of Conservation*

1. On the environmental damage wrought by the mining era, see Brechin, *Imperial San Francisco*, 1999; and Andrew Isenberg, *Mining California: An Ecological History* (New York: Hill and Wang, 2005).

2. On New York conservation, see Samuel Hays, *Conservation and the Gospel of Efficiency: The Progressive Conservation Movement, 1890–1920* (Cambridge, Mass.: Harvard University Press, 1959); Schmitt, *Back to Nature*, 1969; Frank Graham Jr., *The Audubon Ark: A History of the National Audubon Society* (New York: Knopf, 1990); and Gandy, *Concrete and Clay*, 2002. On the intellectual connections between the Northeast and Northern California, see Kevin Starr, *Americans and the California Dream, 1850–1915* (New York: Oxford University Press, 1973).

3. Holway Jones, *John Muir and the Sierra Club: The Battle for Yosemite* (San Francisco: Sierra Club, 1965); Nash, *Wilderness and the American Mind*, 1967; Fox, *American Conservation Movement*, 1981; Douglas Strong, *Dreamers and Defenders: American Conservationists* (Lincoln: University of Nebraska Press, 1988); Michael Cohen, *The History of the Sierra Club, 1892–1970* (San Francisco: Sierra Club Books, 1988). Sierra Club founders were entirely male; the only female board member in the early days was Marion Randall Parsons.

4. On the sensation over Yosemite, and its roots in tourism, nationalism, and John Ruskin, see Sears, *Sacred Places*, 1998. Yosemite and Sequoia were enlarged to their present dimensions in 1906 and 1926, respectively; Crater Lake (in southern Oregon) followed in 1902, Mount Lassen in 1916. On Sequoia, see Larry Dilsaver and William Tweed, *Challenge of the Big Trees: A Resource History of Sequoia and Kings Canyon National Parks* (Three Rivers, Calif.: Sequoia National History Association, 1990). President Clinton added a Giant Sequoia National Monument in 2000.

5. On the delight over California's splendors, see Starr, *California Dream*, 1973. On California's photographic revolution, see Rebecca Solnit, *River of Shadows: Edweard Muybridge and the Technological Wild West* (San Francisco: City Lights Books, 2003). On Western tourism and railroads, see Earl Pomeroy, *In Search of the Golden West: The Tourist in Western America* (New York: Alfred Knopf, 1957); and Marguerite Shaffer, *See America First: Tourism and National Identity, 1880–1940* (Washington, D.C.: Smithsonian Institution, 2001).

6. Hays, *Gospel of Efficiency*, 1959.

7. Hays, *Gospel of Efficiency*, 1959; Nash, *Wilderness and the American Mind*, 1967; Holway Jones, *John Muir and the Sierra Club*, 1965; Fox, *American Conservation Movement*, 1981; Michael Cohen, *History of the Sierra Club*, 1988. By the time of the Hetch Hetchy battle, one-third of the club's members were women.

8. On the myth of wilderness, see Solnit, *Savage Dreams*, 1994; and Cronon, "The Trouble with Wilderness," 1995. The idea of helping the conquerors appreciate their conquest I take from Carl Anthony, "The Civil Rights Movement and Expanding the Boundaries of Environmental Justice in the San Francisco Bay Area, 1960–1999" (tape recording and transcript, Regional Oral History Office, Bancroft Library, University of California, Berkeley, 2003), 40. On John Burroughs, see Schmitt, *Back to Nature*, 1969. The "people's playgrounds" comes from John Muir, *My First Summer in the Sierra* (Boston and New York: Houghton Mifflin, 1911).

9. Nash, *Wilderness and the American Mind*, 1967; McPhee, *Encounters with the Archdruid*, 1971; Michael Cohen, *History of the Sierra Club*, 1988; Gottlieb, *Forcing the Spring*, 1993; Rothman, *Greening of a Nation?* 1998b. Brower was the club's first executive director, from 1952 to 1969. Examples of Exhibit Format books are Eliot Porter, *The Place No One Knew: Glen Canyon on the Colorado*, ed. David Brower (San Francisco: Sierra Club, 1963); Philip Hyde and Francois Leydet, *The Last Redwoods: Photographs and Story of a Vanishing Scenic Resource* (San Francisco: Sierra Club, 1963); and Francois Leydet, *Time and the River Flowing, Grand Canyon* (San Francisco: Sierra Club, 1964).

10. The phrase is Wallace Stegner's, from a letter written to support the campaign for a wilderness act; he refers to wildlands, whereas I refer to people. Sierra Club membership from Ronald Shaiko, *Voices and Echoes for the Environment: Public Interest Representation in the 1990s and Beyond* (New York: Columbia University Press, 1999), 42; and Michael Cohen, *History of the Sierra Club*, 1988, 275, 376. In 1991 one-third of the club's members resided in California.

11. On modern versus antimodern tendencies in American thought, see T. Jackson Lears, *No Place of Grace: Anti-Modernism and the Transformation of American Culture, 1880–1920* (New York: Pantheon, 1981).

12. Ted Wurm and Alvin Graves, *The Crookedest Railroad in the World: A History of the Mount Tamalpais and Muir Woods Railroad of California*, 3rd ed. (Glendale, Calif.: Trans-Anglo Books, 1983); Lincoln Fairley, *Mount Tamalpais: A History* (San Francisco: Scottwall Associates, 1987); Barry Spitz, *Mill Valley: The Early Years* (Mill Valley, Calif.: Potrero Meadow, 1997).

13. On the hiking binge and mountain worship, see Rebecca Solnit, *Wanderlust: A History of Walking* (New York: Viking, 2000) and Susan Schrepfer, *Nature's Altars* (Lawrence: University of Kansas Press, 2005). On the American rage for camping and outdoorsmanship, see Schmitt, *Back to Nature*, 1969; Sears, *Sacred Places*, 1998; and Terence Young, *Heading Out: American Camping from 1869 to 1990* (forthcoming). The first U.S. hiking clubs were Maine's White Mountain Club (1873) and Boston's Appalachian Hiking Club (1876). Hiking and climbing were widespread in the West by the 1870s and included a surprising number of women.

14. Earl Pomeroy, *In Search of the Golden West*, 1957. Cushing and Kent also bought into the Tamalpais Land and Water Company, which developed Mill Valley.

15. Elizabeth Kent, "William Kent, A Biography" (manuscript, Bancroft Library, University of California, Berkeley, 1950); John Hart, *San Francisco's Wilderness Next Door* (San Rafael: Presidio Press, 1979); Jane Futcher and Robert Conover, *Marin: The Place, the People: Profile of a California County* (New York: Holt, Rinehart, and Winston, 1981). The family owned land in Nevada and Nebraska and one of the first vacation homes at Tahoe. Kent's donation to Kelley is from Gottlieb, *Forcing the Spring*, 1993, 62.

16. The Kents had six children, including Roger, a leader in the statewide Dem-

ocratic Party (who married into the Hawaiian Castle and Cooke empire); William Jr., a director of Kern County Land Company and Bank of California; Albert, who managed the family ranchlands; and Elizabeth, an artist who married Robert Howard, son of Berkeley architect John Galen Howard. Roger and Elizabeth were supporters of the Marin Conservation League, while William became a land developer who created Seadrift at Stinson Beach.

17. Spitz, *Mill Valley*, 1997; Willie Yaryan, Denzil Verardo, and Jennie Verardo, *The Sempervirens Story: A Century of Preserving California's Ancient Redwood Forest, 1900–2000* (Los Altos: Sempervirens Fund, 2000). Women Progressives were generally to the left of the men—James Phelan, for example, was a notorious racist.

18. Susan Schrepfer, *The Fight to Save the Redwoods: A History of Environmental Reform, 1917–1978* (Madison: University of Wisconsin Press, 1983). Cf. Sears, *Sacred Places*, 1998.

19. W. W. Bonner, *The Redwoods of California: A Glimpse of the Wonderland of the Golden West* (San Francisco: California Redwood Company, 1884); Fairley, *Mount Tamalpais*, 1987; Sherwood Burgess, "The Forgotten Redwoods of the East Bay," *California Historical Society Quarterly*, March 1951; Gordy Slack, "In the Shadow of Giants," *Bay Nature*, July–September 2004.

20. Frank Stanger, *Sawmills in the Redwoods: Logging on the San Francisco Peninsula, 1849–1967* (San Mateo: San Mateo County Historical Association, 1967); Alan Hynding, *From Frontier to Suburb: The Story of the San Mateo Peninsula* (Belmont, Calif.: Star Books, 1981); Margaret Koch, *Santa Cruz County: Parade of the Past* (Santa Cruz: Western Tanager Press, 1973).

21. Donald Pisani, "Forests and Conservation, 1865–1890," *Journal of American History* 72, no. 2 (1985); Stephanie Pincetl, *Transforming California: A Political History of Land Use and Development* (Baltimore: Johns Hopkins University Press, 1999); Isenberg, *Mining California*, 2005; and David Beesley, *Crow's Range: An Environmental History of the Sierra Nevada* (Reno: University of Nevada Press, 2004).

22. On reinvestment of fortunes into land, see Brechin, *Imperial San Francisco*, 1999; and Richard Walker, "Golden Road to Riches," 2001. On ejection of people from parks, see Solnit, *Savage Dreams*, 1994; and Karl Jacoby, *Crimes Against Nature: Squatters, Poachers, Thieves, and the Hidden History of American Conservation* (Berkeley: University of California Press, 2001).

23. Joseph Engbeck, *State Parks of California, from 1864 to the Present* (Portland, Ore.: Graphic Arts Center Publications, 1980), 29, 32.

24. Yaryan et al., *Sempervirens Story*, 2000. Around 1885 the Bohemian Club rescued Meeker's grove, the last uncut redwoods along the Russian River. This can only partly be attributed to adoration of the sublime qualities of the trees, but their Dionysian theatrics and bare-bottomed cavorting in the forest were a sign of changing sensibilities.

25. On nineteenth-century tourist attractions at scenic wonders, see Sears, *Sacred Places*, 1998.

26. Yaryan et al., *Sempervirens Story*, 2000. Governor George Pardee put the park back under the corrupt hand of the forestry board in 1905, however.

27. Richard Walker and Matthew Williams, "Water from Power: Water Supply and Regional Growth in the Santa Clara Valley," *Economic Geography* 58, no. 2 (1982).

28. Pincetl, *Transforming California*, 1999, 44, calls the Save the Redwoods League a typical Progressive-era organization, but it came late in the day, after the radical edge of Progressivism had been blunted.

29. Schrepfer, *Fight to Save the Redwoods*, 1983, 14–15. On the makeup of the Sierra Club, see Holway Jones, *John Muir and the Sierra Club*, 1965; and Michael Cohen, *History of the Sierra Club*, 1988. See also William Edward Colby, "Early Days of the Sierra Club" (unpublished manuscript, Bancroft Library, University of California, Berkeley, 1959); Newton Drury, "Parks and Redwoods, 1919–1971," interview by Amelia Fry and Susan Schrepfer (Regional Oral History Office, Bancroft Library, University of California, Berkeley, 1972); Stephen Mark, *Preserving the Living Past: John C. Merriam's Legacy in the State and National Parks* (Berkeley: University of California Press, 2005); and Engbeck, *State Parks of California*, 1980. Alston Chase's idea (*In a Dark Wood: The Fight Over Forests and the Rising Tyranny of Ecology*, Boston: Houghton Mifflin, 1995) that the Sierra Club and Save the Redwoods League were made up of different classes of people is absurd; William Badè and Joseph Grant were in both groups, for example. Upper class in this era necessarily meant white and protestant.

30. Figures are from Richard Brewer, *Conservancy: The Land Trust Movement in America* (Dartmouth, Mass.: The University Press of New England, 2003), 28. A call by the Secretary of the Interior in 1879 to reserve 50,000 acres of redwoods fell on deaf ears. Similarly, Congress agreed to a survey for a Redwoods National Park in 1919 but refused to acquire private land for it. Engbeck, *State Parks of California*, 1980.

31. Schrepfer, *Fight to Save the Redwoods*, 1983; Pincetl, *Transforming California*, 1999. On Grant's eugenics, see Brechin, "Conserving the Race," 1996; and Chase, *In a Dark Wood*, 1995.

32. Pincetl, *Transforming California*, 1999, 164–68.

33. On the Northwest timber wars, see Nancy Langston, *Forest Dreams, Forest Nightmares: The Paradox of Old Growth in the Inland West* (Seattle: University of Washington Press, 1995); Paul Hirt, *A Conspiracy of Optimism: Management of the National Forests since World War Two* (Lincoln: University of Nebrasks Press, 1994); and Scott Prudham, *Knock on Wood: Nature as Commodity in Douglas-Fir Country* (New York: Routledge, 2005). There, too, 90 percent of the old growth would be cut before a last-ditch effort to save the remainder.

34. Hyde and Leydet, *Last Redwoods*, 1963; Schrepfer, *Fight to Save the Redwoods*, 1983; Michael Cohen, *History of the Sierra Club*, 1988; Chase, *In a Dark Wood*, 1995.

35. Judi Bari, *Timber Wars* (Monroe, Maine: Common Courage Press, 1994); David Harris, *The Last Stand: The War Between Wall Street and Main Street Over Cal-

ifornia's Ancient Redwoods (New York: Times Books, 1995); A. C. Thompson, "The Judi Bari Bombshell," *San Francisco Bay Guardian*, May 29, 2002; Patrick Beach, *A Good Forest for Dying: The Tragic Death of a Young Man on the Front Lines of the Environmental Wars* (New York: Doubleday, 2004).

2 / FIELDS OF GOLD *Resources at Close Quarters*

1. Henderson, *Fictions of Capital*, 1998, 117.

2. Rodman Paul, "The Beginnings of Agriculture in California: Innovation Versus Continuity," *California Historical Quarterly* 52, no. 1 (1973), 19; Bentham Fabian, *The Agricultural Lands of California* (San Francisco: H. H. Bancroft and Co., 1869), 20. On the geography of California agricultural development, see Richard Walker, *The Conquest of Bread: 150 Years of Agribusiness in California* (New York: The New Press, 2004).

3. The development of intensive agriculture near to cities is a staple of geographic thought that goes back to Johann Von Thünen in the early nineteenth century. David Harvey, *Social Justice and the City* (London: Edward Arnold, 1973), and William Cronon, *Nature's Metropolis: Chicago and the Great West* (Chicago: W. W. Norton, 1991), both use Von Thünen, for example, but Harvey's urban theory takes little notice of rural land uses, and Cronon's long-distance model of mercantile extraction leaps over the nearby hinterlands that are the core of Von Thünen's model. On landholdings, see Ellen Liebman, *California Farmland: A History of Large Agricultural Landholdings* (Totowa, N.J.: Rowman and Allanheld, 1983); and Richard Walker, *Conquest of Bread*, 2004.

4. Brechin, *Imperial San Francisco*, 1999; Richard Walker, "Golden Road to Riches," 2001, and *Conquest of Bread*, 2004. A good example is the buyout of the peninsula; see Hynding, *From Frontier to Suburb*, 1981, 69–73.

5. On the delta, see Robert Kelley, *Battling the Inland Sea: American Political Culture, Public Policy, and the Sacramento Valley, 1850–1986* (Berkeley: University of California Press, 1989), 59–62; John Thompson, "The Settlement Geography of the Sacramento–San Joaquin Delta, California" (PhD dissertation, Stanford University, 1957); Roger Minick, *Delta West: The Land and People of the Sacramento–San Joaquin Delta* (Berkeley: Scrimshaw Press, 1969); and Richard Dillon, *Delta Country* (Novato: Presidio Press, 1982). On exports, see Paul, "Beginnings of Agriculture in California," 1973; and Richard Walker, *Conquest of Bread*, 2004.

6. Richard Walker, *Conquest of Bread*, 2004, 194–97.

7. Paul, "Beginnings of Agriculture in California," 1973.

8. Yvonne Jacobson, *Passing Farms, Enduring Values: California's Santa Clara Valley* (Los Altos: William Kaufman, 1984); Richard Walker, *Conquest of Bread*, 2004.

9. Vincent Carosso, *The California Wine Industry, 1830–1895: A Study of the Form-*

ative Years (Berkeley: University of California Press, 1951, 1976); Thomas Pinney, *A History of Wine in America: From Prohibition to the Present* (Berkeley: University of California Press, 2005).

10. Richard Walker, *Conquest of Bread*, 2004, 228–35.

11. Walker and Williams, "Water from Power," 1982.

12. Figures from People for Open Space, *Endangered Harvest: The Future of Bay Area Farmland* (San Francisco: People for Open Space, 1980a).

13. Alvin Sokolow, *The Williamson Act: Twenty-Five Years of Land Conservation* (Sacramento: California Department of Conservation, 1990). Also, Robert Fellmeth, *The Politics of Land* (New York: Grossman, 1973); American Farmland Trust, *Eroding Choices, Emerging Issues: The Condition of California's Agricultural Land Resources* (San Francisco: American Farmland Trust, 1986).

14. People for Open Space, *Endangered Harvest*, 1980a, and *A Search for Permanence: Farmland Conservation in Marin County, California* (San Francisco: People for Open Space, 1981). Cf. Wendell Berry, *Home Economics* (Berkeley: North Point Press, 1987). Recall that much more farmland still existed in the Bay Area in 1980. Greenbelt Alliance, *The Bay Area's Farmlands* (San Francisco: Greenbelt Alliance, 1991c).

15. For critiques of postwar suburbia, see Raymond Dasmann, *The Destruction of California* (New York: MacMillan, 1965); Fellmeth, *Politics of Land*, 1973; and James Howard Kunstler, *The Geography of Nowhere: The Rise and Decline of America's Man-Made Landscape* (New York: Simon and Schuster, 1993).

16. Brechin, *Imperial San Francisco*, 1999. For maps, see William Kahrl, ed., *The California Water Atlas* (Sacramento: Governor's Office of Planning and Research, 1979). Cf. Gandy, *Concrete and Clay*, 2002.

17. On San Francisco water supply, see U.S. Department of the Interior, National Park Service, *The Top of the Peninsula: A History of Sweeney Ridge and the San Francisco Watershed Lands, San Mateo County, California*, by Marianne Babal (San Francisco: National Park Service, Golden Gate National Recreation Area, 1990); and Brechin, *Imperial San Francisco*, 1999. Spring Valley's reservoirs drowned out farms and two towns, Crystal Springs and Searsville. On Marin's water history, see Jack Mason and Helen Van Cleave Park, *The Making of Marin, 1850–1975* (Inverness, Calif.: North Shore Books, 1975); and Marin Municipal Water District, *A Historical Summary of the Marin Municipal Water District* (San Rafael: MMWD, 1975).

18. Sarah Elkind, *Bay Cities and Water Politics: The Battle for Resources in Boston and Oakland* (Lawrence: University of Kansas Press, 1998); Walker and Williams, "Water from Power," 1982. Chabot also had a hand in San Francisco Water Works (1856) and San Jose Water Works (1866); Sherwood Burgess, *The Water King: Anthony Chabot* (Davis, Calif.: Panorama West Publishing, 1992).

19. Walker and Williams, "Water from Power," 1982; Elkind, *Bay Cities and Water Politics*, 1998; Brechin, *Imperial San Francisco*, 1999; John Wesley Noble and Gayle Montgomery, *Its Name Was M.U.D.* (Oakland: East Bay Municipal Utility District,

1999). On the Progressive era and special districts, see Pincetl, *Transforming California*, 1999; and Louise Nelson Dyble, "Paying the Toll: The Golden Gate Bridge and Highway District and the San Francisco Bay Area, 1919–1971" (PhD dissertation, Department of History, University of California, Berkeley, 2002).

20. Paul McHugh, "Miles to Go: Competing Interests Miss Mark in Debate over Access to Peninsula Watershed Trail," *San Francisco Chronicle*, March 19, 2000, A1.

21. Ann Riley, "Overcoming Federal Water Policies: The Wildcat–San Pablo Creeks Case," *Environment* 31, no. 10 (1989), 12–31. Riley went on to found the Waterways Restoration Institute. On creek restoration, see Ann Riley, *Restoring Streams in Cities: A Guide for Planners, Policymakers, and Citizens* (Washington, D.C.: Island Press, 1998).

22. Richard Register, interview at Cordonices Creek restoration project with author, July 17, 2005; Louise Mozingo, "Community Participation and Creek Restoration in the East Bay," in *(Re)Constructing Communities: Design Participation in the Face of Change*, Jeff Hou, Mark Francis, and Nathan Brightbill, eds. (Davis, Calif.: Center for Design Research, 2005), 249–51.

23. Register, interview, 2005.

24. Brechin, *Imperial San Francisco*, 1999. Cf. Cronon, *Nature's Metropolis*, 1991, and Gandy, *Concrete and Clay*, 2002.

25. California State Bureau of Mines, *The Quicksilver Resources of California*, by Walter Bradley (Sacramento: California State Printing Office, Bulletin No. 78, 1918); David St. Clair, "New Almaden and California Quicksilver in the Pacific Rim Economy," *California History* 73, no. 4 (winter 1994–95); Jimmy Schneider, *The Complete History of Santa Clara County's New Almaden Mine* (San Jose: Zella Schneider, 1992). Figures are from California Division of Mines, *Geologic Guidebook of the San Francisco Bay Counties*, ed. Olaf Jenkins (San Francisco: California Division of Mines, Bulletin No. 154, 1951), 263; and California State Mining Bureau, *Mining in California* (Sacramento: California State Mining Bureau, Vol. 17, 1921), 209.

26. Gray Brechin and Robert Dawson, *Farewell, Promised Land* (Berkeley: University of California Press, 1998); Jane Kay, "Tracking a Toxic Trail," *San Francisco Chronicle*, December 22, 2002.

27. California State Mining Bureau, *Annual Report of the State Mineralogist* (Sacramento: California State Printing Office, 1883–84); California State Mining Bureau, *Mining in California*, 1921; California Division of Mines, *Geologic Guidebook*, 1951.

28. Watson Andrews Goodyear, *The Coal Mines of the Western Coast of the United States* (San Francisco: A. L. Bancroft, 1877); Margaret Ballard, "The History of the Coal Mining Industry in the Mount Diablo Region, 1859–1885" (Regional Oral History Office, Bancroft Library, University of California, Berkeley, 1931); Dave Hull, "America's Preeminent Whaling Port: San Francisco," *Sealetter (Journal of the San Francisco Maritime Park Association)*, no. 61 (winter 2001).

29. But see California State Mining Bureau, *The Structural and Industrial Materials of California*, by Lewis Aubury (Sacramento: California State Printing Office, Bulletin No. 38, 1906); California State Mining Bureau, *Mining in California*, 1921; and California Division of Mines, *Geologic Guidebook*, 1951.

30. Thanks to Waverly Lowell for pointing these out.

31. Vicky Elliott, "Extreme Makeover: Leona Quarry to Get a New Life as Oakland's Biggest Subdivision," *San Francisco Chronicle*, October 30, 2005, K1, 5–6.

32. Matt Kondolf, "Hungry Water: Effects of Dams and Gravel Mining on River Channels," *Environmental Management* 21, no. 4 (1997), 533–51.

33. California State Mining Bureau, *Structural and Industrial Materials*, 1906, and *Mining in California*, 1921.

34. A total of more than $50 million, 1896–1926. California State Mining Bureau, *The Clay Resources and the Ceramic Industry of California*, by Waldemar Dietrich (Sacramento: California State Printing Office, Bulletin No. 99, 1928), 33. Also, see California State Mining Bureau, *Annual Report*, 1883–84, 145, and *Structural and Industrial Materials*, 1906.

35. California State Mining Bureau, *Structural and Industrial Materials*, 1906, 187, 203, and *Mining in California*, 1921, 233; California Division of Mines, *Geologic Guidebook*, 1951, 216, 231. Masonry cement, which came earlier than Portland cement, was manufactured at Olema, Benecia, Davenport, and Redwood City.

3 / MOVING OUTDOORS *Parks for the People*

1. Galen Cranz, *The Politics of Park Design: A History of Urban Parks in America* (Cambridge, Mass.: MIT Press, 1982); David Schuyler, *The New Urban Landscape: The Redefinition of City Form in Nineteenth-Century America* (Baltimore: Johns Hopkins University Press, 1986); Sears, *Sacred Places*, 1989; Daniel Bluestone, *Constructing Chicago* (New Haven, Conn.: Yale University Press, 1991); Roy Rosensweig and Elizabeth Blackmar, *The Park and the People* (Ithaca: Cornell University Press, 1992).

2. Terence Young, *Building San Francisco's Parks, 1850–1930* (Baltimore: Johns Hopkins University Press, 2004); also Marjorie Dobkin, "The Great Sand Park: The Origin of Golden Gate Park" (Master's thesis, Department of Geography, University of California, Berkeley, 1979); Raymond Clary, *The Making of Golden Gate Park: The Early Years, 1865–1906* (San Francisco: California Living Books, 1980); and Brechin, *Imperial San Francisco*, 1999. Local historians are much more attuned to the roles of land speculation and politics in park making than are the national historians.

3. Schuyler, *New Urban Landscape*, 1986, 36. Schuyler argues that the pastoral park was a distinctively American contribution, despite antecedents in nineteenth-century Parisian cemeteries and parks and eighteenth-century gardens and parks of England. Cf. Fishman, *Bourgeois Utopias*, 1987.

4. Young, *Building San Francisco's Parks*, 2004. On similar popularization of Central Park, see Schuyler, *New Urban Landscape*, 1986.

5. On the Shanon playground, see Heath Schenker, "Women's and Children's Quarters in Golden Gate Park, San Francisco," *Gender, Place, and Culture* 3, no. 1 (1996), 293–308. The national playground movement, adapting ideas from Germany, took hold in Boston in the 1890s. Spencer-Wood, "American Playground Movement," 1994; and Mozingo, "Neighborhood Parks," n.d. On Progressivism's linking of parks, sanitation, and recreation, see Cranz, *Politics of Park Design*, 1982; Jon Teaford, *The Unheralded Triumph: City Government in America, 1870–1900* (Baltimore: Johns Hopkins University Press, 1984); and Gottlieb, *Forcing the Spring*, 1993.

6. Mozingo, "Neighborhood Parks," n.d.; Frederick Law Olmsted Jr. and John Nolen, "The Normal Requirements of American Towns and Cities in Respect to Public Open Spaces," *Charities and the Commons* 16, no. 14 (1906).Thanks to Louise Mozingo for bringing neighborhood park history to my attention.

7. San Francisco is well endowed with parks compared with San Jose; in 1970, the latter had only 8 acres of open space per 1,000 residents to 35 acres per 1,000 in San Francisco. Stanford Environmental Law Society, *San Jose: Sprawling City* (Stanford: Stanford Environmental Law Society, 1971).

8. Gunter Barth, "Mountain View: Nature and Culture in an American Park Cemetery," in *The Mirror of History: Essays in Honor of Fritz Fellner*, ed. Solomon Wank (Santa Barbara, Calif.: ABC-CLIO, 1988), 153–69; Paul Covel, *People Are for the Birds* (Oakland: Western Interpretive Press, 1978); Beth Bagwell, *Oakland: Story of a City* (Novato: Presidio Press, 1982). On nineteenth-century wildlife slaughter and protection, see Graham, *Audubon Ark*, 1990; and Jennifer Price, *Flight Maps: Adventures with Nature in Modern America* (New York: Basic Books, 1999). The Audubon Society was founded in 1886. On California's massacres, see Davis, *Ecology of Fear*, 1998.

9. Mel Scott, *The San Francisco Bay Area: A Metropolis in Perspective*, 2nd ed. (Berkeley: University of California Press, 1959), 126–29. On Oakland tax politics, see Chris Rhomberg, *No There There: Class and Community in Oakland* (Berkeley: University of California Press, 2004). On its lack of parks, see Olmsted Brothers and Ansel Hall, *Report on Proposed Park Reservations for East Bay Cities*, reprint (Berkeley: University of California, Bureau of Public Administration, [1930] 1984). On the metropolitan parks in the east, see Schuyler, *New Urban Landscape*, 1986, and Cynthia Zaitzevsky, *Frederick Law Olmsted and the Boston Park System* (Cambridge, Mass.: Harvard University Press, 1982).

10. Gray Brechin, "Living the Dream in Berkeley," *California Monthly*, March–April 1984; Richard Walker, "Landscape and City Life: Four Ecologies of Residence in the San Francisco Bay Area," *Ecumene* 2, no. 1 (1995b); Olmsted and Hall, *Proposed Park Reservations*, [1930] 1984.

11. Disney acknowledged his debt to Children's Fairyland in a letter to William

Penn Mott (no relation to Frank Mott). Mary Ellen Butler, *The Prophet of the Parks: The Story of William Penn Mott Jr.* (Ashburn, Va.: National Parks and Recreation Association, 1999), 47. On the campus plan and the Lawrence labs, see Brechin, *Imperial San Francisco*, 1999.

12. Robert Shankland, *Steve Mather of the National Parks* (New York: Alfred Knopf, 1954); Donald Swain, *Wilderness Defender: Horace M. Albright and Conservation* (Chicago: University of Chicago Press, 1970); Horace Albright, Russell Dickinson, and William Penn Mott Jr., *The National Park Service: The Story Behind the Scenery*, ed. Mary Lu Moore (Las Vegas: KC Publications, 1987); Richard Sellers, *Preserving Nature in the National Parks* (New Haven, Conn.: Yale University Press, 1997). Mather, Albright, Kent, and Lane were all deeply involved in resource-extraction industries, but they could already see the money to be made by turning the West into a place of recreation. The only one to oppose mass recreation was Drury. On the growth of auto recreation, see Paul Sutter, *Driven Wild: How the Fight Against Automobiles Launched the Modern Wilderness Movement* (Seattle: University of Washington Press, 2002).

13. Engbeck, *State Parks of California*, 1980; Schrepfer, *Fight to Save the Redwoods*, 1983; Freeman Tilden, *The State Parks: Their Meaning in American Life* (New York: Alfred Knopf, 1962). Also U.S. Department of the Interior, Works Progress Administration, *History of the California State Parks*, ed. Joseph Neasham (Sacramento: California Resources Agency, Division of Parks, 1937). On McDuffie, see Michael Cohen, *History of the Sierra Club*, 1988, 54–55; Engbeck, *State Parks of California*, 1980, 45–47.

14. Engbeck, *State Parks of California*, 1980, 55–60. The commission's powers were reduced in 1936, in favor of the State Division of Parks (later Parks and Beaches).

15. Engbeck, *State Parks of California*, 1980, 64–67; Gray Brechin, *The Living New Deal: Excavating the Public Domain in California* (Berkeley: University of California Press, forthcoming).

16. Engbeck, *State Parks of California*, 1980, 64–67. William Penn Mott became director in 1967, changing the name again to the State Department of Parks and Recreation.

17. Bond figures and quote by Engbeck, *State Parks of California*, 1980, 121. Oil fee numbers from Daniel Press, *Saving Open Space: The Politics of Local Preservation in California* (Berkeley: University of California Press, 2002), 34. Press also has a complete list of state park bonds in Appendix 4.

18. Norman LaForce, "Creating the Eastshore State Park: An Activist History," interview by Ann Lage (Regional Oral History Office, Bancroft Library, University of California, Berkeley, 2002); Chiori Santiago, "Betting on Point Molate," *Bay Nature*, July–September 2005, 38–42. Dwight Steele and Jean Siri were others active in creating the park.

19. Peter Schrag, *Paradise Lost: California's Experience, America's Future* (New

York: The New Press, 1998); Pincetl, *Transforming California*, 1999; Dyble, "Paying the Toll," 2002.

20. Michael Svanevik and Shirley Burgett, *San Mateo County Parks: A Remarkable Story of Extraordinary Places and the People Who Built Them* (Menlo Park, Calif.: San Mateo County Parks and Recreation Foundation, 2001).

21. Chiori Santiago, "A Modest Majesty: Seventy Years of the East Bay Parks," *Bay Nature*, October–December 2004; Slack, "Shadow of Giants," 2004; www.ebparks.org.

22. Schuyler, *New Urban Landscape*, 1986, calls the Boston "Emerald Necklace" the culmination of naturalistic park development. It no doubt figured in the thinking of Olmsted and Hall. Thanks to Louise Mozingo for alerting me to this.

23. Seth Adams, "Who Was Walter P. Frick? The Creation of Mount Diablo State Park," *Diablo Watch* 37–38 (winter–spring, fall 2004b); David Wallace, "Speak of the Devil," *Bay Nature*, July–September 2006.

24. Mimi Stein, *A Vision Achieved: Fifty Years of the East Bay Regional Parks District* (Oakland: East Bay Regional Parks District, 1984).

25. Olmsted and Hall, *Proposed Park Reservations*, [1930] 1984. Samuel May was a classic Progressive who put in place Berkeley's city manager system of government. He was also well connected to the Sierra Club. See Bernice Hubbard May, "A Native Daughter's Leadership in Public Affairs" (Regional Oral History Office, Bancroft Library, University of California, Berkeley, 1976).

26. Stein, *A Vision Achieved*, 1984; Mason McDuffie Company records, carton 26, F13–32 (Bancroft Library, University of California, Berkeley).

27. By contrast, in Los Angeles in 1930 an ambitious regional parkways plan drawn up by Olmsted and Bartholomew sank without a trace, over fears of excessive public debt and the powers to be conferred on the Parks Board. Davis, *Ecology of Fear*, 1998; Greg Hise and William Deverell, eds., *Eden by Design* (Berkeley: University of California Press, 2000).

28. Stein, *A Vision Achieved*, 1984, 56; M. E. Butler, *Prophet of the Parks*, 1999.

29. Mary Ellen Butler, *Prophet of the Parks*, 1999.

30. Trudeau also established the East Bay Regional Parks Foundation and the California Greenways Foundation and was executive director of the Mott Foundation. Richard Trudeau, "From Skyline to Seashore: Twenty-Two Years of Leadership, Land Acquisition, and Lobbying at the East Bay Regional Parks District, 1964–1986," interview by Ann Lage (Regional Oral History Office, Bancroft Library, University of California, Berkeley, 2003). Also, Robert Doyle, telephone interview with author, December 21, 2005.

31. Jean Siri, interview in El Cerrito with author, December 21, 2005; Stephen Vincent, personal communication with the author, October 4, 2004. In 2005 Siri was still fighting to save marshland west of Parchester Village in North Richmond and add it to Point Pinole Regional Shoreline.

32. On the origins of Save Mount Diablo, see Galen Rowell, *Bay Area Wild*, 1997.

Bob Walker (no relation) was a key activist in the 1980s, working with the EBRPD, Sierra Club and Greenbelt Alliance, before his untimely death in 1992. He was also the founder of Gay and Lesbian Sierrans in the Sierra Club.

33. Mark Evanoff of Greenbelt Alliance and Janet Cobb of the EBRPD deserve special credit for the passage of the 1988 measure. Winning tax increases and bond elections has been tricky in recent years. The district lost one tax increase in 1998, another in 2002, and a parcel tax in Contra Costa County in 2004. Alameda County voters approved a parcel tax in 2004.

34. Doyle, interview, 2005. Essential maintenance is done by the East Bay Conservation Corps, employing inner city youth of color.

35. Phil Patton, *Open Road: A Celebration of the American Highway* (New York: Simon and Schuster, 1986); Samuel Hays, *Beauty, Health, and Permanence: Environmental Politics in the United States, 1955–85* (New York: Cambridge University Press, 1987); Hal Rothman, *Devil's Bargains: Television in the 20th Century American West* (Lawrence: University of Kansas Press, 1998a); Sutter, *Driven Wild*, 2002; Young, "Heading Out," manuscript.

36. Yaryan et al., *Sempervirens Story*, 2000, 51.

37. Sutter, *Driven Wild*, 2002; Michael Cohen, *History of the Sierra Club*, 1988, ch. 3; Sellers, *Preserving Nature*, 1997.

38. On controversies of wilderness management, Laura Watt, "Managing Cultural Landscapes: Reconciling Local Ideology and Institutional Ideology in the National Park Service" (PhD dissertation, Department of Environmental Science, Policy, and Management, University of California, Berkeley, 2001).

4 / THE UPPER WEST SIDE *Suburbia and Conservation*

1. Press, *Saving Open Space*, 2002. Press ranks Los Angeles County highest in the state in land conservation, but this includes Santa Catalina Island.

2. Rome, *Bulldozer*, 2001. Also, Jackson, *Crabgrass Frontier*, 1985; Rothman, *Greening of a Nation?* 1998b. Carson was, in fact, well connected to popular opposition. Rachel Carson, *Silent Spring* (Boston: Houghton Mifflin, 1962); Norwood and Linda Lear, *Rachel Carson: Witness for Nature* (New York: Holt, 1997).

3. Alas, Rome does not connect the critique of the bulldozers to the grassroots rebellion. On California's advanced suburbanization, see Davis, *City of Quartz*, 1990; Richard Walker, "Landscape and City Life," 1995b; Greg Hise, *Magnetic Los Angeles: Planning the Twentieth-Century Metropolis* (Baltimore: Johns Hopkins University Press, 1997); and Marc Weiss, *The Rise of the Community Builders: The American Real Estate Industry and Urban Planning* (New York: Columbia University Press, 1987).

4. David Jones, *California's Freeway Era in Historical Perspective* (Berkeley: University of California Press, Institute for Transportation Studies, 1989).

5. The plan is reproduced in Martin Griffin, *Saving the Marin-Sonoma Coast: The Battles for Audubon Canyon Ranch, Point Reyes, and California's Russian River* (Healdsburg, Calif.: Sweetwater Springs Press, 1998), 100–101. On McCarthy, see Dyble, "Paying the Toll," 2002. On the politics of local growth, see John Logan and Harvey Molotch, *Urban Fortunes: The Political Economy of Place* (Berkeley and Los Angeles: University of California Press, 1986); and Stone, *Regime Politics*, 1989.

6. Mark Gelfand, *A Nation of Cities: The Federal Government and Urban America, 1933–1975* (New York: Oxford University Press, 1975); Jackson, *Crabgrass Frontier*, 1985; Edward Eichler, *The Merchant Builders* (Cambridge, Mass.: MIT Press, 1982).

7. Elizabeth Burns, "The Process of Suburban Residential Development: The San Francisco Peninsula, 1860–1970" (PhD dissertation, Department of Geography, University of California, Berkeley, 1975); Hynding, *From Frontier to Suburb*, 1981; Gray Brechin, "Mr. Leavitt of the Sunset," *San Francisco Focus*, June 1990, 23–26; Richard Walker, "Landscape and City Life," 1995b.

8. On this process in general, see Marion Clawson, *Surburban Land Conversion in the United States: An Economic and Governmental Process* (Baltimore: Johns Hopkins University Press, 1971); and Richard Walker, "A Theory of Suburbanization: Capitalism and the Construction of Urban Space in the United States," in *Urbanization and Urban Planning in Capitalist Societies*, eds. Michael Dear and Allen Scott (New York: Methuen, 1981).

9. Stanford Environmental Law Society, *San Jose*, 1971; Fellmeth, *Politics of Land*, 1973; Kunstler, *Geography of Nowhere*, 1993.

10. Griffin, *Saving the Marin-Sonoma Coast*, 1998, 10; John Hart, *Wilderness Next Door*, 1979; Grace Wellman, interviews by Carla Ehat and Genevieve Martinelli, February 19, 1982 (California Room, Marin Public Library, San Rafael); Nancy Wise, *Marin's Natural Assets: A Historic Look at Marin County* (San Rafael: Marin Conservation League, 1985). On van Pelt, see Winifred Starr Dobyns, *California Gardens* (New York: Macmillan, 1931). On garden-club women, see Shana Cohen, "American Garden Clubs," 2005.

11. Stewart is quoted in Wise, *Marin's Natural Assets*, 1985, 37; Dorothy Ward Erskine, "Environmental Quality and Planning," 1976, 138. Sepha Evers was number two in the league; Helen van Pelt moved out of Marin; Portia Forbes was inactive after the early years.

12. Marin Conservation League, *Marvelous Marin County: San Francisco's Most Unique Suburb* (San Rafael: Marvelous Marin, Inc., MCL, 1936); Hugh Pomeroy, *A County Planning Program for Marin County, California* (San Rafael: Marin Planning Survey Committee, for Marin County Planning Commission, 1935). On Pomeroy's role in L.A., see Hise and Deverell, *Eden by Design*, 2000.

13. Undated Marin Conservation League history flyer, in the file with Wellman, interview, 1982; Bertram Dunshee, interview by Anne Kent and Carla Ehat, Novem-

ber 15, 1979 (California History Room, Marin County Public Library, San Rafael). Barbara Eastman was also active in securing these parks. Hart and Kittle, *Legacy*, 2006, 75.

14. Most notably, Robert Putnam, *Bowling Alone: The Collapse and Revival of American Community* (New York: Simon and Schuster, 2000).

15. Angel Island Association, *Angel Island: Jewel of San Francisco Bay* (Tiburon, Calif.: Angel Island Association, 1999). Livermore worked with many others on these projects, including Verna Dunshee, Marty Griffin, and David Steinhardt.

16. Of the early Livermores, Horatio G. was a Forty-Niner, industrialist, and state senator. His son, Horatio Putnam, built the first hydroelectric dam at Folsom in 1893 and acquired property on the summit of Russian Hill, the original upper-class bohemian enclave. Horatio Putnam's son, Norman Sr., was a director of PG&E, Cal-Pak, and Crocker Bank. Norman Livermore, "Man in the Middle: High Sierra Packer, Timberman, Conservationist, and California Resources Secretary," interviews by Ann Lage and Gabrielle Morris, 1981–82 (Regional Oral History Office, Bancroft Library, University of California, Berkeley, 1983).

17. Helen Baker Reynolds, interview in San Francisco by Carla Ehat and Anne Kent, February 28, 1979 (California Room, Marin Public Library, San Rafael). On the billboard campaign nationally, see Catherine Gudis, *Buyways: Billboards, Automobiles, and the American Landscape* (New York: Routledge, 2004). On saving Lake Tahoe, see Beesley, *Crow's Range*, 2004.

18. Harold Gilliam, *Island in Time: The Point Reyes Peninsula* (San Francisco: Sierra Club Books, 1962); John Hart, *Wilderness Next Door*, 1979; George Collins, "The Art and Politics of Park Planning and Preservation, 1920–1979" (Regional Oral History Office, Bancroft Library, University of California, Berkeley, 1980); Peter Behr, interview by Ann Lage (Regional Oral History Office, Bancroft Library, University of California, Berkeley, 1988); Margaret Azevedo, Peter Behr, Katy Miller Johnson, William Kahrl, Pete McClosky, and Boyd Stewart, "Saving Point Reyes National Seashore: An Oral History of Citizen Action in Conservation, 1969–70," interviews by Ann Lage and William Duddleson (Regional Oral History Office, Bancroft Library, University of California, Berkeley, 1993).

19. Behr went on to the State Senate, where he authored the California Wild and Scenic Rivers Act in 1972. Rancher Boyd Stewart was also crucial in rallying landowner support for the park.

20. Griffin, *Saving the Marin-Sonoma Coast*, 1998, 42. Griffin was a doctor in Kentfield.

21. John Hart, *Wilderness Next Door*, 1979; Wise, *Marin's Natural Assets*, 1985; Martha Alexander Gerbode, "Environmentalist, Philanthropist, and Volunteer in the San Francisco Bay Area and Hawaii," interviews by Huey Johnson, Garland Farmer, Esther Fuller, et al. (Regional Oral History Office, Bancroft Library, University of California, Berkeley, 1995); Griffin, *Saving the Marin-Sonoma Coast*, 1998; Martin Rosen, "Trust for Public Land Founding Member and President, 1972–1997: The

Ethics and Practice of Land Conservation" (Regional Oral History Office, Bancroft Library, University of California, Berkeley, 2000). Johnson, Rosen, Ferguson, and Gerbode all lived in Mill Valley.

22. Behr, interview, 1988; Griffin, *Saving the Marin-Sonoma Coast*, 1998. The defeat of extending Bay Area Rapid Transit (BART) to Marin was also critical; Dyble, "Paying the Toll," 2002. As was San Francisco's freeway battle of the late 1950s; William Issel, "Land Values, Human Values, and the Preservation of the City's Treasured Appearance: Environmentalism, Politics, and the San Francisco Freeway Revolt," *Pacific Historical Review* 68 (1999).

23. The key report was Marin County Planning Department, *Can the Last Place Last? Preserving the Environmental Quality of Marin*, by Albert Solnit (San Rafael: Marin County Planning Department, 1971). The Marin plan was revised in 1989 and 1994, but not significantly altered.

24. Azevedo, Behr, et al., "Saving Point Reyes," 1993. Azevedo later ran afoul of some Marinites for advocating higher densities and mixed housing in the county.

25. Griffin, *Saving the Marin-Sonoma Coast*, 1998, 139–50.

26. Cf. Grant McConnell, *Private Power and American Democracy* (New York: Vintage Books, 1966); Logan and Molotch, *Urban Fortunes*, 1986.

27. Quote is from Hal Rothman, *The New Urban Park: Golden Gate National Recreation Area and Civic Environmentalism* (Lawrence: University of Kansas Press, 2004), 24; Edgar Wayburn, "Sierra Club Statesman, Leader of the Parks and Wilderness Movement, etc.," interviews by Ann Lage and Susan Schrepfer (Regional Oral History Office, Bancroft Library, University of California, Berkeley, 1985); Michael Cohen, *History of the Sierra Club*, 1988; Harold Gilliam, "The Quiet Conservationist," *San Francisco Chronicle Magazine*, April 25, 2004. Another Wayburn legacy is the Alaska Lands Act of 1980. Cf. Edgar Wayburn, "Sierra Club Statesman," 1985. See also the writings of Peggy Wayburn, e.g., P. Wayburn, *Adventuring in the San Francisco Bay Area* (San Francisco: Sierra Club Books, 1987).

28. Amy Meyer, *New Guardians of the Golden Gate* (Berkeley: University of California Press, 2006); Sierra Club History Committee, "The Sierra Club and the Urban Environment II: Labor and the Environment in the San Francisco Bay Area, 1960s–1970s," interviews with David Jenkins, Amy Meyer, Anthony Ramos, and Dwight Steele (Regional Oral History Office, Bancroft Library, University of California, Berkeley, 1983); Rothman, *New Urban Park*, 2004.

29. On the park's origins, see Rothman, *New Urban Park*, 2004. Also Margot Patterson Doss, *Paths of Gold: In and Around Golden Gate National Recreation Area* (San Francisco: Chronicle Books, 1974); John Hart, *Wilderness Next Door*, 1979; U.S. Department of the Interior, National Park Service, *A Civil History of the Golden Gate National Recreation Area and Point Reyes National Seashore, California*, by Anna Toogood (Washington, D.C.: National Park Service, 1980); Sierra Club History Committee, "Sierra Club and the Urban Environment II," 1983; Harold Gilliam and Ann

Lawrence, *Marin Headlands: Portals of Time* (San Francisco: Golden Gate National Park Association, 1993). GGNRA was twinned with the Gateway National Recreation Area in New York City.

30. John Jacobs, *A Rage for Justice: The Passion and Politics of Phillip Burton* (Berkeley: University of California Press, 1995).

31. Jenkins observed, "[Burton] was out of our ranks [and] when Phil became a major advocate of the environmentalists, he was able to do it because his base with us was so secure." Sierra Club History Committee, "Sierra Club and the Urban Environment II," 1983, 29. Executive director McCloskey advocated an urban parks policy, but the club board nixed the idea; Michael Cohen, *History of the Sierra Club*, 1988, 398.

32. Meyer is quoted from Sierra Club History Committee, "Sierra Club and the Urban Environment II," 1983, 3. Jacobs, *Rage for Justice*, 1995. It is a delicious irony that the wilderness area within Point Reyes National Seashore was named after Burton.

33. Rothman, *New Urban Park*, 2004. Only half of GGNRA's lands were newly protected, but the whole has a coherence that the pieces lacked before.

34. Meyer, *New Guardians*, 2006; Rothman *New Urban Park*, 2004. Rothman is dismissive of opposition to the Presidio Trust, but it includes the Sierra Club Bay Chapter and is based on sound concerns about privatization and lack of public control.

35. Figures calculated by Peter Cohen from data at www.openspacecouncil.org, ca. 2002.

36. David Sharkey, "The Conservation and Environmental Movement in San Mateo County," *La Peninsula* 20, no. 3 (1980), 21–31; and Yaryan et al., *Sempervirens Story*, 2000.

37. Yaryan et al., *Sempervirens Story*, 2000; Pincetl, *Transforming California*, 1999, 166–69.

38. Dorothy Varian, *The Inventor and the Pilot: Russell and Sigurd Varian* (Palo Alto: Pacific Books, 1983); Yaryan et al., *Sempervirens Story*, 2000. Richard Leonard worked as a lawyer for Varian Associates.

39. Yaryan et al., *Sempervirens Story*, 2000, 53ff.

40. Martin Kenney, ed., *Understanding Silicon Valley: The Anatomy of an Entrepreneurial Region* (Stanford: Stanford University Press, 2000).

41. For the history of the Committee for Green Foothills, see Phyllis Butler, ed., *20-20 Vision: In Celebration of the Peninsula Hills* (Palo Alto: Western Tanager Press, Committee for Green Foothills, 1982); Ruth Spangenberg, "The Founding of the Committee for Green Foothills," interview by Richard Walker, August 16, 2001 (Regional Oral History Office, Bancroft Library, University of California, Berkeley, 2004); Mary Davey, "Saving Open Space on the San Francisco Peninsula," interview by Richard Walker, August 26, 2001 (Regional Oral History Office, Bancroft Library, University of California, Berkeley, 2004); Lennie Roberts, "Protecting San Mateo

County Open Space," interview by Richard Walker, August 23, 2001 (Regional Oral History Office, Bancroft Library, University of California, Berkeley, 2004). Also, Spangenberg's reminiscences in *Green Footnotes,* spring 2000; Jay Thorwaldson, "Gracious Activist: Lois Hogle's Fight for the Foothills," *Palo Alto Weekly* 94 (August 22, 2001); and time line in *Green Footnotes,* fortieth anniversary issue, 2002.

42. In the early years, Committee for Green Foothills members were almost all Republicans and were desperate to bring a few token Democrats on their board; this all changed by the 1980s. The committee never endorses political candidates or parties. Pete McCloskey is not related to Mike McCloskey of the Sierra Club.

43. On Stegner and Litton, see Michael Cohen, *History of the Sierra Club*, 1988 (quote from Brower, at 157).

44. Lewis Mumford, from letter to Ruth Spangenberg, November 1960, on file at Bancroft Library, University of California, Berkeley. Spangenberg is typical of westside activists in having known practically everyone in the local elite, from Stanford presidents and Fred Terman to the Hewletts, Packards, Varians, and Duvenecks. Spangenberg, "Founding of the Committee for Green Foothills," 2004. Mumford was a frequent visitor to Berkeley. Tresidder had been president of the Yosemite Park and Curry Company and a Sierra Club member.

45. Spangenberg, "Founding of the Committee for Green Foothills," 2004; Davey, "Saving Open Space," 2004; Roberts, "Protecting San Mateo County," 2004; Thorwaldson, "Gracious Activist," 2001.

46. On Hanko, see Rowell, *Bay Area Wild*, 1997, 181–85.

47. Livingston and Blayney, *Foothills Environmental Design Study: Report No. 3 to the City of Palo Alto,* with the assistance of Lawrence Halperin and Associates (San Francisco: Livingston and Blayney, 1970). Larry Livingston had worked as a planner in Los Altos and Saratoga, so he was no stranger to the foothills crowd. Goodwin Steinberg, *From the Ground Up: Building Silicon Valley* (Stanford: Stanford University Press, 2002), 14.

48. Kevin Cool, "This Precious Plot," *Stanford Magazine*, January–February 2001; *Green Foothills,* various issues.

49. Route modifications were also won by Stanford and Los Altos Hills. Boushey learned highway politics from Helen Reynolds of the California Roadside Council. Hart and Kittle, *Legacy*, 2006, 59.

50. On the California League of Conservation Voters, see Robert Duffy, *The Green Agenda in American Politics: New Strategies for the Twenty-first Century* (Lawrence: University of Kansas Press, 2003), 164; www.ecovote.org.

51. Claire Dedrick, interview by Ann Lage (unedited transcript, Regional Oral History Office, Bancroft Library, University of California, Berkeley, 1998); Tom Jordan, "Remembering Claire and Kent Dedrick," *Green Footnotes*, fall 2005, 15.

52. Hynding, *From Frontier to Suburb*, 1981; Svanevik and Burgett, *San Mateo County Parks*, 2001. The state owns 300 acres on Mount San Bruno, which it leases to

the county to operate; 700 acres are private open space, which bristle with broadcasting and microwave towers.

53. Bill Workman, "Environmental Crusader for Mountain Carries On," *San Francisco Chronicle*, March 2, 2000; John King, "Build It and They Will Cringe," *San Francisco Chronicle*, December 13, 2004.

54. This campaign was led by Lennie Roberts and Olive Mayer of CGF and Zoe Kersten-Tucker of Citizens for a Tunnel (and later executive director of Committee for Green Foothills). Roberts, "Protecting San Mateo County," 2004; Hart and Kittle, *Legacy*, 2006, 57; *Green Footnotes*, various issues.

55. A crucial $200,000 donation was given by Ivan Pejcha, a Czech immigrant entrepreneur irritated by the supervisors' progrowth stance. Roberts, "Protecting San Mateo County," 2004.

56. Hynding, *From Frontier to Suburb*, 1981; Roberts, "Protecting San Mateo County," 2004.

57. By the late 1980s, Marin was the most expensive county to live in on the West Coast (second in the country). Richard Walker et al., "Playground of U.S. Capitalism?" 1990. Marin was 96 percent white in 1970 and still 79 percent white in 2000.

58. Dedrick, interview, 1998.

5 / THE GREEN AND THE BLUE *Saving the Bay and the Coast*

1. John Hart and David Sanger, *San Francisco Bay: Portrait of an Estuary* (Berkeley: University of California Press, 2003); Dennis Anderson and Jerry George, *Hidden Treasures of San Francisco Bay* (Berkeley: Heyday Books, 2003).

2. Briton Busch, *The War Against the Seals: A History of the North American Seal Fishery* (Montreal: McGill-Queens University Press, 1985); Arthur McEvoy, *The Fisherman's Problem: Ecology and Law in the California Fisheries, 1850–1980* (New York: Cambridge University Press, 1986); Ronald Yoshiyama, "A History of Salmon and People in the Central Valley Region of California," *Reviews in Fisheries Science* 7, no. 3 (1999). State regulation of fishing goes back to 1872, but it wasn't able to halt the declines.

3. Morgan's oysters were imports from Chesapeake Bay; native oysters had been done in by pollution by the 1900s, though a few survive today. Leslie was owned by the Whitneys, then the Schillings, George Ishiyama, and finally Cargill, ending the Leslie name. On cement and salt, see California Division of Mines, *Salt in California*, by William Ver Plank and Robert Heizer (Sacramento: California State Printing Office, Bulletin No. 175, 1958); and Hynding, *From Frontier to Suburb*, 1981, 86–87, 251.

4. McEvoy, *Fisherman's Problem*, 1986; L. Eve Ma, "Chinese in California's Fishing Industry, 1850–1941," *California History* 60, no. 2 (1981).

5. A typical reaction of the time was that of my high school girlfriend's mother,

who said, as we drove along the bayshore, that it was better to fill the mudflats since they smelled so bad.

6. Mel Scott, *San Francisco Bay Area*, 1959; Gray Brechin, "Progress in San Francisco: It Could Have Been Worse," *San Francisco Magazine*, October 1983; Stuart Cook, "The Unfulfilled Bay: The Dreamers Didn't Always Get Their Way," *California Waterfront Age* 1, no. 4 (1985). Reber's plan had antecedents going back to the 1910s and even as far back as 1879. David Long, "Mistaken Identity: Putting the John Reber Plan for the San Francisco Bay Area into Historical Context," in *American Cities and Towns: Historical Perspectives*, ed. Joseph Rishel (Pittsburgh: Duquesne University Press, 1992).

7. On Foster's use of the public purse for private gain, see Fellmeth, *Politics of Land*, 1973; and Hynding, *From Frontier to Suburb*, 1981, 280–83. Bruce Brugmann, a *Redwood City Tribune* reporter at the time, blew the whistle on the Rockefeller-Crocker plan. Sharkey, "Conservation Movement in San Mateo," 1980.

8. For a map circa 1964 of Bay fill projects, see www.bcdc.ca.gov/archive/sfb csc/fill.

9. Everyone there except the three Berkeley women and Dorothy Erskine was male, which may have had something to do with the reluctance to help out. This is according to Sylvia McLaughlin, as quoted by Briggs Nisbet, telephone interview with author, July 15, 2005.

10. On the origins of Save San Francisco Bay Association, see Mel Scott, *The Future of San Francisco Bay* (Berkeley: University of California, Berkeley, Institute of Governmental Studies, 1963); Harold Gilliam, *Between the Devil and the Deep Blue Bay: The Struggle to Save San Francisco Bay* (San Francisco: Chronicle Books, 1969); Kay Kerr, Sylvia McLaughlin, and Esther Gulick, "Save San Francisco Bay Association, 1961–1986," introduction by Harold Gilliam (Regional Oral History Office, Bancroft Library, University of California, Berkeley, 1987); Rowell, *Bay Area Wild*, 1997, 153–55; Hart and Sanger, *San Francisco Bay*, 2003.

11. San Francisco Bay Conservation and Development Commission, *San Francisco Bay Plan* (San Francisco: SFBCDC, 1979); Joseph E. Bodovitz, Melvin B. Lane, and E. Clement Shute Jr., "The San Francisco Bay Conservation and Development Commission, 1964–1973," interviews by Malca Chall (Regional Oral History Office, Bancroft Library, University of California, Berkeley, 1984); Joseph Bodovitz, "The Shrinking of San Francisco Bay and How It Was Stopped," *California Waterfront Age* 1, no. 4 (1985). The only failure was getting just a 100-foot shoreline review, instead of 1,000 feet back from the bay.

12. Roberts, "Protecting San Mateo County," 2004.

13. Esther Gulick, Catherine Kerr, and Sylvia McLaughlin, "Saving San Francisco Bay: Past, Present, and Future," in *The Horace M. Albright Lectureship in Conservation* (Berkeley: College of Natural Resources, University of California, Berkeley) 28 (April 14, 1988). On alternative solutions for air traffic management, see www.savesfbay.org.

14. On water diversions, see Michael Storper and Richard Walker, "The Expanding California Water System," in *Use and Protection of the San Francisco Bay System*, eds. William Kockelman, Tom Conomos, and Alan Leviton (San Francisco: Pacific Division, American Association for the Advancement of Science, 1982); Fred Nichols, James Cloern, Samuel Luoma, and David Peterson, "The Modification of an Estuary," *Science* 231, no. 4738 (1986); Bay Institute, *From the Sierra to the Sea: The Ecological History of the San Francisco Bay–Delta Watershed* (San Rafael: The Bay Institute, 1998); Jane Kay, "Toward a Healthy Bay," *Bay Nature* magazine, October–December 2003b; and www.tbi.org. On hydraulic mining, see L. Allen James, "Channel Changes Wrought by Gold Mining: Northern Sierra Nevada, California," *Effects of Human-Induced Changes on Hydrologic Systems* (American Water Resources Association) 86 (June 1994); Brechin, *Imperial San Francisco*, 1999.

15. Figures are from the Bay Institute, cited by Kay, "Toward a Healthy Bay," 2003b.

16. On the Sierra Club and rivers, see Michael Cohen, *History of the Sierra Club*, 1988, ch. 4; on the Stanislaus, see Tim Palmer, *Stanislaus: The Struggle for a River* (Berkeley: University of California Press, 1982). Happily, Auburn Dam was defeated.

17. Bill Davoren, interview in Tiburon with author, July 13, 2005. Davoren got crucial support from Hap Dunning of UC Davis, Will Siri of the Sierra Club, Jane Rogers of San Francisco Foundation, and the Packard Foundation.

18. Storper and Walker, "California Water System," 1982; Marc Reisner, *Cadillac Desert: The American West and Its Disappearing Water* (New York: Viking, 1986); Nick Arguimbau, telephone interview with author, July 18, 2004.

19. A key precedent was the Mono Lake decision (1983), which declared a public trust in the state's waters. John Hart, *Storm Over Mono: The Mono Lake Battle and the California Water Future* (Berkeley: University of California Press, 1996). Racanelli later regretted that he didn't impose a solution, given the state's weak resolve. Davoren, interview, 2005. Davoren's sole regret is never having brought a straightforward public trust suit against the delta diverters.

20. For a basic history, see Katherine Jacobs, Samual Luoma, and Kim Taylor, "CALFED: An Experiment in Science and Decisionmaking," *Environment* 45, no. 1 (2003), 31–41. However, these authors miss the Racanelli decision and Peripheral Canal precedents. Moreover, in the meantime, environmentalists had succeeded in having the bay and delta recognized officially as an estuary. The 1987 National Estuaries Act had led to a San Francisco Estuary Project, which issued a conservation and management plan in 1993. Out of that came the San Francisco Estuary Institute and its ongoing research program. See www.sfei.org.

21. Quote from Nisbet, interview, 2005. Davoren, interview, 2005, has a similarly jaundiced view of CALFED, as does former Save the Bay director Barry Nelson. Kay, "Toward a Healthy Bay," 2003b. Under the second Bush administration,

CALFED has gotten no more federal money, and its lofty science program has been sorely underfunded. The naivete of CALFED scientists about the political realities of California water is touching. Compare Jacobs, Luoma, and Taylor, "CALFED: An Experiment," 2003, with Richard Walker, *Conquest of Bread* (New York: The Free Press, 2004), 171–81.

22. For ongoing studies of bay invasive species, see the work of the San Francisco Estuary Institute, www.sfei.org; Kay, "Toward a Healthy Bay," 2003b. The striped bass was also an Atlantic introduction.

23. The first study of wetlands loss was the Bay Institute, *Citizens' Report on the Diked Historical Baylands of San Francisco Bay* (San Rafael: The Bay Institute, 1987). Robin Grossinger of San Francisco Estuary Institute has been reconstructing what the bayshore and its wetlands looked like in the past; see www.sfei.org.

24. Florence and Phil LaRiviere, interview in Palo Alto with author, July 8, 2005; Hart and Kittle, *Legacy,* 2006, 45.

25. Ibid.

26. This leaves the Citizens Committee still unsatisfied over unprotected marshes, operating salt ponds, and the bounty paid to Cargill. LaRivieres, interview, 2005.

27. Mark Holmes, telephone interview with author, July 22, 2005; Hart and Sanger, *San Francisco Bay,* 2003.

28. Holmes, interview, 2005; Nisbet, interview, 2005.

29. Matthew Bettelheim, "The Endemic Nature of the Antioch Dunes," *Bay Nature*, January–March 2005.

30. Nisbet, interview, 2005. Glen Martin, "Invader in the Bay," *San Francisco Chronicle*, October 11, 2005, B1, B12.

31. Hart and Sanger, *San Francisco Bay,* 2003, 44; www.sfei.org. See also Save the Bay, *Protecting Wetlands* (Oakland: Save the Bay), and Cynthia Patton, *Turning Salt into Gold* (Oakland: Save the Bay).

32. Charles Coleman, PG&E *of California: The Centennial Story of Pacific Gas and Electric Company, 1852–1952* (New York: McGraw-Hill, 1952); Thomas Wellock, *Critical Masses: Opposition to Nuclear Power in California, 1958–1978* (Madison: University of Wisconsin Press, 1998).

33. On Berkeley's role in atomic power, see Brechin, *Imperial San Francisco*, 1999.

34. Joel Hedgpethe, "Bodega: A Case History of Intense Controversy," in *Environmental Quality and Water Development*, eds. C. Goldman, J. McEvoy, and P. Richerson (San Francisco: W. H. Freeman, 1973); Fellmeth, *Politics of Land*, 1973; Wellock, *Critical Masses*, 1998. Greens later succeeded in closing down the Vallecitos GE plant for dumping tritium, shuttering the experimental reactor at UC Berkeley, and urging Berkeley and Oakland to declare Nuclear Free Zones. Pesonen went on to the East Bay Regional Parks District and California State Forestry Board.

35. Michael Cohen, *History of the Sierra Club*, 1988, ch. 8; Wellock, *Critical Masses*, 1998.

36. Barbara Epstein, *Political Protest and Cultural Revolution* (Berkeley: University of California Press, 1991). John Gofman of Livermore created the Committee for Nuclear Responsibility in 1971.

37. Wellock, *Critical Masses*, 1998, 4. Taming the larger nuclear beast has proved elusive, however. Although the Berkeley Labs converted to alternative-energy research in the 1970s, there was no such change of heart among the ardent cold warriors of Livermore Labs and the Hungarian Darth Vader, Edward Teller. The disarmament movement drew thousands to the Livermore Action Group's blockade in 1982–83. LAG took up where Abalone Alliance left off, drawing on the same countercultural currents and direct-action tactics but attracting an even wider spectrum of political activists and people worried by Reagan's militarism. Epstein, *Political Protest*, 1991.

38. Harvey Molotch, "Oil in Santa Barbara and Power in America," *Sociological Inquiry* 40 (1970).

39. Quote by Lew Reid, author of Proposition 20, in Rosa Gustaitis, "Lew Reid, Defender of the Coast," *Coast and Ocean* 12, no. 4 (winter 2002–03), 30. On the origins of Proposition 20, see also William Kortum, "Environmentalist Extraordinaire in Sonoma County," interviews by Richard Walker and Martin Bennett, June 9, 2003, and May 5, 2005 (Regional Oral History Office, Bancroft Library, University of California, Berkeley).

40. Legislative leaders Bob Moretti in the California Assembly and James Mills of the State Senate were instrumental in confirming the act in 1976. On the debate over the coastal commission, see Stanley Scott, ed., *Governing California's Coast* (Berkeley: University of California, Berkeley, Institute of Governmental Studies, 1975); Robert Healy, ed., *Protecting the Golden Shore: Lessons from the California Coastal Commission* (Washington, D.C.: Conservation Foundation, 1978).

41. On the performance of the coastal commission, see Michael Heiman, *Coastal Recreation in California* (Berkeley: University of California, Berkeley, Institute of Governmental Studies, 1987); Pincetl, *Transforming California*, 1999. The 1972 federal Coastal Zone Management Act was influenced by bay and coastal protection in California, especially the demand for federal adherence to local regulatory standards. The CZMA clarified federal consistency with state laws, as well as funded state-level coastal planning in the mold of California.

42. Nationally, the Sanctuaries Act has not been terribly effective, and fishing is not limited. Furthermore, upwelling has been altered by global warming, resulting in a marked falloff in the mass of plankton.

43. Peter White, *The Farallon Islands: Sentinels of the Golden Gate* (San Francisco: Scottwall Associates, 1995).

44. Michael Bhargava, "Sanctuaries: Safe Havens for Marine Life," *Coast and Ocean* 19, no. 2 (2003).

1. U.S. Department of Commerce, *Future Development of the San Francisco Bay Area, 1960–2020,* prepared for U.S. Army Corps of Engineers (Washington, D.C.: U.S. Government Printing Office, 1960); Association of Bay Area Governments, *Regional Plan, 1970–1990—San Francisco Bay Region* (Berkeley: ABAG, 1970).

2. For standard accounts, see Gottlieb, *Forcing the Spring,* 1993; Rothman, *Greening of a Nation?* 1998b.

3. Francis Violich, "The Planning Pioneers," *California Living,* Magazine of the Sunday Chronicle and Examiner, February 26, 1978; T. J. Kent Jr., "Professor and Political Activist: A Career in City and Regional Planning in the San Francisco Bay Area," in *Statewide and Regional Land-Use Planning in California,* vol. 2, interview by Malca Chall (Regional Oral History Office, Bancroft Library, University of California, Berkeley, 1983); Marc Treib and Dorothée Imbert, *Garrett Eckbo: Modern Landscapes for Living* (Berkeley: University of California Press, 1997); Pete Allen, "A Space for Living: The Arc of Regional Modernism in the San Francisco Bay Area" (PhD diss., Department of Architecture, University of California, Berkeley, 2007. Jack Kent (no relation to the Kents of Marin) worked in the National Resource Planning Board office in Berkeley, formed an early friendship with Mumford, and was deeply influenced by Modernists he met as a student traveling in Europe (especially the London branch of the Congress of International Architects). He served after the Second World War as planning director of San Francisco, then as chair of the new Department of City and Regional Planning at UC Berkeley, and later on the Berkeley City Council. Mel Scott was inspired by social housing projects he saw in 1930s Europe, then did studies for urban renewal in Los Angeles and San Francisco during the Second World War (to his later regret). He wrote the first urban history of the Bay Area and a history of twentieth-century American planning. Mel Scott, *San Francisco Bay Area,* 2nd ed. [1959] 1985; Mel Scott, *American City Planning Since 1890* (Berkeley: University of California Press, 1969).

4. Some key bureau/institute publications were Stanley Scott and John Bollens, *Special Districts in the Government of California* (Berkeley: University of California, Berkeley, Bureau of Public Administration, 1949); Stanley Scott, *A Golden Gate Authority for the San Francisco Bay Area: The Problem of Representation* (Berkeley: University of California, Berkeley, Bureau of Public Administration, 1960); T. J. Kent Jr., *City and Regional Planning for the Metropolitan San Francisco Bay Area* (Berkeley: University of California, Berkeley, Institute of Governmental Studies, 1963), and *Open Space for Bay Area,* 1970; Stanley Scott and John Bollens, *Governing a Metropolitan Region: The San Francisco Bay Area* (Berkeley: University of California, Berkeley, Institute of Governmental Studies, 1968); Harriet Nathan and Stanley Scott, eds., *Toward a Bay Area*

Regional Organization (Berkeley: University of California, Berkeley, Institute of Governmental Studies, 1969); and Victor Jones, *Government of the San Francisco Bay Area* (Berkeley: University of California, Institute of Government Studies, 1964). (Mel Scott and Stan Scott were not related.) Victor Jones was a University of Chicago product, influenced by Charles Merriam and the urban Progressives there—a contemporary and friend of my father. Victor Jones, "Political Scientist: Observer and Consultant in Metropolitan Governance," in *Statewide and Regional Land-Use Planning in California,* vol. 2, interview by Malca Chall (Regional Oral History Office, Bancroft Library, University of California, Berkeley, 1983).

5. Raymond Dasmann, *The Destruction of California* (New York: Macmillan, 1965); also Dasmann, *Environmental Conservation* (New York: Wiley, 1959); Dasmann, *The Last Horizon* (New York: Macmillan, 1963); Dasmann, *A Different Kind of Country* (New York: Macmillan, 1968); Dasmann, *Planet in Peril? Man and the Biosphere Today* (Harmondsworth, UK: Penguin, 1972); Samuel Wood and Alfred Heller, *California, Going, Going: Our State's Struggle to Remain Beautiful and Productive* (Sacramento: California Tomorrow, 1962); Wood and Heller, *The Phantom Cities of California* (Sacramento: California Tomorrow, 1963); issues of *Cry California,* 1965–1981; William Bronson, *How to Kill a Golden State* (Garden City, N.J.: Doubleday, 1968); Gilliam, *Island in Time,* 1962; Gilliam, *Between the Devil and the Deep Blue Bay,* 1969; Harold Gilliam, *For Better or for Worse: The Ecology of an Urban Area* (San Francisco: Chronicle Books, 1972); Stuart Udall, *The Quiet Crisis* (New York: Holt, Rinehart. and Winston, 1963); Ernest Callenbach, *Ecotopia* (Berkeley: Banyan Tree Books, 1975). Catherine Bauer, William Wurster, and David Brower were key influences on California Tomorrow. Alfred Heller, "California Tomorrow: A Voice for State and Regional Planning," in *Statewide and Regional Land-Use Planning in California,* vol. 1, interview by Malca Chall (Regional Oral History Office, Bancroft Library, University of California, Berkeley, 1983). Harold Gilliam had studied writing with Wallace Stegner at Stanford. Ernest Callenbach was an editor at University of California Press.

6. For the postmodern thesis, see Fred Buttel, "Environmentalism: Origins, Processes, and Implications for Rural Social Change," *Rural Sociology* 57, no. 1 (1992). The mainstream conservationists of the Sierra Club, though not New Dealers, fell out with the Eisenhower Republicans in the 1950s and allied themselves with liberal Democrats such as Stuart Udall and the Great Society Congress of 1964–68. For example, Democrats Hubert Humphrey and John Saylor were the original cosponsors of the Sierra Club–backed Wilderness Act in 1956. Michael Cohen, *History of the Sierra Club,* 1988.

7. Thanks to Stephanie Pincetl for pointing out Mumford's antimodern streak, which inflected Kent and other Bay Area planners and writers—and recalls the contradictions of Muir (see chap. 1); she also observes that mainstream planning has never incorporated environmentalism.

8. On Marshall, see Gottlieb, *Forcing the Spring,* 1993, 15–19; Sutter, *Driven Wild,*

2002. Robert Sterling Yard and Aldo Leopold were antagonistic to the New Deal for both its liberalism and intrusive public-works projects.

9. Dorothy Ward Erskine, interview by Carla Ehat and Anne Kent, February 14, 1979 (California Room, Marin Public Library, San Rafael); T. J. Kent Jr., "Professor and Political Activist," 1983, 107–09; John Erskine, interview with author in San Francisco, July 19, 2005, Janet Thiessen, "Oh, That Vigor: A Life of Dorothy Erskine" (manuscript).

10. John Erskine, interview, 2005.

11. See also Samuel Dana and Myron Krueger, *California Lands: Ownership, Use, and Management* (Washington, D.C.: American Forestry Association, 1958); Outdoor Recreation Resources Review Commission, *Outdoor Recreation for America* (Washington, D.C.: Government Printing Office, 1962). David Pesonen of the Sierra Club worked on the wilderness portion of the ORRRC report.

12. Dorothy Ward Erskine, "Environmental Quality and Planning," 1976, quote at 133. Erskine put out a monthly newsletter, *Regional Exchange*, from 1966 to 1982, which reached thousands of activists and was a key source of information for the Bay Area conservationists. John Erskine, interview, 2005.

13. Erskine, "Environmental Quality and Planning," 1976, quotes at 162, 134, 141. Cf. Karl Polanyi, *The Great Transformation* (New York: Farrar and Rinehart, 1944); Gaventa, *Power and Powerlessness*, 1980.

14. A classic statement of this point of view is by my father, Robert Walker, *The Planning Function in Urban Government*, 2nd ed. (Chicago: University of Chicago Press, [1941] 1950), who studied under Professor Charles Merriam, an old Chicago Progressive. Also see Pincetl, *Transforming California*, 1999.

15. Scott and Bollens, *Government of California*, 1949; John Bollens, *Local Government in California* (Berkeley: University of California Press, 1951); Bollens, *Special District Governments in the United States* (Berkeley: University of California Press, 1957); Stan Scott and John Corzine, *Special Districts in the Bay Area* (Berkeley: University of California, Institute of Governmental Studies, 1963); Wood and Heller, *California, Going, Going*, 1962; Wood and Heller, *Phantom Cities of California*, 1963. Ironically, Progressive legislation under Governor Hiram Johnson strengthened local Home Rule and the proliferation of special districts, worsening government fragmentation. Pincetl, *Transforming California*, 1999; Dyble, "Paying the Toll," 2002.

16. Mel Scott, *San Francisco Bay Area*, 2nd ed. [1959] 1985; Pincetl, *Transforming California*, 1999; Kent, "Professor and Political Activist," 1983; Heller, "California Tomorrow," 1983; Samuel Wood, "Administration, Research, and Analysis in Behalf of Environmental Quality," in *Statewide and Regional Land-Use Planning in California*, vol. 1 (Regional Oral History Office, Bancroft Library, University of California, Berkeley, 1983); Dyble, "Paying the Toll," 2002.

17. Pincetl, *Transforming California*, 1999, 140–42.

18. For example, "Despite the gradual improvement of county plans for open space systems, county legislative bodies can be expected to continue to ignore . . . the

pro–open space, anti-sprawl policies of their own general plans. In other words, they will continue to give lip service to admirable plans, while adopting specific zoning ordinances and capital improvement projects that promote sprawl and destroy open space." Kent, "Open Space for Bay Area," 1970, 41.

19. On the Golden Gate Authority fight (and the maneuvers of the Golden Gate Bridge district to kill it), see Dyble, "Paying the Toll," 2002. The Golden Gate Authority was doomed, as well, by the father of the freeways, state Senator Randolph Collier. The Bay Area Council's report on regional planning—V. B. Stanbery, *Regional Planning Needs of the San Francisco Bay Area* (San Francisco: Bay Area Council, 1954)— was written by another New Dealer and friend of Kent.

20. ABAG was one of the first regional councils of government, before they were generalized by the federal Housing Act of 1966 and Intergovernmental Cooperation Act of 1968. ABAG was promoted by the League of California Cities, led by Berkeley's mayor and city manager. Dyble, "Paying the Toll," 2002; also Victor Jones, *Government of the Bay Area*, 1964; Kent, "Berkeley's First Liberal Democratic Regime, 1961–70," in *Toward a Bay Area Regional Organization*, eds. Harriet Nathan and Stanley Scott (Berkeley: University of California, Berkeley, Institute for Governmental Studies, 1969); Kent, "Professor and Political Activist," 1983.

21. ABAG, *Regional Plan*, 1970; Kent, "Professor and Political Activist," 1983, quote at 59. For open-space sentiment, see ABAG, *An Inventory of Parks and Open Spaces of the San Francisco Bay Region* (Berkeley: ABAG, Regional Recreation Committee, 1963); Eckbo, Dean, Austin, and Williams, *Open Space: The Choices Before California* (San Francisco: Diablo Press, 1969); Eckbo, Dean, Austin, and Williams, *Open Space and Resource Conservation Program for California* (San Francisco: Eckbo, Dean, Austin, and Williams, 1972); Jones and Stokes Associates, *Land Use, Open Space, and the Government Process: The San Francisco Bay Area Experience* (New York: Praeger, 1974). ABAG largely ignored equal access to housing, however. Kristina Ford, "Regional Association and Dissociation in the San Francisco Bay Area" (PhD dissertation, Urban and Regional Planning, University of Michigan, 1976), 63.

22. People for Open Space, *The Case for Open Space in the San Francisco Bay Area* (San Francisco: People for Open Space, 1969a), 2–3; POS, *Economic Impact of a Regional Open Space Program for the San Francisco Bay Area,* by Development Research Associates and consultants Livingston and Blayney (Los Angeles: Development Research Associates 1969b); cf. U.S. Council on Environmental Quality, *The Costs of Sprawl,* by Real Estate Research Corporation (Washington, D.C.: Government Printing Office, 1974). Alfred Heller, ed., *The California Tomorrow Plan* (Los Altos: W. Kaufmann, 1972); California Tomorrow, *Democracy in the Space Age: Regional Government Under a California State Plan* (San Francisco: California Tomorrow, 1973). The POS board in the 1970s included Alf Heller, Barbara Eastman, Mary Wayburn, Lennie Roberts, Joe Bodovitz, Janet Grey Hayes, Lois Hogle, Marty Rosen, Bill Kor-

tum, Sylvia McLaughlin, Jean Siri, and Volker Eisele. Presidents were Jerry Kohl, Dan Luten, Mel Scott, John Sutter, Irwin Luckman, T. J. Kent Jr., and Allen Jacobs.

23. For reflections on the fight, see Mel Scott, *San Francisco Bay Area*, 2nd ed. [1959] 1985, 318; Kent, "Professor and Political Activist," 1983; Larry Orman, telephone interview with author, September 18, 2001. For other views on ABAG, see Victor Jones, "Political Scientist," 1983; Ford, "Regional Association and Dissociation," 1976; Scott and Bollens, *Governing a Metropolitan Region*, 1968; and Dyble, "Paying the Toll," 2002.

24. Nationally, only Minneapolis and Portland ever established metropolitan growth planning. On the perennial clash of U.S. cities and federalism, see Kathy Johnson, "Sovereigns and Subjects: A Geopolitical History of Metropolitan Reform in the U.S.A." *Environment and Planning A*, vol. 38 (2006). The national land-use control act failed in 1975, and the wind went out of the metropolitan thinking with the New Federalism of the Nixon and Reagan Republicans, who favored the suburbs over the Democratic cities. On the national legislation, see Rome, *Bulldozer*, 2001.

25. On the MTC, see Kathy Johnson, "Securing Regional Autonomy: Creation of the Bay Area's Metropolitan Transportation Commission, 1956–70" (paper, Department of Geography, University of California, Berkeley, 2000); Matt Williams, telephone interview with author, August 6, 2004.

26. Dorothy Erskine remained on the POS board until her death in 1982.

27. People for Open Space, *A Greenbelt for the Bay Area: Open Space for People*, (San Francisco: People for Open Space, 1980b); Orman, interview, 2001. Larry Orman, personal communication with author, November 2005.

28. Bay Vision 2020 Commission, *Report of the Bay Vision 2020 Commission* (San Francisco: Bay Vision 2020 Commission, 1991); Peter Lydon, *San Francisco's Bay Vision 2020 Commission: A Civic Initiative for Change* (Berkeley: University of California, Berkeley, Institute of Governmental Studies, 1993); Orman, interview, 2001. Cf. Bay Area Council, *Making Sense of the Region's Growth* (San Francisco: Bay Area Council, 1988); Greenbelt Alliance, *Reviving the Sustainable Metropolis: Guiding Bay Area Conservation and Development into the 21st Century* (San Francisco: Greenbelt Alliance, 1989); Sherman Lewis, *Managing Urban Growth in the San Francisco Bay Region* (Hayward: California State University, Hayward, Center for Public Service Education and Research, 1990).

29. Bay Area Council, *Making Sense of Region's Growth*, 1988, 11.

30. There are, of course, plenty of problems with Modernist planning, which I take seriously. E.g., James Holstan, *The Modernist City: An Anthropological Critique of Brasilia* (Chicago: University of Chicago Press, 1989). A minor revival of regionalism occurred in the early 2000s under Nick Bollman's California Center for Regional Leadership, but the movement lacked any popular base. See www.calregions.org; Elisa Barbour, *Metropolitan Growth Planning in California 1990–2000* (San Francisco: Public Policy Institute of California, 2002).

31. Clawson, *Suburban Land Conversion*, 1971; Fellmeth, *Politics of Land*, 1973; Logan and Molotch, *Urban Fortunes*, 1986; Amy Bridges, *Morning Glories: Municipal Reform in the Southwest* (Princeton, N.J.: Princeton University Press, 1997). Robert Walker, *Planning Function in Urban Government*, 2nd ed. [1941] 1950, was the first to show the capture of local planners by development interests. See also McConnell, *Private Power*, 1966.

32. James Longtin, *150 Years of Land Use: A Brief History of Land Use Regulation* (Berkeley: Local Government Publications, 1999), quote at 3.

33. Richard LeGates and Teresa Selfa, *Growing Old Gracefully: The Petaluma Plan Reaches Middle Age* (San Francisco: San Francisco State University, Public Research Institute, Working Paper 89-14, 1989), show that controls gave Petaluma a more compact city without crimping growth. Rohnert Park, by contrast, has had unchecked growth and now has no town center to speak of.

34. Doug Greenberg, "Growth and Conflict at the Suburban Fringe: The Case of the Livermore-Amador Valley, California" (PhD dissertation, Department of Geography, University of California, Berkeley, 1986).

35. Richard LeGates and Claude Pellerin, *Planning Tomorrowland: The Transformation of Pleasanton, California* (San Francisco: San Francisco State University, Public Research Institute, 1989).

36. In the 1980s, the Legislature finally required all developers to show where they will get water before local agencies can approve their permits. This bill was pushed by the lobbyist for EBMUD.

37. Gary Schoennauer, telephone interview with author, September 27, 2001; Orman, interview, 2001. Because San Jose invested in a huge sewage plant and Santa Clara County got state and federal water, service hookups have never been a tool of growth control in the South Bay. Walker and Williams, "Water from Power," 1982.

38. Santa Clara County Industry and Housing Management Task Force, *Living Within Our Limits* (San Jose: Santa Clara County Planning Department, 1979); Annalee Saxenian, "Genesis of Silicon Valley," in *Silicon Landscapes*, eds. Peter Hall and Ann Markusen (Boston: Allen and Unwin, 1985); Stephen Payne, *Santa Clara County: Harvest of Change* (Northridge: Windsor, in cooperation with County of Santa Clara Historical Heritage Commission, 1987); Schoennauer, interview, 2001. Geraldine Steinberg, who chaired the General Plan Committee, got on the Board of Supervisors during the 1970s because she fought to regulate the Permanente quarry—a perennial sore spot for foothills greens. Steinberg, *From the Ground Up*, 2002. Santa Cruz County voters also enacted a measure in 1978 to limit population growth and protect farmland.

39. On San Diego's planning failures, see Davis, Mayhew, and Miller, *Under the Perfect Sun*, 2003.

40. On the national shift, see Lawrence Burrows, *Growth Management* (New Brunswick, N.J.: Rutgers University, Center for Urban Policy Research, 1978); David J. Brower, David Godschalk, and Douglas Porter, eds., *Understanding Growth Manage-*

ment (Washington, D.C.: The Urban Institute, 1989); Douglas Porter, *Profiles in Growth Management* (Washington, D.C.: The Urban Institute, 1996).

41. Retail malls were favored because they generated sales taxes instead of property taxes. Dean Misczynski, "The Fiscalization of Land Use," in *California Policy Choices*, eds. John Kirlin and Donald Winkler (Los Angeles: University of Southern California, 1986). For a jaundiced view of growth control in L.A., see Davis, *City of Quartz*, 1990.

42. Longtin, *150 Years of Land Use*, 1999, provides a useful summary of growth-control techniques. Of the firmest one-fourth of California cities, 14 percent link construction to infrastructure, 6 percent have annual permit quotas, and 6 percent have green lines. Paul Lewis and Max Neiman, *Cities Under Pressure: Local Growth Controls and Residential Development Policy* (San Francisco: Public Policy Institute of California, 2002).

43. For attempts to evaluate the net effects of growth controls, see Richard LeGates and Sean Nikas, "Growth Management Through Residential Tempo Controls in the San Francisco Bay Area" (San Francisco: San Francisco State University, Public Research Institute, Working Paper 89-11, 1989); John Landis, *Do Growth Controls Work? An Evaluation of Local Growth Control Programs in Seven California Cities* (Berkeley: University of California, Berkeley, California Policy Seminar, 1992); Madelyn Glickfield and Ned Levine, *Regional Growth . . . Local Reaction: The Enactment and Effects of Local Growth Control and Management Measures in California* (Cambridge, Mass.: Lincoln Land Institute, 1992); Governor's Office of Planning and Research, *Local Government Growth Management Survey* (Sacramento: Governor's Office of Planning and Research, 1991); League of California Cities and County Supervisors Association, *Growth Control–Management Survey* (Sacramento: League of California Cities, 1989); Ted Bradshaw, *Is Growth Control a Planning Failure?* (Berkeley: University of California, Berkeley, Institute of Urban and Regional Development, 1993); Lewis and Neiman, *Cities Under Pressure*, 2002.

44. Joel Garreau, *Edge City: Life on the New Frontier* (New York: Doubleday, 1991). Greenbelt Alliance, *At Risk: The Bay Area Greenbelt* (San Francisco: Greenbelt Alliance, periodic reports, 1991a, 1994, 2000, 2006).

45. Landis, *Do Growth Controls Work?* 1992, quote at 45; Pat McGovern, "Contra Costa County Edge Cities: The New Political Economy of Planning" (PhD dissertation, Department of City and Regional Planning, University of California, Berkeley, 1994).

46. Seth Adams, telephone interview with author, August 18, 2004.

47. Gerrit Knapp and Arthur Nelson, *The Regulated Landscape: Lessons on State Land Use Planning from Oregon* (Cambridge, Mass.: Lincoln Institute of Land Policy, 1992).

48. Statewide figure is from Lewis and Neiman, *Cities Under Pressure*, 2002. Greenbelt Alliance, *Beyond Sprawl: New Patterns of Growth to Fit the New California* (San Francisco: Greenbelt Alliance, with Bank of America, California Resources

Agency, and Low Income Housing Fund, 1995). Also, Greenbelt Alliance, *Bound for Success: A Citizen's Guide to Using Urban Growth Boundaries for More Livable Communities and Open Space Protection in California* (San Francisco: Greenbelt Alliance, n.d. [ca. 1997]).

49. Brent Schoradt, telephone interview with author, July 11, 2005.

50. Adams, interview, 2004; *Greenbelt Action*, various issues. Oakley has approved 2,500 homes below sea level in the delta.

51. John Woodbury, interview in Oakland with author, August 16, 2004.

52. Vicky Moore, telephone interview with author, September 24, 2001. Quote from *San Francisco Chronicle*, February 16, 2000, A15.

53. Environmentalists split over tactics. Greenbelt Alliance and the Open Space Authority cut a deal with Cisco and San Jose not to seek legal redress in exchange for $100 million in private and public funds for a Silicon Valley conservation fund—which could increase open space by 50,000 acres. Committee for Green Foothills wanted to stop Cisco entirely. Greenbelt Alliance, *Getting It Right: Preventing Sprawl in Coyote Valley* (San Francisco: Greenbelt Alliance, 2003b); various issues of *Greenbelt Action* and *Green Footnotes*.

54. David Dowall, *The Suburban Squeeze* (Berkeley: University of California Press, 1984), quote at 14; Bernard Frieden, *The Environmental Protection Hustle* (Cambridge, Mass.: MIT Press, 1979). Also Lynne Sagalyn and George Sternlieb, *Zoning and Housing Costs: The Impact of Land Use Controls on Housing Price* (New Brunswick, N.J.: Rutgers University, Center for Urban Policy Research, 1972); Lawrence Katz and Kenneth Rosen, *The Effects of Land-Use Controls on Housing Prices* (Berkeley: University of California, Berkeley, Center for Real Estate Research and Urban Economics, 1980).

55. E.g., Dowall, *Surburban Squeeze*, 1984, 5–7. Several national studies reviewed by Dowall are assumed to apply to the Bay Area. By contrast, LeGates and Selfa, in *Growing Old Gracefully*, 1989, show that housing prices did not rise any faster in Petaluma than in unrestricted Santa Rosa. Cf. People for Open Space, *Greenbelt for the Bay Area*, 1980b; POS, *Room Enough: Housing and Open Space in the Bay Area* (San Francisco: POS, 1983); Lewis and Neiman, *Cities Under Pressure*, 2002.

56. For details, see Walker et al., "Playground of U.S. Capitalism?" 1990; Davis, *City of Quartz*, 1990; Rebecca Solnit, *Hollow City: Gentrification and the Eviction of Urban Culture* (London: Verso Press, 2001); Richard Walker, "The Boom and the Bombshell: The New Economy Bubble and the San Francisco Bay Area," in *The Changing Economic Geography of Globalization*, ed. Giovanna Vertova (London: Routledge, 2006).

57. Because housing markets are commonly supply-restricted by the nature of the commodity—the rates at which new units can be built and old units come on the market are low—any upward pressure from demand leads readily to rising prices.

58. Kee Warner and Harvey Molotch, *Growth Control: Inner Workings and External Effects* (Berkeley: University of California, Berkeley, California Policy Seminar, 1992); Landis, *Do Growth Controls Work?* 1992.

59. Davis, *City of Quartz*, 1990. Contrast Pincetl, "Politics of Growth Control," 1992, who argues that environmentalism is a genuine part of Southern California's local opposition to growth.

60. Longtin, *150 Years of Land Use*, 1999; Schrag, *Paradise Lost*, 1998; Pincetl, *Transforming California*, 1999.

61. People for Open Space, *Room Enough*, 1983.

62. Woodbury, interview, 2004.

63. Christopher Alexander, Hajo Nils, Artemis Anninon, and Ingrid King, *A New Theory of Urban Design* (New York: Oxford University Press, 1987); Daniel Solomon, *Re Build* (New York: Princeton Architectural Press, 1992); Peter Calthorpe, *The Next American Metropolis* (New York: Princeton Architectural Press, 1993); Michael Bernick and Robert Cervero, *Transit Villages in the 21st Century* (New York: McGraw-Hill, 1997); Peter Calthorpe and William Fulton, *The Regional City: Planning for the End of Sprawl* (Washington, D.C.: Island Press, 2001).

64. Orman, interview, 2001. John Holtzclaw, telephone interview with author, August 20, 2004. Greenbelt Alliance, *Beyond Sprawl*, 1995; Greenbelt Alliance and Silicon Valley Manufacturers' Group, *Housing Solutions for Silicon Valley* (San Francisco: Greenbelt Alliance, Livable Communities Program, 1999); Greenbelt Alliance, *Smart Infill: Creating More Livable Communities in the Bay Area* (San Francisco: Greenbelt Alliance, 2003a). Sierra Club, *Solving Sprawl* (San Francisco: Sierra Club Books, 1999). Cf. Andres Duany, Elizabeth Plate-Zyberk, and Jeff Speck, *Suburban Nation: The Rise of Sprawl and the Decline of the American Dream* (San Francisco: North Point Press, 2000).

65. See S. Bhargava, B. Brownstein, A. Dean, and S. Zimmerman, *Everyone's Valley: Inclusion and Affordable Housing in Silicon Valley* (San Jose: Working Partnerships, 2001); Pascale Joassart-Marcelli, William Fulton, and Juliet Musso, "Smart Growth or Growth Machine?" in *Up Against the Sprawl*, eds. J. Wolch, M. Pastor, and P. Drier (Minneapolis: University of Minnesota Press, 2004), Mike Davis, "Gentrifying Disaster," *Mother Jones*, October 25, 2005.

66. Vicky Moore, interview, 2001; various issues of *Greenbelt Action*.

67. Greenbelt Alliance and Sonoma County Farm Bureau, *Preventing Sprawl: Farmers and Environmentalists Working Together* (San Francisco: Greenbelt Alliance, 2004).

68. Schoradt, interview, 2005; various issues of *Greenbelt Action*. Also see Greenbelt Alliance, *Balanced Transportation: Achieving Congestion Relief and Meeting Transportation Needs in Solano County* (San Francisco: Greenbelt Alliance, 2002).

69. Schoradt, interview, 2005.

70. Orman, interview, 2001; People for Open Space, *The Greenbelt's Public Lands: Public Lands Database for the San Francisco Bay Area* (San Francisco: People for Open Space, Greenbelt Congress, 1988); Greenbelt Alliance, *At Risk* (1991a, 1994, 2000, 2006); www.openspacecouncil.org.

71. Craig Britton, general manager of Midpeninsula Regional Open Space District, in *San Francisco Chronicle*, July 9, 2001, A13.

72. Olmsted and Hall, *Proposed Park Reservations*, [1930] 1984, 29; Jean Rusmore, *The Bay Area Ridge Trail* (Berkeley: Wilderness Press, 1995); Rowell, *Bay Area Wild*, 1997, 105, 162; www.ridgetrail.org.

73. www.baytrail.org; five-part 2003 *San Francisco Chronicle* series, sfgate.com/baytrail/; cost figures from Carl Hall, "Slowly Closing the Gaps for a Bay Trail Loop," *San Francisco Chronicle*, December 18, 2005, B1, B6.

7 / FASTEN YOUR GREENBELT *Triumph and Trust Funds*

1. The Bay Area was ahead of the national cycle in real estate, which peaked about five years later, ca. 1989. Cynthia Kroll and Linda Kimball, "The R&D Dilemma: The Real Estate Industry and High-Tech Growth," Working Paper 86-116 (Berkeley: University of California, Berkeley, Center for Real Estate and Urban Economics, 1986); Helga Leitner, "Capital Markets, the Development Industry, and Urban Office Market Dynamics: Rethinking Building Cycles," *Environment and Planning A* 26 (1994). California and New York accounted for one-half of all real estate lending in 1984, and by 1989, California banks led all states, with $104 billion outstanding in real estate loans. Moira Johnston, *Roller Coaster: The Bank of America and the Future of American Banking* (New York: Ticknor and Fields, 1990); Sheshunoff and Co., *Banks of California* (Austin, Texas: Sheshunoff and Co., 1989), 21–22; Barney Warf and Joseph Cox, "U.S. Bank Failures and Regional Economic Structure," *Professional Geographer* 47, no. 1 (1995).

2. *Washington Post*, "San Jose's Choice: Tax Retrenchment Hurts Needy Most," September 18, 1983, A1. Santa Clara County Housing Task Force, *Housing: A Call for Action* (San Jose: Santa Clara County Planning Department, 1977); Santa Clara County Industry and Housing Management Task Force, *Living Within Our Limits*, 1979; Saxenian, "Genesis of Silicon Valley," 1985; Lenny Siegel and John Markoff, *The High Cost of High Tech* (New York: Harper and Row, 1985); John Findlay, "Stanford Industrial Park: Downtown for Silicon Valley," in *Magic Lands: Western Cityscapes and American Culture after 1940* (Berkeley: University of California Press, 1992). Santa Clara County completed key remaining highways—237, 101, and 85—by executing a nifty end run around CalTrans: financing them with a local sales tax approved by the voters in 1984. A San Jose light-rail system was also built, using federal mass-transit funds.

3. For a good retrospective on the cost of suburban sprawl at this time, and the reaction it spawned, see Rome, *Bulldozer*, 2001. The list of woes in Santa Clara County is not exactly the same as elsewhere, however.

4. Richard Walker, "Boom and the Bombshell," 2006.

5. Ibid.

6. Greg Miller, "Will the Party Last? In Tech, Some See Signs of a Bubble," *Los*

Angeles Times, December 28, 1997, Business and Technology sec., 1; www.boe.ca.gov/news/sp022301att. Market valuation would be higher than assessed valuation, but no estimates exist.

7. Quote is from Sara Miles, *How to Hack a Party Line: The Democrats and Silicon Valley*, rev. ed. (Berkeley: University of California Press, 2001), 9; Richard Walker, "Boom and the Bombshell," 2006.

8. John Micklethwait, "The Valley of the Money's Delight," *Economist* 324 (March 29, 1997); Michael Lewis, *The New New Thing: A Silicon Valley Story* (Chicago: W. W. Norton, 2000); Miles, *Hack a Party Line*, 2001.

9. Spitz, *Mill Valley*, 1997. Statewide there were 180,000 acres protected by special districts. Press, *Saving Open Space*, 2002, 18.

10. Davey, "Saving Open Space," 2004. Roberts, "Protecting San Mateo County," 2004. The East Bay parks model had been suggested by Livingston and Blayney, *Foothills Environmental Design Study*, 1970.

11. Rowell, *Bay Area Wild*, 1997; Yaryan et al., *Sempervirens Story*, 2000, 63; Davey, "Saving Open Space," 2004. MROSD got preliminary approval in 1998 for a parcel tax to fund expansion to the coast but had not gone back to the voters for the tax as of 2005.

12. Doyle, interview, 2005.

13. Unequal access to open space is well documented in Los Angeles, and the same pattern surely holds in the Bay Area. Jennifer Wolch, Steve Brachman, Jed Fehrenbach, and Jamin Johnson, "Parks and Park Funding in Los Angeles: An Equity Mapping Analysis" (working paper, Department of Geography, University of Southern California, 2001).

14. In Marin, with its more relaxed upper-class atmosphere, there have been repeated skirmishes between those who want light use and habitat protection versus a younger generation that favors trail biking, jogging, and paragliding. Rothman, *New Urban Park*, 2004, 86–92.

15. Clarence Lo, *Small Property Versus Big Government: Social Origins of the Property Tax Revolt* (Berkeley: University of California Press, 1990); Schrag, *Paradise Lost*, 1998; Richard Walker, "California Rages," 1995a; Pincetl, *Transforming California*, 1999.

16. Davey, "Saving Open Space," 2004; Joan Cardellino, telephone interview with author, August 10, 2004.

17. Woodbury, interview, 2004. Proposition 40 in 2002 passed with the support of 77 percent of African Americans, 74 percent of Latinos, 60 percent of Asian Americans, and 56 percent of Euro Americans. www.clipi.org/pdf/forestbrief.pdf.

18. Data provided by Dick Wayman of the Coastal Conservancy staff; also Cardellino, interview, 2004. Similar park bonds and taxes were passed elsewhere around the state and the country in those economically flush years. Pincetl, *Transforming California*, 1999; Press, *Saving Open Space*, 2002, 2; Brewer, *Conservancy*, 2003.

19. The best review of land trusts is Brewer, *Conservancy*, 2003; quote at 11. See also Sallie Fairfax and Darla Guenzler, *Conservation Trusts: Institutional Alternatives for a New Era in Land and Resource Conservation* (Lawrence: University of Kansas Press, 2001); John Wright, *Rocky Mountain Divide: Selling and Saving the West* (Austin: University of Texas Press, 1993).

20. Brewer, *Conservancy*, 2003, 11, 186, 224.

21. Robert Duffy, *Green Agenda in American Politics*, 2003, 9; David Ottaway and Joe Stephens, "Nonprofit Land Bank Amasses Billions," *Washington Post*, May 4, 2004, A1.

22. Ottaway and Stephens, "Nonprofit Land Bank," 2004; it manages another seven million acres.

23. Brewer, *Conservancy*, 2003.

24. A practical guidebook is Elizabeth Byers and Karin Ponti, *The Conservation Easement Handbook* (San Francisco and Alexandria, Va.: Land Trust Alliance, Trust for Public Land, 2005).

25. On the pros and cons of easements, and the lack of good data on them, see A. Merenlender, L. Huntsinger, G. Guthey, and S. Fairfax, "Land Trusts and Conservation Easements: Who Is Conserving What for Whom?" *Conservation Biology* 18, no. 1 (2003), 65–75.

26. This is the judgment of Brewer, in *Conservancy*, 2003, ch. 1, who also rescues Charles Elliott from obscurity.

27. Press, *Saving Open Space*, 2002, 34, 40.

28. Ordway was, like so many conservation women, a veteran of the women's clubs. Brewer, *Conservancy*, 2003, ch. 10. For a critique, see Ottaway and Stephens, "Nonprofit Land Bank," 2004.

29. Brewer, *Conservancy*, 2003, chapter 9. On TNC's income, see Ottaway and Stephens, "Nonprofit Land Bank," 2004.

30. John Hart, "Private Land, Public Good: Taking Stock of Conservation Easements," *Bay Nature*, January–March 2006, 18–32; Woodbury, interview, 2004; www .openspacecouncil.org.

31. Figures are from www.tpl.org and http://nature.org/wherewework/north america/states/california.

32. Figures from John Hart, "Private Land, Public Good," 2006; Web sites of the various trusts; and www.openspacecouncil.org. These sources are not entirely consistent!

33. Rosen, "Trust for Public Land Founding Member," 2000; Huey Johnson interview in Gerbode, "Environmentalist, Philanthropist, and Volunteer," 1995; Brewer, *Conservancy*, 2003, ch. 11; Ralph Benson, telephone interview with author, July 25, 2006.

34. Brewer, *Conservancy*, 2003, 223. TPL played a part in the creation of the Golden Gate National Recreation Area but was not primarily responsible, as claimed by Press, *Saving Open Space*, 2002, 39. TPL also had a hand in local campaigns for taxes to support public parks.

35. Santa Cruz County is covered by Sempervirens Fund and Save the Redwoods League.

36. The Packard Foundation had a $175 million program, 1995–2000, for open space up and down the West Coast. Davey, "Saving Open Space," 2004.

37. On the hazards of poorly protected trust land, see Brewer, *Conservancy*, 2003, 170–75.

38. *San Francisco Chronicle*, July 9, 2001, A13. Directors' bios are on the POST Web page: www.openspacetrust.org.

39. John Hart, *Farming on the Edge: Saving Family Farms in Marin County, California* (Berkeley: University of California Press, 1991). Ellen Straus, a Jewish refugee from wartime Amsterdam, also served on the boards of Marin Conservation League, Greenbelt Alliance, and Environmental Action and cofounded Marin Organic.

40. People for Open Space, *Search for Permanence*, 1981; John Hart, *Farming on the Edge*, 1991; U.S. Department of the Interior, National Park Service, *A Good Life: Dairy Farming in the Olema Valley*, by Dewey Livingston (San Francisco: National Park Service, 1995). Letting diary ranches continue to operate had been tried first on Point Reyes. Marin dairying is enjoying a rebound due to the popularity of organic milk.

41. Gianfrano Gorgoni, *Christo: Running Fence, Sonoma and Marin Counties, California, 1972–76* (New York: Abrams, 1978).

42. Brewer, *Conservancy*, 2003, 241–43; Nisbet, interview, 2005.

43. Cardellino, interview, 2004.

44. Data is from Dick Wayman of the Coastal Conservancy.

45. Woodbury, interview, 2004. Administratively, the council is a project of Greenbelt Alliance, its fiscal agent, and operates happily without a central office. See www.openspacecouncil.org for a list of members. The inventory and maps are done for the council by Green Info Network, headed by Larry Orman.

46. Woodbury, interview, 2004.

47. Orman, interview, 2001.

48. On urban densities, see www.geograpfia.com. Mountain lions are endemic in the East Bay parks, and bears are recolonizing the mountains to the north and the south of the Bay. *San Francisco Chronicle,* June 11, 2001. Also Jane Kay, "Nature Moves Back In," *San Francisco Chronicle*, July 13, 2003a; various issues of *Bay Nature*. Cf. Davis, *Ecology of Fear*, 1998.

49. The Nature Conservancy has come under fire for cozy relations with corporations and wealthy benefactors, resulting in questionable tax, science, and management policies. Ottaway and Stephens, "Nonprofit Land Bank," 2004.

50. "I fear that the erosion of public prerogative in favor of private capital and individual focus is slowly leaving us to the mercies of only money in the conservation world—pragmatism is great, but it can lead down an ever steeper slope" (Larry Orman, personal communication with the author, December 12, 2005).

1. The ten largest wineries in the state still accounted for 81 percent of total California shipments in 1980, Gallo alone more than 50 percent of the total. Kirby Moulton, "The Economics of Wine in California," in *The Book of California Wine*, eds. D. Muscatine, M. Amerine, and B. Thompson (Berkeley: University of California Press, 1984), 390.

2. Sonoma was the principal grape producer before Prohibition, but the Bay Area's share of output fell from 30 percent of the state to 10 percent after Prohibition. Carosso, *California Wine Industry*, 1951; Pinney, *History of Wine in America*, 2005.

3. James Conaway, *Napa: The Story of an American Eden* (New York: Avon Books, 1990); James Lapsley, *Bottled Poetry: Napa Winemaking From Prohibition to the Modern Era* (Berkeley: University of California Press, 1996); Pinney, *History of Wine in America*, 2005. Bank of America provided much of the finance. Moira Johnston, "A Very Civil War," *California* 14, no. 1 (January 1989).

4. For figures on consumption, see Cees Eysberg, *The California Wine Economy* (Utrecht, The Netherlands: Rijkuniversiteit Utrecht, Geografisch Instituut, 1990), 89, 91. Export figures are from www.wineinstitute.org.

5. Eysberg, *California Wine Economy*, 1990, 128–39, 157; Motto, Kryla, and Fisher, *The Economic Impact of California Wine* (St. Helena: Motto, Kryla, and Fisher, 2000).

6. Conaway, *Napa*, 1990; Dan Berger and Richard Hinkle, *Beyond the Grapes: An Inside Look at Napa Valley* (Wilmington, Del.: Atomium Books, 1991).

7. Greig Guthey, "Terroir and the Politics of the Agro-Industry in California's North Coast Wine District" (PhD dissertation, Department of Geography, University of California, Berkeley, 2004); Pinney, *History of Wine in America*, 2005.

8. Conaway, *Napa*, 1990; James Conaway, *The Far Side of Eden: New Money, Old Land, and the Battle for Napa Valley* (Boston: Houghton Mifflin, 2002); Guthey, "Terroir and Politics of Agro-Industry," 2004.

9. Guthey, "Terroir and Politics of Agro-Industry," 2004; Pinney, *History of Wine in America*, 2005.

10. Gallo, the Wine Group (which bought Glen Ellen), and Constellation Blend (which bought CK Mondavi) still control 60 percent of U.S. wine sales.

11. Eysberg, *California Wine Economy*, 1990, 113.

12. Richard Walker, *Conquest of Bread*, 2004.

13. Cf. Sayer, *Moral Significance of Class*, 2005.

14. On labor conditions in the Wine Country, see Lauren Coodley, *Napa: The Transformation of an American Town* (Charleston, S.C.: Arcadia, 2003); Sandra Nichols, "Saints, Peaches, and Wine: Mexican Migrants and the Transformation of Los Haro, Zacatecas, and Napa, California" (PhD dissertation, Department of Geography, University of California, Berkeley, 2002). Cf. Richard Walker, *Conquest of Bread*, 2004.

15. Erskine, "Environmental Quality and Planning," 1976; Dorothy Erskine, interview, 1979; Conaway, *Napa*, 1990, 82–92; Volker Eisele, "Twenty-five Years of Farmland Protection in the Napa Valley," in *California Farmland and Urban Pressures: Statewide and Regional Perspectives*, eds. Albert Medvitz, Alvin Sokolow, and Cathy Lemp (Davis: University of California, Davis, Agricultural Issues Center, 1999). John Erskine, interview, 2005.

16. On the evolution of Napa agricultural-land controls, see Conaway, *Napa*, 1990; Mary Handel and Alvin Sokolow, *Farmland and Open Space Preservation in the Four North Bay Counties* (Davis: University of California, Davis, Cooperative Extension, 1994); Eisele, "Twenty-five Years of Farmland Protection," 1999; Guthey, "Terroir and Politics of Agro-Industry," 2004; Fairfax and Guenzler, *Conservation Trusts*, 2001, ch. 11.

17. This is consistent with Napa's general neglect of its farmworkers. See Coodley, *Napa*, 2003.

18. John Woodbury, telephone interview with author, December 16, 2005. Data from Napa County Land Trust Web site, www.napalandtrust.org.

19. Woodbury, interview, 2005. An extreme property rights measure, to gain compensation for any regulation lowering land value, went down to defeat in 2006, opposed by vintners, growers, and greens, alike.

20. Conaway, *Napa*, 1990; Guthey, "Terroir and Politics of Agro-Industry," 2004; Pinney, *History of Wine in America*, 2005.

21. Conaway, *Napa*, 1990; Guthey, "Terroir and Politics of Agro-Industry," 2004.

22. Johnston, "Very Civil War," 1989, quote at 60.

23. Johnston, "Very Civil War," 1989; Conaway, *Far Side of Eden*, 2002; Conaway, *Napa*, 1990; Eisele, "Twenty-five Years of Farmland Protection," 1999. Cf. Rothman, *Devil's Bargains*, 1998a.

24. Kortum was able to prove that PG&E's Humboldt pilot nuclear generator had contaminated cows there with iodine 131. Kortum, "Environmentalist Extraordinaire," 2005.

25. Kortum, "Environmentalist Extraordinaire," 2005. The Sonoma state park sites had been surveyed by George Collins at the same time as Point Reyes.

26. John Crevelli, *Twenty-five Years of COAAST* (Forestville, Calif.: Walkabout Press, 1993); John Crevelli, interview in Healdsburg with author, July 16, 2004; Chuck Rhinehart, interview near Santa Rosa with author, June 9, 2003; Kortum, "Environmentalist Extraordinaire," 2005. Hinkle, Crevelli, and Rhinehart all came from working-class backgrounds.

27. Kortum, "Environmentalist Extraordinaire," 2005; Marty Bennett, interview in Sonoma with author, May 5, 2005; Crevelli, interview, 2004. John Crevelli's path to activism was similar to Bill Kortum's: he grew up in Santa Rosa and developed a real love for the Sonoma countryside; he watched the U.S. Highway 101 freeway rip through his neighborhood and destroy the nearby creek; then he witnessed the ravages of lumbering on the North Coast.

28. Kortum, "Environmentalist Extraordinaire," 2005.

29. Kortum, "Environmentalist Extraordinaire," 2005; Rudee, interview, 2004; Crevelli, interview, 2004. Rudee was more a feminist than an environmentalist but very much a political moderate (though she did shift to the Democratic Party along the way). Putnam was another pioneering woman in Sonoma County politics, and greener yet, but she died in 1984.

30. Crevelli, *Twenty-five Years of* COAAST, 1993; Crevelli, interview, 2004; Kortum, "Environmentalist Extraordinaire," 2005; Griffin, *Saving Marin-Sonoma Coast*, 1993; Martin Griffin, interview in Sonoma with author and Marty Bennett, July 22, 2004. An earlier ballot measure against the dam was turned back in 1972. The Sonoma County Water Agency was created expressly to pursue the Warm Springs project; the U.S. Army Corps of Engineers designed and built it.

31. On the changes in Sonoma's economy and class structure, see Nari Rhee and Dan Acland, *The Limits of Prosperity: Growth, Inequality, and Poverty in the North Bay* (Berkeley and Santa Rosa: Center for Labor Research and Education, New Economy–Working Solutions, 2005).

32. Katie Scarborough and Scot Stegeman, *Farmlands Worth Saving: The Present Value and Preservation of Sonoma County Agriculture* (Santa Rosa: Sonoma County Farmlands Group, 1989); Kortum, "Environmentalist Extraordinaire," 2005.

33. Handel and Sokolow, *Farmland and Open Space Preservation*, 1994, 19–23.

34. Guthey, "Terroir and Politics of Agro-Industry," 2004, chapter 5.

35. Ibid. Of special note was the creation of the west county "Town Hall Coalition," led by former Sebastopol mayor Lynn Hamilton, to fight forest clearance, spraying, and hillside erosion.

36. Rudee, interview, 2004; Kortum, "Environmentalist Extraordinaire," 2005; Benson interview, 2006; www.sonomalandtrust.org.

37. Figures from sonomaopenspace.org. Together, the Sonoma Land Trust and the open space district protect more than 40,000 acres of farmland, including vineyards— more farm acreage than is protected in Marin. Both counties rank in the top ten in the United States in farmland preservation. www.farmlandpreservationreport.com.

38. Caryl Hart, interview in Berkeley with author, December 13, 2005. Hart, who grew up near Santa Barbara, is a graduate of University of San Francisco law school and a former public defender who combines a strong affinity for open space with a firm belief in social justice and making public land available to all the people. She serves on the advisory committee of the Sonoma County Agricultural Preservation and   Open Space District and the California State Parks Commission.

39. Kortum, "Environmentalist Extraordinaire," 2005; Bennett, interview, 2005. The growth boundaries passed with an average 70 percent margin because there is widespread support across class and racial lines in the county, according to Bennett.

40. Griffin, *Saving Marin-Sonoma Coast*, 1998; Griffin, interview, 2004.

41. Bennett, interview, 2005.

42. Orman, interview, 2001, Bennett, interview, 2005.

43. Bennett, interview, 2005. Michael Allen, president of the North Bay Labor Council, has been an important ally in moving working people in the green direction.

9 / TOXIC LANDSCAPES *Beyond Open Spaces*

1. Martin Melosi, *Garbage in the Cities: Refuse, Reform, and the Environment, 1880–1980* (College Station: Texas A&M University Press, 1981); Gottlieb, *Forcing the Spring*, 1993, chapter 2; Susan Strasser, *Waste and Want: A Social History of Trash* (New York: Metropolitan Books, 1999).

2. Melosi, *Garbage in the Cities*, 1981; Louis Blumberg and Robert Gottlieb, *War on Waste: Can American Win Its Battle with Garbage?* (Washington, D.C.: Island Press, 1989); Rome, *Bulldozer*, 2001; Heather Rogers, *Gone Tomorrow: The Hidden Life of Garbage* (New York: The New Press, 2005). Inexplicably, Rome hardly mentions garbage in his otherwise comprehensive discussion of postwar suburbanization.

3. Leonard Dworsky, ed., *Pollution* (New York: Chelsea House, 1971); Matthew Crenson, *The Unpolitics of Air Pollution: A Study of Non-Decisionmaking in the Cities* (Baltimore: Johns Hopkins University Press, 1971); Clarence Davies and Barbara Davies, *The Politics of Pollution*, 2nd ed. (Indianapolis: Bobbs-Merrill, [1970] 1975); Walt Westman, *Ecology, Impact Assessment, and Environmental Planning* (New York: John Wiley and Sons, 1985); Roger Findley and Daniel Farber, *Environmental Law in a Nutshell*, 3rd ed. (St. Paul, Minn.: West Publishing, 1992).

4. Gottlieb, *Forcing the Spring*, 1993, 127. The big environmental groups had not yet set up Washington, D.C., lobbying offices, so they played only an indirect part in the legislation. Cf. Robert Duffy, *Green Agenda in American Politics*, 2003, 57.

5. The Clean Water Act was complemented by the Safe Drinking Water Act of 1974. On industry reaction to regulation, see Richard Walker, "Erosion of the Clean Air Act of 1970: A Study in the Failure of Government Regulation and Planning," *Boston College Environmental Affairs Law Review* 7, no. 2 (1978).

6. On solid-waste laws, see Melosi, *Garbage in the Cities*, 1981; Blumberg and Gottlieb, *War on Waste*, 1989; Rogers, *Gone Tomorrow*, 2005; and Anne Scheinberg, "The Proof of the Pudding: Urban Recycling in North America as a Process of Ecological Modernization," *Environmental Politics* 12, no. 4 (2003), 49–75. California's history of solid-waste management has been checkered. It took a long time to get recycling going; after many failed tries in the Legislature in the 1970s, a bottle-deposit initiative was smashed by industry and Republican hostility in 1982; a weak law finally passed in 1986. Things began to improve, thanks especially to local initiatives in places such as Berkeley, San Jose, and Marin County, and today about half the waste stream is recycled or recovered. But recycling is still a poor alternative to source reduction. Rogers, *Gone Tomorrow*, 2005.

7. EDF is now Environmental Defense. Sierra Club Legal Defense Fund, now Earthjustice, has always been an autonomous branch of the Sierra Club. Its headquarters is in Oakland.

8. Robert Duffy, *Green Agenda in American Politics*, 2003. For examples of the lawsuits, see Richard Walker, "Erosion of the Clean Air Act," 1978.

9. Shaiko, *Voices and Echoes for Environment*, 1999, figures at 42.

10. Christopher Manes, *Green Rage: Radical Environmentalism and the Unmaking of Civilization* (Boston: Little, Brown, 1990); Susan Zakin, *Coyotes and Dog Towns: Earth First! and the Environmental Movement* (New York: Viking, 1993). For the critical view, see Gottlieb, *Forcing the Spring*, 1993; Jim Schwab, *Deeper Shades of Green: The Rise of Blue-Collar and Minority Environmentalism in America* (San Francisco: Sierra Club Books, 1994); Andrew Szasz, *Ecopopulism: Toxic Waste and the Movement for Environmental Justice* (Minneapolis: University of Minnesota Press, 1994); Luke Cole and Sheila Foster, *From the Ground Up: Environmental Racism and the Rise of the Environmental Justice Movement* (New York: New York University Press, 2001).

11. The ARB sought zero-emission vehicles (electric cars) in 1990 but has since relented; the emphasis now is on hybrids and other ultralow-emission vehicles. It should be said that in 1949, Pittsburgh and Pennsylvania introduced the earliest legislation to reduce air pollution, by limiting coal usage.

12. Jean Siri reports that she was the first to hassle the BAAQMD in the 1960s, after she discovered the high asthma rate among schoolkids she was working with in North Richmond. She calls the board of that day "a real Irish mafia" and says that it had no public phone number to report complaints. Siri, interview, 2005.

13. Arguimbau, interview, 2004. After 1984, CBE's primary air-pollution work was in the L.A. basin.

14. Julia May, telephone interview with author, August 25, 2004. CBE lawyers won Attorney of the Year from the trade journal *California Lawyer* three times over the last twenty years, an incredible record. Richard Toshiyuki Drury, telephone interview with author, August 11, 2004.

15. Arguimbau, interview, 2004; Richard Walker, Michael Storper, and Ellen Widess, "The Limits of Environmental Control: The Saga of Dow in the Delta," *Antipode* 11, no. 2 (1979).

16. Julia May, interview, 2004; Arguimbau, interview, 2004. Shell (Martinez) and Tesoro (Rodeo) still ranked in the top ten facilities nationally in total environmental releases. See www.scorecard.org/rank-facilities-in-country.tci.

17. Julia May, interview, 2004.

18. Ibid. In 2005, CBE and its allies got BAAQMD to halt flaring altogether—for the first time anywhere in the United States.

19. Richard T. Drury, interview, 2004; Julia May, interview, 2004.

20. Will Rostov, quoted in *San Francisco Chronicle*, August 16, 2004, A8.

21. Holtzclaw, interview, 2004.

22. Arguimbau, interview, 2004. The Sierra Club and others bought into the 1990 Clean Air Act amendments because they won more controls over coal-fired power plants in the east and feared worse things if a settlement wasn't reached. CBE was the only group to buck that agreement.

23. Holtzclaw, interview, 2004.

24. Matt Williams, interview, 2004; Holtzclaw, interview, 2004. The alliance alternative plans were done in 1994 and 2005. TALC began as the Transportation Issues Forum. See www.transcoalition.org.

25. Matt Williams, interview, 2004; Holtzclaw, interview, 2004; Bhavna Shamasunder, telephone interview with author, August 20, 2004. See also www.mtcwatch .org.

26. Chris Carlsson, ed. *Critical Mass: Bicycling's Defiant Celebration* (Oakland, Calif.: AK Press, 2002).

27. Cf. Rome, *Bulldozer*, 2001. In Sonoma County, however, septic tanks were at issue in growth-control fights south of Santa Rosa. See chapter 8.

28. Greg Karras, telephone interview with author, August 17, 2004.

29. Ibid.

30. Mike Belliveau, telephone interview with author, July 21, 2004.

31. Greg Karras, "Pollution Prevention: The Chevron Story," *Environment* 30 (1989); Karras, interview, 2004. On the shortcomings of emission regulation and the need to address production directly, see Robert Gottlieb, ed., *Reducing Toxics: A New Approach to Policy and Industrial Decision-Making* (Washington, D.C., and Covelo, Calif.: Island Press, 1995).

32. Karras, interview, 2004; Ted Smith, telephone interview with author, June 11, 2004.

33. In the East Bay, on the other hand, the problem is the separation of storm and sanitary sewers, so that street effluvium goes directly into creeks and the bay without treatment. No one has solved this mess.

34. Barry Commoner, *The Closing Circle* (New York: Knopf, 1971); Gottlieb, *Reducing Toxics*, 1995; Robert Gottlieb, *Environmentalism Unbound: Exploring New Pathways for Change* (Cambridge, Mass.: MIT Press, 2001).

35. The one area of toxics regulation was food and cosmetics, under the Pure Food and Drug Act of 1906 and the Delaney clause, added in 1958. On resistance to Carson's ideas among the Sierra Club old guard such as Alex Hildebrand, head of research for Standard Oil of California (Chevron), and Tom Jukes, chemist for American Cyanamid, see Michael Cohen, *History of the Sierra Club*, 1988. I recall that during a pesticide-spraying controversy in the Bay Area in the 1970s, Jukes drank a spoonful of malathion to prove it was safe.

36. On the failures of pesticide regulation, see Sandra Steingraber, *Living Downstream: An Ecologist Looks at Cancer and the Environment* (Reading, Mass.: Addison-Wesley, 1997).

37. Samuel Epstein, Lester Brown, and Carl Pope, *Hazardous Waste in America* (San Francisco: Sierra Club Books, 1982); Blumberg and Gottlieb, *War on Waste*, 1989; Gottlieb, *Reducing Toxics*, 1995.

38. David McCaffrey, *OSHA and the Politics of Health Regulation* (New York: Plenum Press, 1982).

39. Lois Gibbs, *Love Canal: My Story* (Albany: State University of New York Press, 1982); Szasz, *Ecopopulism*, 1994; Schwab, *Deeper Shades of Green*, 1994; Cole and Foster, *From the Ground Up*, 2001. On the other hand, the Reagan EPA ended all work on solid-waste management to focus on toxics.

40. Szasz, *Ecopopulism*, 1994; Schwab, *Deeper Shades of Green*, 1994. On environmental health, see Richard Hofrichter, ed., *Reclaiming the Environmental Debate: The Politics of Health in a Toxic Culture* (Cambridge, Mass.: MIT Press, 2000).

41. California enacted legislation in 1972 creating a Waste Management Board and dump permit program (four years ahead of Congress). The board was long enamored of megadumps.

42. Robert Van den Bosch, *The Pesticide Conspiracy* (Garden City, N.J.: Doubleday, 1978); Pulido, "Critical Review of Methodology," 1996a.

43. David Weir and Marc Shapiro, *The Circle of Poison* (San Francisco: Center for Investigative Reporting, Food First Institute, 1979).

44. Monica Moore, interview in Berkeley with author, July 26, 2005.

45. Ibid. On Californians for Pesticide Reform, see cpr.radicaldesigns.org. On body burden of chemicals, see, e.g., Kristin Shafer, Margaret Reeves, Skip Spitzer, and Susan Kegley, *Chemical Trespass: Pesticides in Our Bodies and Corporate Responsibility* (San Francisco: Pesticide Action Network, 2004).

46. Glen Martin, "Our Poisoned Bay: Despite End to Direct Piping of Sewage, Pollution Worse Now than Thirty Years Ago," *San Francisco Chronicle*, August 2, 1999, A1, 6–7; Kay, "Nature Moves Back In," 2003a.

47. Mike Belliveau, *Toxics in the Bay* (Oakland: Citizens for a Better Environment, 1981); Belliveau, interview, 2004; Nichols et al., "Modification of an Estuary," 1986, 571.

48. Mike Belliveau, *Toxic Hot Spots* (Oakland: Citizens for a Better Environment, 1987); Belliveau, interview, 2004; Karras, interview, 2004.

49. Popular protest would stop virtually all incinerator projects in California. Blumberg and Gottlieb, *War on Waste*, 1989, 70–71.

50. Henry Clark, "The West County Toxics Coalition" (tape recording and transcript, Regional Oral History Office, Bancroft Library, University of California, Berkeley, 2003); Belliveau, interview, 2004; Erin Reding, "Environmental Justice in Contra Costa County" (senior honors thesis, Department of Geography, University of California, Berkeley, 2004). Belliveau and Clark served on each other's boards, but they parted ways later and there's still tension between the organizations.

51. Figures from www.scorecard.org. Release data show no clear downward trend over the last twenty years, despite better regulation and two recessions. Reding, "Envi-

ronmental Justice," 2004. On Larsen, see Gar Smith, "Toxic Tour: Driving Through One of the West Coast's Deadliest Neighborhoods," *Earth Island Journal* 20, no. 3 (autumn 2005).

52. Ted Smith, "Pioneer Activist for Environmental Justice in Silicon Valley, 1967–2000," interview by Carl Wilmsen, 2000 (Regional Oral History Office, Bancroft Library, University of California, Berkeley, 2003); see also www.svtc.org. SVTC began as a project of Santa Clara Committee for Occupational Safety and Health (COSH), then hived off in 1984. The Santa Clara Central Labor Council and Peter Cervantes-Gautschi were helpful, as well.

53. The companies had shifted to CFCs from more-conventional cleaning solvents—TCE (trichloroethylene) and TCA (trichloroethane)—when the latter were recognized as carcinogens in the 1970s. The corporate denial model is one Smith recounts first hearing from Ralph Nader. The last stage is the company taking credit for what it was forced to do in the first place. Ted Smith, "Pioneer Activist," 2003, 56–57, 105–7.

54. Cliff Rechtstaffen, "The Lead Poisoning Challenge," *Harvard Law Review* 21 (1997).

10 / GREEN JUSTICE   *Reclaiming the Inner City*

1. Richard Moore, for example, came out of Reis Tijerina's Poor People's Campaign and Pam Tau Lee from the Chinatown Red Guards. Cole and Foster, *From the Ground Up*, 2001, 20–29, give too much credit to civil rights and not enough to toxics.

2. See, e.g., Bullard, *Dumping in Dixie*, 1990; Gottlieb, *Forcing the Spring*, 1993; Schwab, *Deeper Shades of Green*, 1994; Szasz, *Ecopopulism*, 1994; Cole and Foster, *From the Ground Up*, 2001.

3. Charles Lee, *Toxic Wastes and Race in the United States* (New York: United Church of Christ, Committee for Racial Justice, 1987). For a review of subsequent studies, see Pulido, "Critical Review of Methodology," 1996a.

4. See, e.g., Bullard, *Dumping in Dixie*, 1990; Gottlieb, *Forcing the Spring*, 1993; Schwab, *Deeper Shades of Green*, 1994; Cole and Foster, *From the Ground Up*, 2001.

5. Luke Cole preface to Anthony, "Expanding the Boundaries," 2003, i. Cole's statement stands in odd contrast to the lack of evidence about Bay Area activism in his book. Cole and Foster, *From the Ground Up*, 2001. In the judgment of several younger activists I talked to, the overall state of environmental justice organizing in Los Angeles is better than that in the Bay Area.

6. Anthony, "Expanding the Boundaries," 2003. Anthony and Linn knew each other from Philadelphia, where Linn was a professor of landscape architecture at the University of Pennsylvania and active in setting up neighborhood gardens to bring poor communities together. Karl Linn, interview in Berkeley with author, July 6,

2004. See also Richard Register, *Ecocity Berkeley: Building Cities for a Healthy Future* (Berkeley: North Atlantic Books, 1987).

7. Anthony, "Expanding the Boundaries," 2003, 53.

8. Pam Tau Lee, "Community and Union Organizing, and Environmental Justice in the San Francisco Bay Area, 1967–2000" (Regional Oral History Office, Bancroft Library, University of California, Berkeley, 2003); Vivian Chang, interview in Oakland with author, August 9, 2004. There are other Bay Area Asian community groups whose work touches on environment, such as Asian Immigrant Women's Advocates and the Chinese Progressive Association.

9. Henry Clark, "West County Toxics Coalition," 2003; Belliveau, interview, 2004; Reding, "Environmental Justice," 2004.

10. www.peopleunited.org; Dawn Phillips, interview with author, April 3, 2006.

11. Bradley Angel, interview with author, April 3, 2006; Shamasunder interview, 2004.

12. Steve Costa, Meena Palaniappan, Arlene Wong, Jeremy Hays, Clara Landeiro, and Jane Rongerude, *Neighborhood Knowledge for Change: The West Oakland Indicators Project* (Oakland: Pacific Institute for Studies in Development, Environment, and Security, 2002); Douglas Fisher, "Tour Targets Poor Neighborhoods," *Oakland Tribune*, February 19, 2004; Shamasunder, interview, 2004. Angel interview, 2006.

13. Phillips interview, 2006; Shamasunder, interview, 2004; Williams, interview, 2004.

14. Angel interview, 2006; www.lejyouth.org; Wanda Sabir, "Growing a Greenway in Hunter's Point," *Bay Nature* 6 (2006).

15. Angel, interview 2006; www.greenaction.org; arceology.org.; sfenvironment.org.

16. Jeff Miller, interview in San Francisco with author, July 24, 2004.

17. Pam Tau Lee, "Community and Union Organizing," 2003, 72–73.

18. Anthony, "Expanding the Boundaries," 2003, 56–57.

19. Belliveau, interview, 2004; Arguimbau, interview, 2004; Karras, interview, 2004; Marta Segura, telephone interview by Juan DeLara (author's research assistant), August 6, 2004; Julia May, interview, 2004; Richard T. Drury, interview, 2004. I heard the story from Stephanie Pincetl in 1998.

20. Smith admits it was a struggle to diversify staff and board, with frequent setbacks. Ted Smith, interview, 2004.

21. Anthony, "Expanding the Boundaries," 2003, 37, quote at 44; Karras, interview, 2004.

22. Douglas Massey and Nancy Denton, *American Apartheid: Segregation and the Making of the Underclass* (Cambridge, Mass.: Harvard University Press, 1993). The literature on urban segregation and unequal exposure to hazards is well summarized in Pulido, "Rethinking Environmental Racism," 2000. It should be noted that in studies such as that by Massey and Denton, the Bay Area always shows up as by far the least segregated of all large American cities.

23. For example, the Sierra Club was embroiled in an attempt by anti-immigrant forces nationally to secure control of the board in 2004–05; they were barely turned back by the national leadership in San Francisco. See Pope, "Virus of Hate," 2004, 14.

24. Henry Clark, "West County Toxics Coalition," 2003, 65–66.

25. Ted Smith, interview, 2004. On race and environment more generally, see Donald Moore, Jake Kosek, and Anand Padian, eds., *Race, Nature, and the Politics of Difference* (Durham, NC: Duke University Press, 2003).

26. Rosen, "Trust for Public Land Founding Member," 2000; Anthony, "Expanding the Boundaries," 2003; Chang, interview, 2004; Ted Smith, "Pioneer Activist," 2003; Belliveau, interview, 2004. The best published account that sees both sides is Hurley, *Environmental Inequalities*, 1995.

27. Clark is quoted in Reding, "Environmental Justice," 2004, 24.

28. Henry Clark, "West County Toxics Coaltion," 2003; Belliveau, interview, 2004; Chang, interview, 2004; Ellis, interview, 2004; Segura, interview, 2004; Angel interview, 2006. I have, in my research, expressly stayed away from trying to understand the ins and outs of such disputes within and among environmental justice groups. If this gives my narrative an overly positive bent, so be it.

29. Juliet Ellis, interview in Oakland with author, July 26, 2004.

30. Daniel Faber and Deborah McCarthy, "Green of Another Color: Building Effective Partnership Between Foundations and the Environmental Justice Movement" (report for the Philanthropy and Environmental Justice Resesarch Project, Northwestern University, Chicago, 2001); Pam Tau Lee, *Community and Union Organizing*, 2003, 70; Anthony, "Expanding the Boundaries," 2003, 50; Ted Smith, "Pioneer Activist," 2003, 18, 114; Henry Clark, "West County Toxics Coalition," 2003; Ellis, interview, 2004; Chang, interview, 2004; Shamasunder, interview, 2004; Reding, "Environmental Justice," 2004.

31. Ellis, interview, 2004.

32. Eliza Strickland, "The New Face of Environmentalism," *East Bay Express*, November 2–8, 2005, 15–26.

33. Lizette Hernandez, Torri Estrada, and Cataline Garzón, *Building upon Our Strengths, a Community Guide to Brownfields Redevelopment in the San Francisco Bay Area* (San Francisco: Urban Habitat, 1999); Cameron Yee, *Crash Course in Bay Area Transportation Investment* (San Francisco: Urban Habitat, 1999); also Gottlieb, *Reducing Toxics*, 1995. Urban Habitat joined in the Clean Air Task Force lawsuit against the MTC over worsening regional air pollution. The Alameda base conversion was highly controversial and not very successful from the point of view of community activists.

34. Anthony, "Expanding the Boundaries," 2003, quotes at 70, 78–79.

35. Marta Matsuoka, *Building Healthy Communities from the Ground Up: Environmental Justice in California* (Oakland: Asian Pacific Islander Environmental Network, Communities for a Better Environment, Environmental Health Coalition,

People Organized to Defend Economic and Environmental Rights, Silicon Valley Toxics Coalition, 2003).

36. Community benefits agreements originated in Los Angeles in the late 1990s, then came north—although Berkeley activists got the same kind of deal in the 1980s from the Bayer Corporation.

37. Ellis, interview, 2004; Shamasunder, interview, 2004.

38. Ellis, interview, 2004. Also Anthony, "Expanding the Boundaries," 2003, 82–84.

39. Ellis, interview, 2004.

40. Segura, interview, 2004.

CONCLUSION *City and Country Reconciled?*

1. Cf. Fishman, *Bourgeois Utopias*, 1987.

2. Gottlieb, *Forcing the Spring*, 1993. Richard White made his case in a paper, "Whole Earth," presented at the Berkeley Workshop on Environmental Politics, Institute of International Studies, University of California, Berkeley, February 21, 2003. Michael Shellenberger and Ted Nordhaus, *The Death of Environmentalism: Global Warming Politics in a Post-Environmental World* (Berkeley: The Breakthrough Institute, 2004).

3. Shaiko, *Voices and Echoes for Environment*, 1999; Robert Duffy, *Green Agenda in American Politics*, 2003. The Sierra Club, in particular, took a renewed interest in membership involvement and led the way toward electoral campaigning to turn back the conservative Republican tide in the 1990s.

4. On fights with the Republican Congress of the 1990s, see Robert Duffy, *Green Agenda in American Politics*, 2003, 74–79. For a vigorous critique of the George W. Bush record, see Carl Pope and Paul Rauber, *Strategic Ignorance: Why the Bush Administration Is Recklessly Destroying a Century of Environmental Progress* (San Francisco: Sierra Club Books, 2004).

5. Figures showing a long slowdown in Bay Area growth are misleading because so much of it now occurs outside the nine-county region. As of 2006, 40,000 homes were planned for the southern delta alone—all below sea level behind weak levees. On Central Valley growth, see Brian Muller, "Local Growth Strategy and the Transformation of Exurbia: The Central Valley of California, 1960–1996" (PhD dissertation, Department of City and Regional Planning, University of California, Berkeley, 2001).

6. Richard Register, *Ecocities: Building Cities in Balance with Nature* (Berkeley: Berkeley Hills Books, 2002); Peter Berg, Beryl Magilavy, and Seth Zuckerman, *A Green City Program for the San Francisco Bay Area and Beyond* (San Francisco: Planet Drum Books, 1990).

BIBLIOGRAPHY

Adams, Seth. 2004a. "Building Bridges: Save Mount Diablo and Unusual Allies." *Diablo Watch* 38 (fall).

———. 2004b. "Who Was Walter P. Frick? The Creation of Mount Diablo State Park." *Diablo Watch* 37–38 (winter–spring, fall).

———. 2004c. Telephone interview with author. August 18. Director of land programs, Save Mount Diablo, Walnut Creek, Calif.

Agnew, John. 1987. *Place and Politics: The Geographical Mediation of State and Society.* Boston: Allen and Gavin.

Albright, Horace, Russell Dickinson, and William Penn Mott Jr. 1987. *The National Park Service: The Story Behind the Scenery.* Edited by Mary Lu Moore. Las Vegas: KC Publications.

Alexander, Christopher, Hajo Nils, Artemis Anninon, and Ingrid King. 1987. *A New Theory of Urban Design.* New York: Oxford University Press.

Allen, Pete. 2007. "A Space for Living: The Arc of Regional Modernism in the SF Bay Area, 1939–1969." PhD diss., Department of Architecture, University of California, Berkeley.

Alpers, Paul. 1996. *What Is Pastoral?* Chicago: University of Chicago Press.

American Farmland Trust. 1986. *Eroding Choices, Emerging Issues: The Condition of California's Agricultural Land Resources.* San Francisco: American Farmland Trust.

Anderson, Dennis, and Jerry George. 2003. *Hidden Treasures of San Francisco Bay.* Berkeley: Heyday Books.

Andersen, Benedict. 1983. *Imagined Communities: Reflections on the Origins and Spread of Nationalism.* London: Verso.

Angel, Bradley. 2006. Telephone interview with author, April 3. Founder and director of Greenaction for Environmental Health and Justice (1998–present), San Francisco.

Angel Island Association. 1999. *Angel Island: Jewel of San Francisco Bay*. Tiburon, Calif.: Angel Island Association.

Anthony, Carl. 2003. "The Civil Rights Movement and Expanding the Boundaries of Environmental Justice in the San Francisco Bay Area, 1960–1999." Tape recording and transcript. Regional Oral History Office, Bancroft Library, University of California, Berkeley.

Arguimbau, Nick. 2004. Telephone interview with author. July 18. Former attorney, air pollution program director, member of board of directors (1984–97), and chair of board of directors, Citizens for a Better Environment; past chair of Northern California air quality committee, Sierra Club; Fairfax.

Association of Bay Area Governments. 1963. *An Inventory of Parks and Open Spaces of the San Francisco Bay Region*. Berkeley: Association of Bay Area Governments, Regional Recreation Committee.

———. 1970. *Regional Plan, 1970–1990—San Francisco Bay Region*. Berkeley: Association of Bay Area Governments.

Azevedo, Margaret. 1984. Interviews with Carla Ehat and Anne Kent. California History Room, Marin County Public Library, San Rafael.

Azevedo, Margaret, Peter Behr, Katy Miller Johnson, William Kahrl, Pete McClosky, and Boyd Stewart. 1993. "Saving Point Reyes National Seashore: An Oral History of Citizen Action in Conservation, 1969–70." Interviews by Ann Lage and William Duddleson. Tape recording and transcript. Regional Oral History Office, Bancroft Library, University of California, Berkeley.

Babal, Marianne. *See* U.S. Department of the Interior.

Bagwell, Beth. 1982. *Oakland: Story of a City*. Novato: Presidio Press.

Ballard, Margaret. 1931. "The History of the Coal Mining Industry in the Mount Diablo Region, 1859–1885." Tape recording and transcript. Regional Oral History Office, Bancroft Library, University of California, Berkeley.

Barbour, Elisa. 2002. *Metropolitan Growth Planning in California 1990–2000*. San Francisco: Public Policy Institute of California.

Bari, Judi. 1994. *Timber Wars*. Monroe, Maine: Common Courage Press.

Barth, Gunter. 1975. *Instant Cities: Urbanization and the Rise of San Francisco and Denver*. New York: Oxford University Press.

———. 1988. "Mountain View: Nature and Culture in an American Park Cemetery." In *The Mirror of History: Essays in Honor of Fritz Fellner*, 153–69, edited by Solomon Wank. Santa Barbara, Calif.: ABC-CLIO.

Bay Area Council. 1988. *Making Sense of the Region's Growth*. San Francisco: Bay Area Council.

Bay Area Open Space Council. 2004. *Parks, People, and Change: Ethnic Diversity and*

Its Significance for Parks, Recreation, and Open Space Conservation in the San Francisco Bay Area. Oakland: Bay Area Open Space Council.

Bay Institute. 1987. *Citizens' Report on the Diked Historical Baylands of San Francisco Bay.* San Rafael: The Bay Institute.

———. 1998. *From the Sierra to the Sea: The Ecological History of the San Francisco Bay–Delta Watershed.* San Rafael: The Bay Institute.

Bay Vision 2020 Commission. 1991. *Report of the Bay Vision 2020 Commission.* San Francisco: Bay Vision 2020 Commission.

Beach, Patrick. 2004. *A Good Forest for Dying: The Tragic Death of a Young Man on the Front Lines of the Environmental Wars.* New York: Doubleday.

Beckert, Sven. 2001. *The Monied Metropolis.* New York: Cambridge University Press.

Beesley, David. 2004. *Crow's Range: An Environmental History of the Sierra Nevada.* Reno: University of Nevada Press.

Behr, Peter. 1988. Interview by Ann Lage. Tape recording and transcript. Regional Oral History Office, Bancroft Library, University of California, Berkeley.

Belliveau, Mike. 1981. *Toxics in the Bay.* Oakland: Citizens for a Better Environment.

———. 1987. *Toxics Hot Spots.* Oakland: Citizens for a Better Environment.

———. 2004. Telephone interview with author. July 21. Hazardous waste program director (1980–90) and executive director (1990–98), Citizens for a Better Environment, Oakland; director (2000–present), Environmental Health Strategies Center, Orono, Maine.

Bennett, Marty. 2005. Interview in Sonoma with author. May 5. Professor, Santa Rosa Junior College (1990–); executive director, New Economy–Working Solutions (2002–); board member, North Bay Labor Council (2002–) board, Sonoma County Conservation Associates (2003–).

Benson, Ralph. 2006. Telephone interview with author, July 25. Director, Sonoma Land Trust (2003–), Petaluma; former General Counsel and Chief Operating Officer, Trust for Public Land, San Francisco.

Benton, Lisa. 1998. *The Presidio: From Army Post to National Park.* Boston: Northeastern University Press.

Berg, Peter, Beryl Magilavy, and Seth Zuckerman. 1990. *A Green City Program for the San Francisco Bay Area and Beyond.* San Francisco: Planet Drum Books.

Berger, Dan, and Richard Hinkle. 1991. *Beyond the Grapes: An Inside Look at Napa Valley.* Wilmington, Del.: Atomium Books.

Bernick, Michael, and Robert Cervero. 1997. *Transit Villages in the 21st Century.* New York: McGraw-Hill.

Berry, Wendell. 1987. *Home Economics.* Berkeley: North Point Press.

Bettelheim, Matthew. 2005. "The Endemic Nature of the Antioch Dunes." *Bay Nature,* January–March, 8–11.

Bhargava, Michael. 2003. "Sanctuaries: Safe Havens for Marine Life." *Coast and Ocean* 19, no. 2. www.coastalconservancy.ca.gov/coast&ocean/2003.

Bhargava, S., B. Brownstein, A. Dean, and S. Zimmerman. 2001. *Everyone's Valley: Inclusion and Affordable Housing in Silicon Valley*. San Jose: Working Partnerships.

Blackburn, Robin. 1988. *The Overthrow of Colonial Slavery*. London: Verso Press.

Bluestone, Daniel. 1991. *Constructing Chicago*. New Haven, Conn.: Yale University Press.

Blumberg, Louis, and Robert Gottlieb. 1989. *War on Waste: Can America Win Its Battle with Garbage?* Washington, D.C.: Island Press.

Bodovitz, Joseph. 1985. "The Shrinking of San Francisco Bay and How It Was Stopped." *California Waterfront Age* 1, no. 4:21–27.

Bodovitz, Joseph E., Melvin B. Lane, and E. Clement Shute Jr. 1984. "The San Francisco Bay Conservation and Development Commission, 1964–1973." Interviews by Malca Chall. Tape recording and transcript. Regional Oral History Office, Bancroft Library, University of California, Berkeley.

Bollens, John. 1951. *Local Government in California*. Berkeley: University of California Press.

———. 1957. *Special District Governments in the United States*. Berkeley: University of California Press.

Bonner, W. G. 1884. *The Redwoods of California: A Glimpse of the Wonderland of the Golden West*. San Francisco: California Redwood Company.

Bookman, Ann, and Sandra Morgan, eds. 1988. *Women and the Politics of Empowerment*. Philadelphia: Temple University Press.

Bourdieu, Pierre. 1984. *Distinction*. Translated by Richard Nice. Cambridge, Mass.: Harvard University Press.

———. 2000. *Pascalian Meditations*. Translated by Richard Nice. Stanford: Stanford University Press.

Bradshaw, Ted. 1993. *Is Growth Control a Planning Failure?* Berkeley: University of California, Berkeley, Institute of Urban and Regional Development.

Branaman, Marybeth. 1955. *Growth of the San Francisco Bay Area Urban Core*. Berkeley: University of California, Berkeley, Bureau of Business and Economic Research.

Brechin, Gray. 1983. "Progress in San Francisco: It Could Have Been Worse." *San Francisco Magazine*, October, 58–63.

———. 1984. "Living the Dream in Berkeley." *California Monthly*, Mar.–Apr. 24–25.

———. 1990. "Mr. Levitt of the Sunset." *San Francisco Focus*, June, 23–26.

———. 1996. "Conserving the Race: Natural Aristocracies, Eugenics, and the U.S. Conservation Movement." *Antipode* 28, no. 3:229–45.

———. 1999. *Imperial San Francisco: Urban Power, Earthly Ruin*. Berkeley: University of California Press.

———. In press. *The Living New Deal: Excavating the Public Domain in California*. Berkeley: University of California Press.

Brechin, Gray, and Robert Dawson. 1998. *Farewell, Promised Land*. Berkeley: University of California Press.

Breton, Mary Joy. 1998. *Women Pioneers for the Environment.* Boston: Northeastern University Press.

Brewer, Richard. 2003. *Conservancy: The Land Trust Movement in America.* Dartmouth, Mass.: The University Press of New England.

Bridges, Amy. 1997. *Morning Glories: Municipal Reform in the Southwest.* Princeton, N.J.: Princeton University Press.

Bronson, William. 1968. *How to Kill a Golden State.* Garden City, N.J.: Doubleday.

Brower, David J., David Godschalk, and Douglas Porter, eds. 1989. *Understanding Growth Management.* Washington, D.C.: The Urban Institute.

Brower, David R., ed. 1969. *Not Man Apart: Lines from Robinson Jeffers, Photographs of the Big Sur Coast.* New York: Ballantine Books.

Bullard, Robert. 1990. *Dumping in Dixie: Race, Class and Environmental Quality.* Boulder, Colo.: Westview Press.

Bullard, Robert, ed. 2005. *The Fight for Environmental Justice: Human Rights and the Politics of Pollution.* San Francisco: Sierra Club Books.

Burgess, Sherwood. 1951. "The Forgotten Redwoods of the East Bay." *California Historical Society Quarterly,* March, 1–14.

———. 1992. *The Water King: Anthony Chabot.* Davis, Calif.: Panorama West Publishing.

Burns, Elizabeth. 1975. "The Process of Suburban Residential Development: The San Francisco Peninsula, 1860–1970." PhD diss., Department of Geography, University of California, Berkeley.

Burrows, Lawrence. 1978. *Growth Management.* New Brunswick, N.J.: Rutgers University, Center for Urban Policy Research.

Busch, Briton. 1985. *The War Against the Seals: A History of the North American Seal Fishery.* Montreal: McGill-Queens University Press.

Butler, Mary Ellen. 1999. *The Prophet of the Parks: The Story of William Penn Mott Jr.* Ashburn, Va: National Parks and Recreation Association.

Butler, Phyllis, ed. 1982. *20-20 Vision: In Celebration of the Peninsula Hills.* Palo Alto: Western Tanager Press, Committee for Green Foothills.

Buttel, Fred. 1992. "Environmentalism: Origins, Processes, and Implications for Rural Social Change." *Rural Sociology* 57, no. 1:1–27.

Byers, Elizabeth, and Karin Ponti. 2005. *The Conservation Easement Handbook.* San Francisco and Alexandria, Va.: Land Trust Alliance, Trust for Public Land.

California Division of Mines. 1951. *Geologic Guidebook of the San Francisco Bay Counties,* edited by Olaf Jenkins. Bulletin No. 154. San Francisco: California Division of Mines.

———. 1958. *Salt in California,* by William Ver Plank and Robert Heizer. Bulletin No. 175. Sacramento: California State Printing Office.

California State Bureau of Mines. 1918. *The Quicksilver Resources of California,* by Walter Bradley. Bulletin No. 78. Sacramento: California State Printing Office.

California State Mining Bureau. 1883–84. *Annual Report of the State Mineralogist.* Sacramento: California State Printing Office.

⸺. 1906. *The Structural and Industrial Materials of California,* by Lewis Aubury. Bulletin No. 38. Sacramento: California State Printing Office.

⸺. 1921. *Mining in California.* Vol. 17. Sacramento: California State Mining Bureau.

⸺. 1928. *The Clay Resources and the Ceramic Industry of California,* by Waldemar Dietrich. Bulletin No. 99. Sacramento: California State Printing Office.

California Tomorrow. 1973. *Democracy in the Space Age: Regional Government Under a California State Plan.* San Francisco: California Tomorrow.

⸺. 1988. *Crossing the Schoolhouse Border: Immigrant Students and the California Public Schools.* San Francisco: California Tomorrow.

Callenbach, Ernest. 1975. *Ecotopia.* Berkeley: Banyan Tree Books.

Calthorpe, Peter. 1993. *The Next American Metropolis.* New York: Princeton Architectural Press.

Calthorpe, Peter, and William Fulton. 2001. *The Regional City: Planning for the End of Sprawl.* Washington, D.C.: Island Press.

Cammett, Melani. 2005. "Fats Cats and Self-made Men: Globalization and the Paradoxes of Collective Action." *Comparative Politics* 37, no 4:379–400.

Cardellino, Joan. 2004. Telephone interview with author. August 10. Coastal access program director, California Coastal Conservancy, Oakland.

Carlsson, Chris, ed. 2002. *Critical Mass: Bicycling's Defiant Celebration.* Oakland, Calif.: AK Press.

Carosso, Vincent. 1951. *The California Wine Industry, 1830–1895: A Study of the Formative Years.* Reprint, Berkeley: University of California Press, 1976.

Carson, Rachel. 1962. *Silent Spring.* Boston: Houghton Mifflin.

Castells, Manuel. 1983. *The City and the Grassroots.* Berkeley: University of California Press.

Chang, Vivian. 2004. Interview in Oakland with author. August 9. Organizer (1999–2000), organizing director (2001–03), and executive director (2004–), Asian–Pacific Islander Environmental Network (APEN), Oakland.

Chase, Alston. 1995. *In a Dark Wood: The Fight Over Forests and the Rising Tyranny of Ecology.* Boston: Houghton Mifflin.

Clark, Henry. 2003. "The West County Toxics Coalition." Tape recording and transcript. Regional Oral History Office, Bancroft Library, University of California, Berkeley.

Clark, Jesse, ed. 2005. "Moving the Environment: Transportation Justice." *Race, Poverty, and Environment* 16, no. 4 (special issue).

Clary, Raymond. 1980. *The Making of Golden Gate Park: The Early Years, 1865–1906.* San Francisco: California Living Books.

Clawson, Marion. 1971. *Suburban Land Conversion in the United States: An Economic and Governmental Process.* Baltimore: Johns Hopkins University Press.

Cohen, Michael. 1988. *The History of the Sierra Club, 1892–1970*. San Francisco: Sierra Club Books.

Cohen, Shana. 2005. "American Garden Clubs and the Fight for Nature Preservation, 1890–1980." PhD diss., Department of Environmental Science, Policy, and Management, University of California, Berkeley.

Colby, William Edward. 1959. "Early Days of the Sierra Club." Unpublished manuscript. Bancroft Library, University of California, Berkeley.

Cole, Luke, and Sheila Foster. 2001. *From the Ground Up: Environmental Racism and the Rise of the Environmental Justice Movement*. New York: New York University Press.

Coleman, Charles. 1952. *PG&E of California: The Centennial Story of Pacific Gas and Electric Company, 1852–1952*. New York: McGraw-Hill.

Collins, George. 1980. "The Art and Politics of Park Planning and Preservation, 1920–1979." Tape recording and transcript. Regional Oral History Office, Bancroft Library, University of California, Berkeley.

Commoner, Barry. 1971. *The Closing Circle*. New York: Knopf.

Conaway, James. 1990. *Napa: The Story of an American Eden*. New York: Avon Books.

———. 2002. *The Far Side of Eden: New Money, Old Land, and the Battle for Napa Valley*. Boston: Houghton Mifflin.

Coodley, Lauren. 2003. *Napa: The Transformation of an American Town*. Charleston, S.C.: Arcadia.

Cook, Stuart. 1985. "The Unfulfilled Bay: The Dreamers Didn't Always Get Their Way." *California Waterfront Age* 1, no. 4:6–20.

Cool, Kevin. 2001. "This Precious Plot." *Stanford Magazine*, January–February, 60–71.

Cosgrove, Denis. 1984. *Social Formation and Symbolic Landscape*. London: Croom Helm.

Costa, Steve, Meena Palaniappan, Arlene Wong, Jeremy Hays, Clara Landeiro, and Jane Rongerude. 2002. *Neighborhood Knowledge for Change: The West Oakland Indicators Project*. Oakland: Pacific Institute for Studies in Development, Environment, and Security.

Covel, Paul. 1978. *People Are for the Birds*. Oakland: Western Interpretive Press.

Cox, Kevin, ed. 1997. *Spaces of Globalization: Reasserting the Power of the Local*. New York: Guilford Press.

Cranz, Galen. 1982. *The Politics of Park Design: A History of Urban Parks in America*. Cambridge, Mass.: MIT Press.

Crenson, Matthew. 1971. *The Unpolitics of Air Pollution: A Study of Non-Decisionmaking in the Cities*. Baltimore: Johns Hopkins University Press.

Crevelli, John. 1993. *Twenty-five Years of COAAST*. Forestville, Calif.: Walkabout Press.

———. 2004. Interview in Healdsburg with author. July 16. Cofounder, Californians Organized to Acquire Access to State Tidelands (COAAST); former professor of history, Santa Rosa Junior College; board of directors, Conservation Action.

Cronon, William. 1991. *Nature's Metropolis: Chicago and the Great West*. Chicago: W. W. Norton.

———. 1995. "The Trouble with Wilderness; or, Getting Back to the Wrong Nature." In *Uncommon Ground: Toward Reinventing Nature*, 69–90, edited by William Cronon. New York: W. W. Norton.

Cry California (journal), 1965–1982. San Francisco: California Tomorrow.

Dana, Samuel, and Myron Krueger. 1958. *California Lands: Ownership, Use, and Management*. Washington, D.C.: American Forestry Association.

Dasmann, Raymond. 1959. *Environmental Conservation*. New York: Wiley.

———. 1963. *The Last Horizon*. New York: Macmillan.

———. 1965. *The Destruction of California*. New York: Macmillan.

———. 1968. *A Different Kind of Country*. New York: Macmillan.

———. 1972. *Planet in Peril? Man and the Biosphere Today*. Harmondsworth, UK: Penguin.

Davey, Mary. 2004. "Saving Open Space on the San Francisco Peninsula." Interview by Richard Walker. August 26, 2001. Tape recording and transcript. Regional Oral History Office, Bancroft Library, University of California, Berkeley.

Davies, Clarence, and Barbara Davies. [1970] 1975. *The Politics of Pollution*. 2nd ed. Indianapolis: Bobbs-Merrill.

Davis, Mike. 1986. *Prisoners of the American Dream*. London: Verso Press.

———. 1990. *City of Quartz: Excavating the Future in Los Angeles*. London: Verso Press.

———. 1998. *Ecology of Fear: Los Angeles and the Imagination of Disaster*. New York: Henry Holt, Metropolitan.

———. 2005. "Gentrifying Disaster: In New Orleans, Ethnic Cleansing, GOP-Style." *Mother Jones*, October 25.

Davis, Mike, Kelly Mayhew, and Jim Miller. 2003. *Under the Perfect Sun: The San Diego Tourists Never See*. New York: The New Press.

Davoren, Bill. 2005. Interview in Tiburon with author. July 13. Founder, the Bay Institute.

Dedrick, Claire. 1998. Interview by Ann Lage. Unedited transcript. Regional Oral History Office, Bancroft Library, University of California, Berkeley.

DeLeon, Richard. 1992. *Left Coast City: Progressive Politics in San Francisco, 1975–1991*. Lawrence: University of Kansas Press.

DeVault, Marjorie. 1991. *Feeding the Family: The Social Organization of Caring as Gendered Work*. Chicago: University of Chicago Press.

Deverell, William, and Tom Sitton, eds. 1994. *California Progressivism Revisited*. Berkeley: University of California Press.

Deverell, William, and Greg Hise, eds. 2005. *Land of Sunshine: An Environmental History of Metropolitan Los Angeles*. Pittsburgh, Penn.: University of Pittsburgh Press.

Dillon, Richard. 1982. *Delta Country*. Novato: Presidio Press. Photography by Steve Simmons; foreword by Harold Gilliam.

Dilsaver, Larry, and William Tweed. 1990. *Challenge of the Big Trees: A Resource History of Sequoia and Kings Canyon National Parks*. Three Rivers, Calif.: Sequoia National History Association.

Dobkin, Marjorie. 1979. "The Great Sand Park: The Origin of Golden Gate Park." Master's thesis, Department of Geography, University of California, Berkeley.

Dobyns, Winifred Starr. 1931. *California Gardens*. New York: Macmillan.

Doss, Margot Patterson. 1974. *Paths of Gold: In and Around Golden Gate National Recreation Area*. San Francisco: Chronicle Books.

Dowall, David. 1984. *The Suburban Squeeze*. Berkeley: University of California Press.

Doyle, Robert. 2005. Telephone interview with author. December 21. Assistant general manager, East Bay Regional Parks District, Oakland (c. 1989–); chair of board of directors, Save Mount Diablo (1978–89).

Drury, Newton. 1972. "Parks and Redwoods, 1919–1971." Interview by Amelia Fry and Susan Schrepfer. Tape recording and transcript. Regional Oral History Office, Bancroft Library, University of California, Berkeley.

Drury, Richard Toshiyuki. 2004. Telephone interview with author. August 11. Staff attorney (1993–2003) and legal director (1997–2003) and board of directors (2003–), Communities for a Better Environment, Oakland; senior attorney, Adams, Broadwell, Joseph and Cardoso, South San Francisco.

Duany, Andres, Elizabeth Plate-Zyberk, and Jeff Speck. 2000. *Suburban Nation: The Rise of Sprawl and the Decline of the American Dream*. San Francisco: North Point Press.

Duffy, Robert. 2003. *The Green Agenda in American Politics: New Strategies for the Twenty-First Century*. Lawrence: University of Kansas Press.

Duffy, Rosaleen. 2002. *A Trip Too Far: Ecotourism, Politics and Exploitation*. London: Earthscan Publications.

Duncan, James, and Nancy Duncan. 2004. *Landscapes of Privilege: The Politics of the Aesthetic in an American Suburb*. New York: Routledge.

Dunshee, Bertram. 1979. Interview by Anne Kent and Carla Ehat. November 15. California History Room, Marin County Public Library, San Rafael.

Duveneck, Josephine Whitney. 1978. *Life on Two Levels: An Autobiography*. Los Altos: W. Kaufmann. Introduction by Wallace Stegner.

Dworsky, Leonard. ed. 1971. *Pollution*. New York: Chelsea House.

Dyble, Louise Nelson. 2001. "North of the Golden Gate: A Historical Perspective on Land Use Policy in Marin and Sonoma Counties." Report for the NRI Project group, March 15, 2001. Berkeley: College of Natural Resources, University of California, Berkeley.

———. 2002. "Paying the Toll: The Golden Gate Bridge and Highway District and the San Francisco Bay Area, 1919–1971." PhD diss., Department of History, University of California, Berkeley.

Eckbo, Dean, Austin, & Williams. 1969. *Open Space: The Choices Before California*. San Francisco: Diablo Press.

————. 1972. *Open Space and Resource Conservation Program for California*. San Francisco: Eckbo, Dean, Austin, & Williams.

Eichler, Edward. 1982. *The Merchant Builders*. Cambridge, Mass.: MIT Press.

Eisele, Volker. 1999. "Twenty-five Years of Farmland Protection in the Napa Valley." In *California Farmland and Urban Pressures: Statewide and Regional Perspectives*, edited by Albert Medvitz, Alvin Sokolow, and Cathy Lemp. Davis: University of California, Davis, Agricultural Issues Center.

Elkind, Sarah. 1998. *Bay Cities and Water Politics: The Battle for Resources in Boston and Oakland*. Lawrence: University of Kansas Press.

Elliott, Vicky. 2005. "Extreme Makeover: Leona Quarry to Get a New Life as Oakland's Biggest Subdivision." *San Francisco Chronicle*, October 30, K1, 5–6.

Ellis, Juliet. 2004. Interview in Oakland with author. July 26. Executive director (2001–), Urban Habitat, Oakland.

Engbeck, Joseph. 1980. *State Parks of California, from 1864 to the Present*. Portland, Ore.: C. H. Belding.

Epstein, Barbara. 1991. *Political Protest and Cultural Revolution*. Berkeley: University of California Press.

Epstein, Samuel, Lester Brown, and Carl Pope. 1982. *Hazardous Waste in America*. San Francisco: Sierra Club Books.

Erskine, Dorothy Ward. 1976. "Environmental Quality and Planning: Continuity of Volunteer Leadership." In *Bay Area Foundation History Series*, vol. 3. Berkeley: Regional Oral History Office, Bancroft Library, University of California, Berkeley.

————. 1979. Interview by Carla Ehat and Anne Kent. February 14. California Room, Marin Public Library, San Rafael.

Erskine, John. 2005. Interview in San Francisco with author. July 19. Son of Dorothy and Morris Erskine; board of directors (1982–1995), People for Open Space–Greenbelt Alliance.

Ethington, Philip. 1994. *The Public City: The Political Construction of Urban Life in San Francisco, 1850–1900*. New York: Cambridge University Press.

————. 2001. "Mapping the Local State." *Journal of Urban History* 27, no. 5:686–702.

Eysberg, Cees. 1990. *The Californian Wine Economy*. Utrecht, The Netherlands: Rijkuniversiteit Utrecht, Geografisch Instituut.

Faber, Daniel, and Deborah McCarthy. 2001. "Green of Another Color: Building Effective Partnership Between Foundations and the Environmental Justice Movement." Report for the Philanthropy and Environmental Justice Research Project, Northwestern University, Chicago. www.socant.neu.edu.

Fabian, Bentham. 1869. *The Agricultural Lands of California*. San Francisco: H. H. Bancroft and Co.

Fairfax, Sallie, and Darla Guenzler. 2001. *Conservation Trusts: Institutional Alternatives for a New Era in Land and Resource Conservation*. Lawrence: University Press of Kansas.

Fairley, Lincoln. 1987. *Mount Tamalpais: A History*. San Francisco: Scottwall Associates.

Fellmeth, Robert. 1973. *The Politics of Land*. New York: Grossman.

Findlay, John. 1992. "Stanford Industrial Park: Downtown for Silicon Valley." In *Magic Lands: Western Cityscapes and American Culture after 1940*, 117–59. Berkeley: University of California Press.

Findley, Roger, and Daniel Farber. 1992. *Environmental Law in a Nutshell*. 3d ed. St. Paul, Minn.: West Publishing.

Fischer, Douglas. 2004. "Tour Targets Poor Neighborhoods." *Oakland Tribune*, February 19.

Fishman, Robert. 1987. *Bourgeois Utopias: The Rise and Fall of Suburbia*. New York: Basic Books.

Flanagan, Maureen. 1996. "The City Profitable, the City Livable: Environmental Policy, Gender, and Power in Chicago in the 1910s." *Journal of Urban History* 22, no. 2:163–90.

Ford, Kristina. 1976. "Regional Association and Dissociation in the San Francisco Bay Area." PhD diss., Urban and Regional Planning, University of Michigan.

Fox, Stephen. 1981. *The American Conservation Movement: John Muir and His Legacy*. Madison: University of Wisconsin Press.

Frieden, Bernard. 1979. *The Environmental Protection Hustle*. Cambridge, Mass.: MIT Press.

Futcher, Jane, and Robert Conover. 1981. *Marin: The Place, the People: Profile of a California County*. New York: Holt, Rinehart, and Winston.

Gandy, Matthew. 2002. *Concrete and Clay: Reworking Nature in New York City*. Cambridge, Mass.: MIT Press.

Garreau, Joel. 1991. *Edge City: Life on the New Frontier*. New York: Doubleday.

Gaventa, John. 1980. *Power and Powerlessness: Quiescence and Rebellion in an Appalachian Valley*. Urbana: University of Illinois Press.

Gelfand, Mark. 1975. *A Nation of Cities: The Federal Government and Urban America, 1933–1975*. New York: Oxford University Press.

Gerbode, Martha Alexander. 1995. "Environmentalist, Philanthropist, and Volunteer in the San Francisco Bay Area and Hawaii." Interviews with Huey Johnson, Garland Farmer, Esther Fuller, et al. Tape recording and transcript. Regional Oral History Office, Bancroft Library, University of California, Berkeley.

Gibbs, Lois. 1982. *Love Canal: My Story*. Albany: State University of New York Press.

Gilliam, Harold. 1957. *San Francisco Bay*. Garden City, N.J.: Doubleday.

———. 1962. *Island in Time: The Point Reyes Peninsula*. San Francisco: Sierra Club Books.

———. 1969. *Between the Devil and the Deep Blue Bay: The Struggle to Save San Francisco Bay*. San Francisco: Chronicle Books.

———. 1972. *For Better or for Worse: The Ecology of an Urban Area*. San Francisco: Chronicle Books.

———. 2004. "The Quiet Conservationist." *San Francisco Chronicle Magazine,* April 25, 17–21.

Gilliam, Harold, and Ann Lawrence. 1993. *Marin Headlands: Portals of Time.* San Francisco: Golden Gate National Park Association.

Glickfeld, Madelyn, and Ned Levine. 1992. *Regional Growth . . . Local Reaction: The Enactment and Effects of Local Growth Control and Management Measures in California.* Cambridge, Mass.: Lincoln Land Institute.

Goodyear, Watson Andrews. 1877. *The Coal Mines of the Western Coast of the United States.* San Francisco: A. L. Bancroft.

Gorgoni, Gianfrano. 1978. *Christo: Running Fence, Sonoma and Marin Counties, California, 1972–76.* Chronicle by Calvin Tomkins; narrative text by David Bourdon. New York: Abrams.

Gottlieb, Robert. 1993. *Forcing the Spring: The Transformation of the American Environmental Movement.* Washington, D.C.: Island Press.

———. 2001. *Environmentalism Unbound: Exploring New Pathways for Change.* Cambridge, Mass.: MIT Press.

Gottlieb, Robert, ed. 1995. *Reducing Toxics: A New Approach to Policy and Industrial Decision-Making.* Washington, D.C., and Covelo, Calif.: Island Press.

Governor's Office of Planning and Research. 1991. *Local Government Growth Management Survey.* Sacramento: Governor's Office of Planning and Research.

Graham, Frank Jr. 1990. *The Audubon Ark: A History of the National Audubon Society.* New York: Knopf.

Gramsci, Antonio. 1971. *Prison Notebooks.* New York: International Publishers.

Greenbelt Action. 1986–. San Francisco: People for Open Space, Greenbelt Alliance.

Greenbelt Alliance. 1989. *Reviving the Sustainable Metropolis: Guiding Bay Area Conservation and Development into the 21st Century.* San Francisco: Greenbelt Alliance.

———. 1991a, 1994, 2000, 2006. *At Risk: The Bay Area Greenbelt.* Periodic reports. San Francisco: Greenbelt Alliance.

———. 1991b. *Making the Greenbelt a Reality: Results of the Greenbelt Campaign: 1987–1991.* San Francisco: Greenbelt Alliance.

———. 1991c. *The Bay Area's Farmlands.* San Francisco: Greenbelt Alliance.

———. 1995. *Beyond Sprawl: New Patterns of Growth to Fit the New California.* San Francisco: Greenbelt Alliance, with Bank of America, California Resources Agency, and Low Income Housing Fund.

———. n.d. (ca. 1997). *Bound for Success: A Citizen's Guide to Using Urban Growth Boundaries for More Livable Communities and Open Space Protection in California.* San Francisco: Greenbelt Alliance.

———. 2000. "Wall of Sprawl Threatens the Greenbelt." Special report, *Greenbelt Action.* Winter.

———. 2002. *Balanced Transportation: Achieving Congestion Relief and Meeting Transportation Needs in Solano County.* San Francisco: Greenbelt Alliance.

————. 2003a. *Smart Infill: Creating More Livable Communities in the Bay Area*. San Francisco: Greenbelt Alliance.

————. 2003b. *Getting It Right: Preventing Sprawl in Coyote Valley*. San Francisco: Greenbelt Alliance.

Greenbelt Alliance, ed. 1995. *Reader on Urban Growth Boundaries*. San Francisco: Greenbelt Alliance.

Greenbelt Alliance and Silicon Valley Manufacturers' Group. 1999. *Housing Solutions for Silicon Valley*. San Francisco: Greenbelt Alliance, Livable Communities Program.

Greenbelt Alliance and Sonoma County Farm Bureau. 2004. *Preventing Sprawl: Farmers and Environmentalists Working Together*. San Francisco: Greenbelt Alliance.

Greenberg, Doug. 1986. "Growth and Conflict at the Suburban Fringe: The Case of the Livermore-Amador Valley, California." PhD diss., Department of Geography, University of California, Berkeley.

Green Footnotes. 1967–. Palo Alto: Committee for Green Foothills.

Griffin, Martin. 1998. *Saving the Marin-Sonoma Coast: The Battles for Audubon Canyon Ranch, Point Reyes, and California's Russian River*. Healdsburg, Calif.: Sweetwater Springs Press.

————. 2004. Interview in Healdsburg with author and Marty Bennett. July 22. Former director, Marin chapter of Audubon Society; founder, Audubon Ranch; proprietor, Hop Kiln Winery, Healdsburg, Calif.

Gudis, Catherine. 2004. *Buyways: Billboards, Automobiles, and the American Landscape*. New York: Routledge.

Gulick, Esther, Catherine Kerr, and Sylvia McLaughlin. 1988. "Saving San Francisco Bay: Past, Present, and Future." In *The Horace M. Albright Lectureship in Conservation* (Berkeley: College of Natural Resources, University of California, Berkeley) 28 (April 14): 1–10.

Gustaitis, Rosa. 2002–03. "Lew Reid, Defender of the Coast." *Coast and Ocean* 12, no. 4 (winter): 30–32.

Guthey, Greig. 2004. "Terroir and the Politics of the Agro-Industry in California's North Coast Wine District." PhD diss., Department of Geography, University of California, Berkeley.

Guthman, Julie. 2004. *Agrarian Dreams: The Paradox of Organic Farming in California*. Berkeley: University of California Press.

Hall, Carl. 2005. "Slowly Closing the Gaps for a Bay Trail Loop." *San Francisco Chronicle*, December 18, B1, 6.

Halper, Jon, ed. 1991. *Gary Snyder: Dimensions of a Life*. San Francisco: Sierra Club Books.

Hamalian, Linda. 1991. *A Life of Kenneth Rexroth*. New York: W. W. Norton.

Handel, Mary, and Alvin Sokolow. 1994. *Farmland and Open Space Preservation in the Four North Bay Counties*. Davis: University of California, Cooperative Extension.

Harris, David. 1995. *The Last Stand: The War Between Wall Street and Main Street Over California's Ancient Redwoods*. New York: Times Books.

Hart, Caryl. 2005. Interview in Berkeley with author. December 13. Founder, Sonoma County LandPaths; member, State Parks Commission (2000–); member, Advisory Committee of Sonoma County Agricultural Preservation and Open Space District (1993–).

Hart, John. 1979. *San Francisco's Wilderness Next Door*. San Rafael: Presidio Press.

———. 1991. *Farming on the Edge: Saving Family Farms in Marin County, California*. Berkeley: University of California Press.

———. 1996. *Storm Over Mono: The Mono Lake Battle and the California Water Future*. Berkeley: University of California Press

———. 2006. "Private Land, Public Good: Taking Stock of Conservation Easements." *Bay Nature*, January–March, 18–32.

Hart, John, and David Sanger, 2003. *San Francisco Bay: Portrait of an Estuary*. Berkeley: University of California Press.

Hart, John, and Nancy Kittle. 2006. *Legacy: Portraits Of 50 Bay Area Environmental Elders*. Photographs by Nancy Kittle; text by John Hart. San Francisco: Sierra Club Books.

Harvey, David. 1973. *Social Justice and the City*. London: Edward Arnold.

———. 1997. *Justice, Nature, and the Geography of Difference*. Oxford: Basil Blackwell.

———. 2001. *Spaces of Hope*. Berkeley: University of California Press.

Hayden, Delores. 1985. *The Grand Domestic Revolution: A History of Feminist Designs for American Homes, Neighborhoods, and Cities*. Cambridge, Mass.: Harvard University Press.

Hays, Samuel. 1959. *Conservation and the Gospel of Efficiency: The Progressive Conservation Movement, 1890–1920*. Cambridge, Mass.: Harvard University Press.

———. 1987. *Beauty, Health, and Permanence: Environmental Politics in the United States, 1955–85*. New York: Cambridge University Press.

Healy, Robert, ed. 1978. *Protecting the Golden Shore: Lessons from the California Coastal Commission*. Washington, D.C.: Conservation Foundation.

Hedgpethe, Joel. 1973. "Bodega: A Case History of Intense Controversy." In *Environmental Quality and Water Development*, 439–54, edited by C. Goldman, J. McEvoy, and P. Richerson. San Francisco: W. H. Freeman.

Heiman, Michael. 1987. *Coastal Recreation in California*. Berkeley: University of California, Berkeley, Institute of Governmental Studies.

———. 1988. *The Quiet Evolution: Power, Planning, and Profits in New York State*. New York: Praeger.

———. 1989. "Production Confronts Consumption: Landscape Perception and Social Conflict in the Hudson Valley." *Society and Space* 7, no. 2:165–78.

Heller, Alfred. 1983. "California Tomorrow: A Voice for State and Regional Planning." In *Statewide and Regional Land-Use Planning in California*, vol. 1, 360–436, interview by Malca Chall. Regional Oral History Office, Bancroft Library, University of California, Berkeley.

Heller, Alfred, ed. 1972. *The California Tomorrow Plan*. Los Altos: W. Kaufmann.

Henderson, George. 1998. *California and the Fictions of Capital*. New York: Oxford University Press.

Henwood, Doug. 2003. *After the New Economy*. New York: The New Press.

Hernandez, Lizette, Torri Estrada, and Cataline Garzón. 1999. *Building upon Our Strengths: A Community Guide to Brownfields Redevelopment in the San Francisco Bay Area*. San Francisco: Urban Habitat.

Hirt, Paul. 1994. *A Conspiracy of Optimism: Management of the National Forests since World War Two*. Lincoln: University of Nebraska Press.

Hise, Greg. 1997. *Magnetic Los Angeles: Planning the Twentieth-Century Metropolis*. Baltimore: Johns Hopkins University Press.

Hise, Greg, and William Deverell, eds. 2000. *Eden by Design: The 1930 Olmsted-Bartholomew Plan for the Los Angeles Regions*. Berkeley: University of California Press.

———, eds. 2005. *Land of Sunshine: An Environmental History of Metropolitan Los Angeles*. Pittsburgh: University of Pittsburgh Press.

Hochschild, Arlie. 1989. *The Second Shift: Inside the Two-Job Marriage*. New York: Viking.

Hofrichter, Richard, editor. 2000. *Reclaiming the Environmental Debate: The Politics of Health in a Toxic Culture*. Cambridge, Mass.: MIT Press.

Holbrook, Deirdre. 2004. Telephone interview with author. August 16. Information director, Peninsula Open Space Trust, Menlo Park, Calif.

Hollingsworth, Buckner. 1962. *Her Garden Was Her Delight*. New York: Macmillan.

Holmes, Mark. 2005. Telephone interview with author. July 22. Bay restoration program director (1999–present), the Bay Institute; former restoration officer (1986–99), Save the Bay Association, Oakland.

Holstan, James. 1989. *The Modernist City: An Anthropological Critique of Brasilia*. Chicago: University of Chicago Press.

Holtzclaw, John. 2004. Telephone interview with author. August 20. Former chair, Bay Chapter; former chair, Urban Environment Committee; chair, National Transportation Committee; chair, National Sprawl Campaign, Sierra Club, San Francisco.

Hornbeck, David. 1983. *California Patterns: A Geographical and Historical Atlas*. Palo Alto: Mayfield Publishing.

Hull, Dave. 2001. "America's Preeminent Whaling Port: San Francisco." *Sealetter (Journal of the San Francisco Maritime Park Association)* 61 (winter): 26–31.

Hurley, Andrew. 1995. *Environmental Inequalities: Class, Race, and Industrial Pollution in Gary, Indiana, 1945–1980*. Chapel Hill: University of North Carolina Press.

Hyde, Philip, and Francois Leydet. 1963. *The Last Redwoods: Photographs and Story of a Vanishing Scenic Resource.* San Francisco: Sierra Club. Foreword by Stewart L. Udall.

Hynding, Alan. 1981. *From Frontier to Suburb: The Story of the San Mateo Peninsula.* Belmont, Calif.: Star Books.

Isenberg, Andrew. 2005. *Mining California: An Ecological History.* New York: Hill and Wang.

Issel, William. 1999. "Land Values, Human Values, and the Preservation of the City's Treasured Appearance: Environmentalism, Politics, and the San Francisco Freeway Revolt." *Pacific Historical Review* 68:611–46.

Jackson, Kenneth. 1985. *The Crabgrass Frontier: The Suburbanization of the United States.* New York: Oxford University Press.

Jacobs, John. 1995. *A Rage for Justice: The Passion and Politics of Phillip Burton.* Berkeley: University of California Press.

Jacobs, Katherine, Samual Luoma, and Kim Taylor. 2003. "CALFED: An Experiment in Science and Decisionmaking." *Environment* 45, no. 1:31–41.

Jacobson, Yvonne. 1984. *Passing Farms, Enduring Values: California's Santa Clara Valley.* Los Altos: William Kaufman.

Jacoby, Karl. 2001. *Crimes Against Nature: Squatters, Poachers, Thieves, and the Hidden History of American Conservation.* Berkeley: University of California Press.

James, L. Allen. 1994. "Channel Changes Wrought by Gold Mining: Northern Sierra Nevada, California." *Effects of Human-Induced Changes on Hydrologic Systems* (American Water Resources Association) 86 (June): 629–37.

Joassart-Marcelli, Pascale, William Fulton, and Juliet Musso. 2004. "Smart Growth or Growth Machine?" In *Up Against the Sprawl,* 75–95, edited by J. Wolch, M. Pastor, and P. Drier. Minneapolis: University of Minnesota Press.

Johns, Michael. 1997. *The City of Mexico in the Age of Diaz.* Austin: University of Texas Press.

Johnson, Kathy. 1996. "Securing Regional Autonomy: Creation of the Bay Area's Metropolitan Transportation Commission, 1956–70." Paper, Department of Geography, University of California, Berkeley.

———. 2006. "Sovereigns and Subjects: A Geopolitical History of Metropolitan Reform in the U.S.A." *Environment and Planning A.* 38:149–68.

Johnston, Moira. 1989. "A Very Civil War." *California* 14, no. 1 (January): 54–61.

———. 1990. *Roller Coaster: The Bank of America and the Future of American Banking.* New York: Ticknor and Fields.

Jones, David. 1989. *California's Freeway Era in Historical Perspective.* Berkeley: University of California, Berkeley, Institute for Transportation Studies.

Jones, Holway. 1965. *John Muir and the Sierra Club: The Battle for Yosemite.* San Francisco: Sierra Club.

Jones, Victor. 1964. *Government of the San Francisco Bay Area.* Berkeley: University of California, Institute of Governmental Studies.

————. 1983. "Political Scientist: Observer and Consultant in Metropolitan Governance." In *Statewide and Regional Land-Use Planning in California*, vol. 2, 126–225, interview by Malca Chall. Regional Oral History Office, Bancroft Library, University of California, Berkeley.

Jones and Stokes Associates. 1974. *Land Use, Open Space, and the Government Process: The San Francisco Bay Area Experience*. New York: Praeger.

Jordan, Tom. 2005. "Remembering Claire and Kent Dedrick." *Green Footnotes* (fall), 15.

Kahrl, William, ed. 1979. *The California Water Atlas*. Sacramento: Governor's Office of Planning and Research.

Karras, Greg. 1989. "Pollution Prevention: The Chevron Story." *Environment* 31, no. 8 (October): 4–5, 45.

————. 2004. Telephone interview with author. August 17. Chief staff scientist, Communities for a Better Environment, Oakland.

Katz, Lawrence, and Kenneth Rosen. 1980. *The Effects of Land-Use Controls on Housing Prices*. Berkeley: University of California, Berkeley, Center for Real Estate Research and Urban Economics.

Kay, Jane. 2002. "Tracking a Toxic Trail." *San Francisco Chronicle*, December 22: A23, 26–27.

————. 2003a. "Nature Moves Back In." *San Francisco Chronicle*, July 13, 2003: 21–22.

————. 2003b. "Toward a Healthy Bay." *Bay Nature* Magazine, October–December: 18–30.

Kelley, Robert. 1989. *Battling the Inland Sea: American Political Culture, Public Policy, and the Sacramento Valley, 1850–1986*. Berkeley: University of California Press.

Kenney, Martin, ed. 2000. *Understanding Silicon Valley: The Anatomy of an Entrepreneurial Region*. Stanford: Stanford University Press.

Kent, Elizabeth. 1950. "William Kent, A Biography." Manuscript, Bancroft Library, University of California, Berkeley.

Kent, T. J. Jr. 1963. *City and Regional Planning for the Metropolitan San Francisco Bay Area*. Berkeley: University of California, Berkeley, Institute of Governmental Studies.

————. 1969. "Berkeley's First Liberal Democratic Regime, 1961–70." In *Toward a Bay Area Regional Organization*, 110–25, edited by Harriet Nathan and Stanley Scott. Berkeley: University of California, Berkeley, Institute of Governmental Studies.

————. 1970. *Open Space for the San Francisco Bay Area: Organizing to Guide Metropolitan Growth*. Appendix 3, A Brief History of People for Open Space. Berkeley: University of California, Berkeley, Institute of Governmental Studies.

————. 1983. "Professor and Political Activist: A Career in City and Regional Planning in the San Francisco Bay Area." In *Statewide and Regional Land-Use Planning in California,* vol. 2, 1–125, interview by Malca Chall. Regional Oral History Office, Bancroft Library, University of California, Berkeley.

Kent, William. 1929. *Reminiscences of Outdoor Life.* San Francisco: A. M. Robertson. Foreword by Stewart Edward White.

Kerr, Kay, Sylvia McLaughlin, and Esther Gulick. 1987. "Save San Francisco Bay Association, 1961–1986." Tape recording and transcript, with introduction by Harold Gilliam. Regional Oral History Office, Bancroft Library, University of California, Berkeley.

King, John. 2004. "Build It and They Will Cringe." *San Francisco Chronicle,* December 13: B1–2.

Knapp, Gerrit, and Arthur Nelson. 1992. *The Regulated Landscape: Lessons on State Land Use Planning from Oregon.* Cambridge, Mass.: Lincoln Institute of Land Policy.

Koch, Margaret. 1973. *Santa Cruz County: Parade of the Past.* Santa Cruz: Western Tanager Press.

Kockelman, William, John Conomos, and Alan Leviton, eds. 1982. *Use and Protection of the San Francisco Bay System.* San Francisco: Pacific Division, American Academy for the Advancement of Science.

Kondolf, Matt. 1997. "Hungry Water: Effects of Dams and Gravel Mining on River Channels." *Environmental Management* 21, no. 4:533–51.

Kortum, William. 2005. "Environmentalist Extraordinaire in Sonoma County." Interviews by Richard Walker and Martin Bennett, June 9, 2003, and May 5, 2005. Tape recording and transcript. Regional Oral History Office, Bancroft Library, University of California, Berkeley.

Kroll, Cynthia, and Linda Kimball. 1986. "The R&D Dilemma: The Real Estate Industry and High-Tech Growth." Working Paper 86-116. Berkeley: University of California, Berkeley, Center for Real Estate and Urban Economics.

Krubiner, Todd. 1999. "Comparison and Analysis of San Francisco Bay Area Regional Open-space Systems: East Bay Regional Park District and Midpeninsula Regional Open Space District." PhD diss., University of Washington.

Kunstler, James Howard. 1993. *The Geography of Nowhere: The Rise and Decline of America's Man-Made Landscape.* New York: Simon and Schuster.

LaForce, Norman. 2002. "Creating the Eastshore State Park: An Activist History." Interview by Ann Lage. Tape recording and transcript. Regional Oral History Office, Bancroft Library, University of California, Berkeley.

Landis, John. 1992. *Do Growth Controls Work? An Evaluation of Local Growth Control Programs in Seven California Cities.* Berkeley: University of California, Berkeley, California Policy Seminar.

Land Trust Alliance. 1988. *Conservation Easement Handbook.* Washington, D.C., and San Francisco: Land Trust Association, Trust for Public Land.

Langston, Nancy. 1995. *Forest Dreams, Forest Nightmares: The Paradox of Old Growth in the Inland West.* Seattle: University of Washington Press.

Lapsley, James. 1996. *Bottled Poetry: Napa Winemaking From Prohibition to the Modern Era*. Berkeley: University of California Press.

LaRiviere, Florence, and Phil LaRiviere. 2005. Interview in Palo Alto with author. July 8. Organizers, Citizens to Complete the Refuge (1981–), Palo Alto.

Lawrence, Elizabeth. 1987. *Gardening for Love: The Market Bulletins*. Edited by Allen Lacy. Durham, N.C.: Duke University Press.

League of California Cities and County Supervisors Association. 1989. *Growth Control–Management Survey*. Sacramento: League of California Cities.

Lear, Norwood, and Linda Lear. 1997. *Rachel Carson: Witness for Nature*. New York: Henry Holt and Company.

Lears, T. Jackson. 1981. *No Place of Grace: Anti-Modernism and the Transformation of American Culture, 1880–1920*. New York: Pantheon.

Lee, Charles. 1987. *Toxic Wastes and Race in the United States*. New York: United Church of Christ, Commission for Racial Justice.

Lee, Pamela Tau. 2003. "Community and Union Organizing, and Environmental Justice in the San Francisco Bay Area, 1967–2000." Interview by Carl Wilmsen. Tape recording and transcript. Regional Oral History Office, Bancroft Library, University of California, Berkeley.

LeGates, Richard, and Sean Nikas. 1989. "Growth Management Through Residential Tempo Controls in the San Francisco Bay Area." Working Paper 89-11. San Francisco: San Francisco State University, Public Research Institute.

LeGates, Richard, and Claude Pellerin. 1989. *Planning Tomorrowland: The Transformation of Pleasanton, California*. San Francisco: San Francisco State University, Public Research Institute.

LeGates, Richard, and Teresa Selfa. 1989. *Growing Old Gracefully: The Petaluma Plan Reaches Middle Age*. Working Paper 89-14. San Francisco: San Francisco State University, Public Research Institute.

Leitner, Helga. 1994. "Capital Markets, the Development Industry, and Urban Office Market Dynamics: Rethinking Building Cycles." *Environment and Planning A* 26:779–802.

Lewis, Michael. 2000. *The New New Thing: A Silicon Valley Story*. Chicago: W. W. Norton.

Lewis, Paul, and Max Neiman. 2002. *Cities Under Pressure: Local Growth Controls and Residential Development Policy*. San Francisco: Public Policy Institute of California.

Lewis, Sherman. 1990. *Managing Urban Growth in the San Francisco Bay Region*. Hayward: California State University, Hayward, Center for Public Service Education and Research.

Leydet, Francois. 1964. *Time and the River Flowing, Grand Canyon*. Edited by David Brower. Exhibit Format series no. 8. San Francisco: Sierra Club.

Liebman, Ellen. 1983. *California Farmland: A History of Large Agricultural Landholdings*. Totowa, N.J.: Rowman and Allanheld.

Limerick, Patricia. 1992. "Disorientation and Reorientation: The American Landscape Discovered from the West." *Journal of American History* 79, no. 3: 1021–49.

Linn, Karl. 2004. Interview in Berkeley with author. July 6. Professor emeritus of landscape architecture, University of Pennsylvania (1959–89); board of directors, Earth Island Institute (1988–2005); board of directors, San Francisco League of Urban Gardeners; co-founder, Urban Habitat.

Livermore, Norman. 1983. "Man in the Middle: High Sierra Packer, Timberman, Conservationist, and California Resources Secretary." Interviews by Ann Lage and Gabrielle Morris, 1981–82. Tape recording and transcript. Regional Oral History Office, Bancroft Library, University of California, Berkeley.

Livingston, Dewey. *See* U.S. Department of the Interior.

Livingston and Blayney. 1970. *Foothills Environmental Design Study: Report No. 3 to the City of Palo Alto.* With assistance of Lawrence Halperin and Associates. San Francisco: Livingston and Blayney.

Lo, Clarence. 1990. *Small Property Versus Big Government: Social Origins of the Property Tax Revolt.* Berkeley: University of California Press.

Logan, John, and Harvey Molotch. 1986. *Urban Fortunes: The Political Economy of Place.* Berkeley and Los Angeles: University of California Press.

Long, David. 1992. "Mistaken Identity: Putting the John Reber Plan for the San Francisco Bay Area into Historical Context." In *American Cities and Towns: Historical Perspectives,* 123–42, edited by Joseph Rishel. Pittsburgh: Duquesne University Press.

Longstreth, Richard. 1997. *City Center to Regional Mall: Architecture, the Automobile, and Retailing in Los Angeles, 1920–1950.* Cambridge, Mass.: MIT Press.

Longtin, James. 1999. *150 Years of Land Use: A Brief History of Land Use Regulation.* Berkeley: Local Government Publications.

Lydon, Peter. 1993. *San Francisco's Bay Vision 2020 Commission: A Civic Initiative for Change.* Berkeley: University of California, Berkeley, Institute of Governmental Studies.

Ma, L. Eve. 1981. "Chinese in California's Fishing Industry, 1850–1941." *California History* 60, no. 2:142–57.

MacIntyre, Alisdair. 1984. *After Virtue.* Bloomington: Indiana University Press.

Mahmood, Saba. 2001. "Rehearsed Spontaneity and the Conventionality of Ritual: Disciplines of Salat." *American Ethnologist* 28, no. 4:827–53.

Manes, Christopher. 1990. *Green Rage: Radical Environmentalism and the Unmaking of Civilization.* Boston: Little, Brown.

Manning, Robert. 1984. "Men and Mountains Meet: Journal of the Appalachian Mountain Club, 1876–1984." *Journal of Forest History,* January, 24–33.

Marin Conservation League. 1936. *Marvelous Marin County: San Francisco's Most Unique Suburb.* San Rafael: Marvelous Marin, Inc., Marin Conservation League.

Marin County Planning Department. 1971. *Can the Last Place Last? Preserving the Environmental Quality of Marin,* by Albert Solnit. San Rafael: Marin County Planning Department.

Marin Municipal Water District. 1975. *A Historical Summary of the Marin Municipal Water District.* San Rafael: Marin Municipal Water District.

Mark, Stephen. 2005. *Preserving the Living Past: John C. Merriam's Legacy in the State and National Parks.* Berkeley: University of California Press.

Martin, Glen. 1999. "Our Poisoned Bay: Despite End to Direct Piping of Sewage, Pollution Worse Now than Thirty Years Ago." *San Francisco Chronicle,* August 2, A1, 6, 7.

———. 2005. "Invader in the Bay." *San Francisco Chronicle,* October 11, B1, B12.

Marx, Leo. 1964. *The Machine in the Garden: Technology and the Pastoral Ideal in America.* New York: Oxford University Press.

Mason, Jack, and Helen Van Cleave Park. 1975. *The Making of Marin, 1850–1975.* Inverness, Calif.: North Shore Books.

Massey, Douglas, and Nancy Denton. 1993. *American Apartheid: Segregation and the Making of the Underclass.* Cambridge, Mass.: Harvard University Press.

Matsuoka, Marta. 2003. *Building Healthy Communities from the Ground Up: Environmental Justice in California.* Oakland: Asian Pacific Islander Environmental Network, Communities for a Better Environment, Environmental Health Coalition, People Organized to Defend Economic and Environmental Rights, Silicon Valley Toxics Coalition.

May, Bernice Hubbard. 1976. "A Native Daughter's Leadership in Public Affairs." Tape recording and transcript. Regional Oral History Office, Bancroft Library, University of California, Berkeley.

May, Julia. 2004. Telephone interview with author. August 25. Staff scientist (1987–2003) and air pollution program director (1995–2003), Communities for a Better Environment, Oakland.

McAdam, Doug. 1982. *Political Process and the Development of Black Insurgency, 1930–1970.* Chicago: University of Chicago Press.

McCaffrey, David. 1982. *OSHA and the Politics of Health Regulation.* New York: Plenum Press.

McCarthy, James. 1999. "The Political and Moral Economies of Wise Use." PhD diss., Department of Geography, University of California, Berkeley.

McConnell, Grant. 1966. *Private Power and American Democracy.* New York: Vintage Books.

McEvoy, Arthur. 1986. *The Fisherman's Problem: Ecology and Law in the California Fisheries, 1850–1980.* New York: Cambridge University Press.

McGirr, Lisa. 2001. *Suburban Warriors: The Origins of the New American Right.* Princeton, N.J.: Princeton University Press.

McGovern, Pat. 1994. "Contra Costa County Edge Cities: The New Political Economy of Planning." PhD diss., Department of City and Regional Planning, University of California, Berkeley.

McHugh, Paul. 2000. "Miles to Go: Competing Interests Miss Mark in Debate over Access to Peninsula Watershed Trail." *San Francisco Chronicle*, March 19, A1.

McPhee, John. 1971. *Encounters with the Archdruid.* New York: Farrar, Straus and Giroux.

Melosi, Martin. 1981. *Garbage in the Cities: Refuse, Reform, and the Environment, 1880–1980.* College Station: Texas A&M University Press.

Merchant, Carolyn. 1996. *Earthcare: Women and the Environment.* New York: Routledge.

Merchant, Carolyn, ed. 1998. *Green Versus Gold: Sources in California Environmental History.* Washington, D.C.: Island Press.

Merenlender, A., L. Huntsinger, G. Guthey, and S. Fairfax. 2003. "Land Trusts and Conservation Easements: Who Is Conserving What for Whom?" *Conservation Biology* 18, no. 1:65–75.

Meyer, Amy. 2006. *New Guardians for the Golden Gate: How America Got a Great National Park.* Berkeley: University of California Press.

Micklethwait, John. 1997. "The Valley of the Money's Delight." *Economist* 342 (March 29).

Miles, Sara. 2001. *How to Hack a Party Line: The Democrats and Silicon Valley.* Rev. ed. Berkeley: University of California Press.

Miller, Greg. 1997. "Will the Party Last? In Tech, Some See Signs of a Bubble." *Los Angeles Times,* December 28, Business and Technology sec., 1.

Miller, Jeffrey. 2004. Interview in San Francisco with author. July 24. Board of directors (1990–2001), board president (1994–97). San Francisco League of Urban Gardeners.

Minick, Roger. 1969. *Delta West: The Land and People of the Sacramento–San Joaquin Delta.* With historical essay by Dave Bohn. Berkeley: Scrimshaw Press.

Misczynski, Dean. 1986. "The Fiscalization of Land Use." In *California Policy Choices,* 73–106, edited by John Kirlin and Donald Winkler. Los Angeles: University of Southern California.

Mitchell, Don. 2003. *The Right to the City: Social Justice and the Fight for Public Space.* New York: Guilford Press.

Molotch, Harvey. 1970. "Oil in Santa Barbara and Power in America." *Sociological Inquiry* 40:131–44.

———. 1996. "LA as Product: How Design Works in a Regional Economy." In *The City: Los Angeles and Urban Theory at the End of the Twentieth Century,* 125–75, edited by Allen Scott and Edward Soja. Los Angeles: University of California Press.

Molotch, Harvey, William Freudenberg, and Krista Paulsen. 2000. "History Repeats Itself, But How? City Character, Urban Tradition, and the Accomplishment of Place." *American Sociological Review* 65:791–823.

Moody, Kim. 1989. *An Injury to All: The Decline of American Unionism*. London: Verso Press.

Moore, Donald, Jake Kosek, and Anand Padian, eds. 2003. *Race, Nature and the Politics of Difference*. Durham, N.C.: Duke University Press.

Moore, Monica. 2005. Interview in Berkeley with author. July 26. Director (1982–present), Pesticide Action Network North America.

Moore, Vicky. 2001. Telephone interview with author. September 24. South Bay field representative, Greenbelt Alliance, San Jose.

Motto, Kryla, and Fisher. 2000. *The Economic Impact of California Wine*. St. Helena: Motto, Kryla, and Fisher.

Moulton, Kirby. 1984. "The Economics of Wine in California." In *The Book of California Wine*, 380–405, edited by D. Muscatine, M. Amerine, and B. Thompson. Berkeley: University of California Press.

Mozingo, Louise. 2005. "Community Participation and Creek Restoration in the East Bay." In *(Re)Constructing Communities: Design Participation in the Face of Change*, 249–51, edited by Jeff Hou, Mark Francis, and Nathan Brightbill. Davis, Calif.: Center for Design Research.

———. N.d. "A Century of Neighborhood Parks." Manuscript, Department of Landscape Architecture, University of California, Berkeley.

Muir, John. 1911. *My First Summer in the Sierra*. Boston and New York: Houghton Mifflin.

Muller, Brian. 2001. "Local Growth Strategy and the Transformation of Exurbia: The Central Valley of California, 1960–1996." PhD diss., Department of City and Regional Planning, University of California, Berkeley.

Nash, Roderick. 1967. *Wilderness and the American Mind*. New Haven, Conn.: Yale University Press.

Nathan, Harriet, and Stanley Scott, eds. 1969. *Toward a Bay Area Regional Organization*. Berkeley: University of California, Berkeley, Institute of Governmental Studies.

Neasham, Joseph. *See* U.S. Department of the Interior.

Nichols, Fred, James Cloern, Samuel Luoma, and David Peterson. 1986. "The Modification of an Estuary." *Science* 231, no. 4738:567–73.

Nichols, Sandra. 2002. "Saints, Peaches, and Wine: Mexican Migrants and the Transformation of Los Haro, Zacatecas, and Napa, California." PhD diss., Department of Geography, University of California, Berkeley.

Nisbet, Briggs. 2005. Telephone interview with author. July 15. Restoration campaigns manager (2001–2005), Save San Francisco Bay Association.

Noble, John Wesley, and Gayle Montgomery. 1999. *Its Name Was M.U.D.* Oakland: East Bay Municipal Utility District.

O'Connor, James. 1998. *Natural Causes: Essays in Ecological Marxism*. New York: Guilford Press.

Ogden, Kate. 1990. "Sublime Vistas and Scenic Backdrops: Nineteenth-century Painters and Photographers at Yosemite." *California History* 69, no. 2:134–53.

Olmsted, Frederick Law Jr., and John Nolen. 1906. "The Normal Requirements of American Towns and Cities in Respect to Public Open Spaces." *Charities and the Commons* 16, no. 14 (July 7): 411–26.

Olmsted Brothers and Ansel Hall. 1930. *Report on Proposed Park Reservations for East Bay Cities.* Repr., 1984. Berkeley: University of California, Bureau of Public Administration.

Olsen, Laurie. 1988. *Crossing the Schoolhouse Border: Immigrant Students and the California Public Schools: A California Tomorrow Policy Research Report.* San Francisco: California Tomorrow.

———. 1999. *Turning the Tides of Exclusion: A Guide for Educators and Advocates for Immigrant Students.* Oakland: California Tomorrow.

Olson, Keith. 1979. *Biography of a Progressive: Franklin K. Lane, 1864–1921.* Westport, Conn.: Greenwood Press.

Olwig, Kenneth. 2002. *Landscape, Nature, and the Body Politic: From Britain's Renaissance to America's New World.* Madison: University of Wisconsin Press.

Orman, Larry. 2001. Telephone interview with author. September 18. Executive director (1976–95), Greenbelt Alliance; founder and executive director (1995–), GreenInfo Network, San Francisco.

———. 2005. Personal communication with author, November 7.

Ottaway, David, and Joe Stephens. 2004. "Nonprofit Land Bank Amasses Billions." *Washington Post,* May 4: A1. First of three article series on The Nature Conservancy, May 4–6.

Outdoor Recreation Resources Review Commission. 1962. *Outdoor Recreation for America.* Washington, D.C.: Government Printing Office.

Overview Corporation. 1973. *How to Implement Open Space Plans for the San Francisco Bay Area,* by Frank Broadhead and Roselyn Rosenfeld. 3 vols. Berkeley: Association of Bay Area Governments.

Page, Brian, and Richard Walker. 1991. "From Settlement to Fordism: The Agro-Industrial Revolution in the American Midwest." *Economic Geography* 67, no. 4:281–315.

———. 1994. "Nature's Metropolis: The Ghost Dance of Christaller and Von Thünen." *Antipode* 26, no. 2:152–62.

Palmer, Tim. 1982. *Stanislaus: The Struggle for a River.* Berkeley: University of California Press.

Patton, Cynthia. 2002. *Turning Salt into Environmental Gold: Wetland Restoration in the South San Francisco Bay Salt Ponds.* Oakland: Save the Bay Association.

Patton, Phil. 1986. *Open Road: A Celebration of the American Highway.* New York: Simon and Schuster.

Paul, Rodman. 1973. "The Beginnings of Agriculture in California: Innovation Versus Continuity." *California Historical Quarterly* 52, no. 1:16–27.

Payne, Stephen. 1987. *Santa Clara County: Harvest of Change.* Northridge: Windsor, in cooperation with County of Santa Clara Historical Heritage Commission.

People for Open Space. 1969a. *The Case for Open Space in the San Francisco Bay Area*. San Francisco: People for Open Space.

———. 1969b. *Economic Impact of a Regional Open Space Program for the San Francisco Bay Area*, by Development Research Associates, associated consultants Livingston and Blayney. Los Angeles: Development Research Associates.

———. 1980a. *Endangered Harvest: The Future of Bay Area Farmland*. Report of the Farmlands Conservation Project. San Francisco: People for Open Space.

———. 1980b. *A Greenbelt for the Bay Area: Open Space for People*. San Francisco: People for Open Space.

———. 1981. *A Search for Permanence: Farmland Conservation in Marin County, California*. San Francisco: People for Open Space.

———. 1983. *Room Enough: Housing and Open Space in the Bay Area*. San Francisco: People for Open Space.

———. 1988. *The Greenbelt's Public Lands: Public Lands Database for the San Francisco Bay Area*. San Francisco: People for Open Space, Greenbelt Congress.

Phillips, Dawn. 2006. Telephone interview with author, April 3. Executive Director, PUEBLO (1999–2004); organizer, Just Cause (2004–), Oakland.

Pile, Stephen. 1998. *Geographics of Resistance*. London: Routledge.

Pincetl, Stephanie. 1992. "The Politics of Growth Control: Struggles in Pasadena, California." *Urban Geography* 13, no. 5:450–67.

———. 1999. *Transforming California: A Political History of Land Use and Development*. Baltimore: Johns Hopkins University Press.

———. 2001. "The Preservation of Nature at the Urban Fringe." Working paper. Los Angeles: University of Southern California, Department of Geography.

Pinney, Thomas. 2005. *A History of Wine in America: From Prohibition to the Present*. Berkeley: University of California Press.

Pisani, Donald. 1985. "Forests and Conservation, 1865–1890." *Journal of American History* 72, no. 2:340–59.

Polanyi, Karl. 1944. *The Great Transformation*. New York: Farrar and Rinehart.

Pomeroy, Earl. 1957. *In Search of the Golden West: The Tourist in Western America*. New York: Alfred Knopf.

———. 1965. *The Pacific Slope*. New York: Alfred Knopf.

Pomeroy, Hugh. 1935. *A County Planning Program for Marin County, California*. San Rafael: Marin Planning Survey Committee, for Marin County Planning Commission.

Pope, Carl. 2004. "The Virus of Hate: The Sierra Club and the Immigration Debate." *Sierra* 89 (May–June): 14.

Pope, Carl, and Paul Rauber. 2004. *Strategic Ignorance: Why the Bush Administration Is Recklessly Destroying a Century of Environmental Progress*. San Francisco: Sierra Club Books.

Porter, Douglas. 1996. *Profiles in Growth Management*. Washington, D.C.: The Urban Institute.

Porter, Eliot. 1963. *The Place No One Knew: Glen Canyon on the Colorado.* Edited by David Brower. San Francisco: Sierra Club.

Poulantzas, Nicos. 1973. *Political Power and Social Classes.* London: New Left Books.

Pred, Allan. 1986. *Place, Practice, and Structure.* Cambridge, UK: Polity Press.

Press, Daniel. 2002. *Saving Open Space: The Politics of Local Preservation in California.* Berkeley: University of California Press.

Price, Jennifer. 1999. *Flight Maps: Adventures with Nature in Modern America.* New York: Basic Books.

Price, Patricia. 2004. *Dry Place: Landscapes of Belonging and Exclusion.* Minneapolis: University of Minnesota Press.

Prudham, Scott. 2005. *Knock on Wood: Nature as Commodity in Douglas-Fir Country.* New York: Routledge.

Przeworski, Adam. 1985. *Capitalism and Social Democracy.* New York: Cambridge University Press.

Pulido, Laura. 1996a. "A Critical Review of the Methodology of Environmental Racism Research." *Antipode* 28, no. 2:142–59.

———. 1996b. *Environmentalism and Economic Justice: Two Chicano Struggles in the Southwest.* Tucson: University of Arizona Press.

———. 2000. "Rethinking Environmental Racism: White Privilege and Urban Development in Southern California." *Annals of the Association of American Geographers* 90, no. 1:12–40.

Putnam, Robert. 2000. *Bowling Alone: The Collapse and Revival of American Community.* New York: Simon and Schuster.

Rechstaffen, Cliff. 1997. "The Lead Poisoning Challenge: An Approach for California and Other States." *Harvard Law Review* 21:387.

Reding, Erin. 2004. "Environmental Justice in Contra Costa County." Senior honors thesis, Department of Geography, University of California, Berkeley.

Register, Richard. 1987. *Ecocity Berkeley: Building Cities for a Healthy Future.* Berkeley: North Atlantic Books.

———. 2002. *Ecocities: Building Cities in Balance with Nature.* Berkeley: Berkeley Hills Books.

———. 2005. Interview at Cordonices Creek restoration project with author. July 17. Director, Eco-City Builders, Berkeley.

Reisner, Marc. 1986. *Cadillac Desert: The American West and Its Disappearing Water.* New York: Viking.

Reynolds, Helen Baker. 1979. Interview in San Francisco by Carla Ehat and Anne Kent. February 28. California Room, Marin Public Library, San Rafael.

Rhee, Nari, and Dan Acland. 2005. *The Limits of Prosperity: Growth, Inequality, and Poverty in the North Bay.* Berkeley and Santa Rosa: Center for Labor Research and Education, New Economy–Working Solutions.

Rhinehart, Chuck. 2003. Interview near Santa Rosa with author and Marty Bennett.

June 9. Chair (ca. 1980–1995), Californians Organized to Acquire Access to State Tidelands (COAAST); former professor of biology, Santa Rosa Junior College.

Rhomberg, Chris. 2004. *No There There: Race, Class, and Community in Oakland.* Berkeley: University of California Press.

Riley, Ann. 1989. "Overcoming Federal Water Policies: The Wildcat–San Pablo Creeks Case." *Environment* 31, no. 10:12–31.

———. 1998. *Restoring Streams in Cities: A Guide for Planners, Policymakers, and Citizens.* Washington, D.C.: Island Press.

Riley, Glenda. 1999. *Women and Nature: Saving the Wild West.* Lincoln: University of Nebraska Press.

Roberts, Lennie. 2004. "Protecting San Mateo County Open Space." Interviews by Richard Walker, August 23, 2001. Tape recording and transcript. Regional Oral History Office, Bancroft Library, University of California, Berkeley.

Rogers, Heather. 2005. *Gone Tomorrow: The Hidden Life of Garbage.* New York: The New Press.

Rogin, Michael. 1987. *"Ronald Reagan," the Movie, and Other Episodes in Political Demonology.* Berkeley: University of California Press.

Rome, Adam. 2001. *The Bulldozer in the Countryside: Suburban Sprawl and the Rise of the American Environmental Movement.* New York: Cambridge University Press.

Rosen, Martin. 2000. "Trust for Public Land Founding Member and President, 1972–1997: The Ethics and Practice of Land Conservation." Interview by Carl Wilmsen. Tape recording and transcript. Regional Oral History Office, Bancroft Library, University of California, Berkeley.

Rosenzweig, Roy, and Elizabeth Blackmar. 1992. *The Park and the People.* Ithaca, N.Y.: Cornell University Press.

Rothman, Hal. 1998a. *Devil's Bargains: Tourism in the Twentieth-Century American West.* Lawrence: University Press of Kansas.

———. 1998b. *The Greening of a Nation? Environmentalism in the United States Since 1945.* Fort Worth, Texas: Harcourt Brace College Publishers.

———. 2004. *The New Urban Park: Golden Gate National Recreation Area and Civic Environmentalism.* Lawrence: University of Kansas Press.

Rowell, Galen, with Michael Sowell. 1997. *Bay Area Wild: A Celebration of the Natural Heritage of the San Francisco Bay Area.* San Francisco: Sierra Club Books, Mountain Light Press.

Rudee, Helen. 2004. Interview in Santa Rosa with author and Marty Bennett. July 16. Board of Supervisors (1976–88), Sonoma County.

Rusmore, Jean. 1995. *The Bay Area Ridge Trail.* Berkeley: Wilderness Press.

Ryan, Mary. 1997. *Civic Wars: Democracy and Public Life in the American City During the Nineteenth Century.* Berkeley: University of California Press.

Sabir, Wanda. 2006. "Growing a Greenway in Hunters Point." *Bay Nature* (April–June): 8–11.

Sagalyn, Lynne, and George Sternlieb. 1972. *Zoning and Housing Costs: The Impact of Land Use Controls on Housing Price*. New Brunswick, N.J.: Rutgers University, Center for Urban Policy Research.

San Francisco Bay Conservation and Development Commission. 1979. *San Francisco Bay Plan*. San Francisco: San Francisco Bay Conservation and Development Commission.

Santa Clara County Housing Task Force. 1977. *Housing: A Call for Action*. San Jose: Santa Clara County Planning Department.

Santa Clara County Industry and Housing Management Task Force. 1979. *Living Within Our Limits*. San Jose: Santa Clara County Planning Department.

Santiago, Chiori. 2004. "A Modest Majesty: Seventy Years of the East Bay Parks." *Bay Nature*, October–December, 12–16.

———. 2005. "Betting on Point Molate." *Bay Nature*, July–September, 38–42.

Save the Bay. 2000. *Protecting Wetlands: A Toolbox for Your Community*. Oakland: Save the Bay Association.

Saxenian, Annalee. 1985. "The Genesis of Silicon Valley." In *Silicon Landscapes*, 20–48, edited by Peter Hall and Ann Markusen. Boston: Allen and Unwin.

———. 1994. *Regional Advantage: Silicon Valley and Route 128 in Comparative Perspective*. Cambridge, Mass.: Harvard University Press.

Sayer, Andrew. 2005. *The Moral Significance of Class*. Cambridge, UK: Cambridge University Press.

Scarborough, Katie, and Scot Stegeman. 1989. *Farmland Worth Saving: The Present Value and Preservation of Sonoma County Agriculture*. Santa Rosa: Sonoma County Farmlands Group.

Scheinberg, Anne. 2003. "The Proof of the Pudding: Urban Recycling in North America as a Process of Ecological Modernization." *Environmental Politics* 12, no. 4:49–75.

Schenker, Heath. 1996. "Women's and Children's Quarters in Golden Gate Park, San Francisco." *Gender, Place, and Culture* 3, no. 1:293–308.

Schmitt, Peter. 1969. *Back to Nature: The Arcadian Myth in Urban America*. New York: Oxford University Press.

Schneider, Jimmy. 1992. *The Complete History of Santa Clara County's New Almaden Mine*. San Jose: Zella Schneider.

Schoennauer, Gary. 2001. Telephone interview with author. September 27. Planning director (1980–97), City of San Jose.

Schoradt, Brent. 2005. Telephone interview with author. July 11. Field representative, Solano and Napa counties, Greenbelt Alliance, Fairfield.

Schrag, Peter. 1998. *Paradise Lost: California's Experience, America's Future*. New York: The New Press.

Schrepfer, Susan. 1983. *The Fight to Save the Redwoods: A History of Environmental Reform, 1917–1978*. Madison: University of Wisconsin Press.

———. 2005. *Nature's Altars: Mountains, Gender and American Environmentalism.* Lawrence: University of Kansas Press.

Schuyler, David. 1986. *The New Urban Landscape: The Redefinition of City Form in Nineteenth-Century America.* Baltimore: Johns Hopkins University Press.

Schwab, Jim. 1994. *Deeper Shades of Green: The Rise of Blue-Collar and Minority Environmentalism in America.* San Francisco: Sierra Club Books.

Scott, Anne. 1991. *Natural Allies: Women's Associations in American History.* Urbana: University of Illinois Press.

Scott, Mel. 1959. *The San Francisco Bay Area: A Metropolis in Perspective.* 2nd ed., 1985. Berkeley: University of California Press.

———. 1963. *The Future of San Francisco Bay.* Berkeley: University of California, Berkeley, Institute of Governmental Studies.

———. 1969. *American City Planning Since 1890.* Berkeley: University of California Press.

Scott, Stanley. 1960. *A Golden Gate Authority for the San Francisco Bay Area: The Problem of Representation.* Berkeley: University of California, Berkeley, Bureau of Public Administration.

Scott, Stanley, ed. 1975. *Governing California's Coast.* Berkeley: University of California, Berkeley, Institute of Governmental Studies.

Scott, Stanley, and John Bollens. 1949. *Special Districts in the Government of California.* Berkeley: University of California, Berkeley, Bureau of Public Administration.

———. 1968. *Governing a Metropolitan Region: The San Francisco Bay Area.* Berkeley: University of California, Berkeley, Institute of Governmental Studies.

Scott, Stan, and John Corzine. 1963. *Special Districts in the Bay Area.* Berkeley: University of California, Berkeley, Institute of Governmental Studies.

Sears, John. [1989] 1998. *Sacred Places: American Tourist Attractions in the Nineteenth Century.* Amherst: University of Massachusetts Press.

Segura, Marta. 2004. Telephone interview by Juan DeLara (author's research assistant). August 6. Southern California director, Communities for a Better Environment, Los Angeles.

Sellers, Richard. 1997. *Preserving Nature in the National Parks.* New Haven, Conn.: Yale University Press.

Shafer, Kristin, Margaret Reeves, Skip Spitzer, and Susan Kegley. 2004. *Chemical Trespass: Pesticides in Our Bodies and Corporate Responsibility.* San Francisco: Pesticide Action Network.

Shaffer, Marguerite. 2001. *See America First: Tourism and National Identity, 1880–1940.* Washington, D.C.: Smithsonian Institution

Shaiko, Ronald. 1999. *Voices and Echoes for the Environment: Public Interest Representation in the 1990s and Beyond.* New York: Columbia University Press.

Shamasunder, Bhavna. 2004. Telephone interview with author. August 20. Environmental justice advocate, Urban Habitat, Oakland.

Shankland, Robert. 1954. *Steve Mather of the National Parks*. New York: Alfred Knopf.

Sharkey, David. 1980. "The Conservation and Environmental Movement in San Mateo County." *La Peninsula* 20, no. 3:21–31.

Shellenberger, Michael, and Ted Norhaus. 2004. *The Death of Environmentalism: Global Warming Politics in a Post-Environmental World*. Berkeley: The Breakthrough Institute.

Sheshunoff and Co. 1985–97. *Banks of California*. Austin, Texas: Sheshunoff and Co.

Siegel, Lenny, and John Markoff. 1985. *The High Cost of High Tech*. New York: Harper and Row.

Sierra Club. 1999. *Solving Sprawl*. San Francisco: Sierra Club Books.

Sierra Club History Committee. 1983. "The Sierra Club and the Urban Environment II: Labor and the Environment in the San Francisco Bay Area, 1960s–1970s." Interviews with David Jenkins, Amy Meyer, Anthony Ramos, and Dwight Steele. Tape recording and transcript. Regional Oral History Office, Bancroft Library, University of California, Berkeley.

Siri, Jean. 2005. Interview in El Cerrito with author. December 21. Environmental advocate and gadfly; city council (1980–85, 1987–91) and mayor (1982–83, 1988–89), El Cerrito; California Solid Waste Board (c. 1970s); Stege Sanitary District Board (1975–79); East Bay Regional Parks Board (1993–2006).

Slack, Gordy. 2004. "In the Shadow of Giants." *Bay Nature*, July–September, 11–15.

Sloan, Doris. 2006. *Geology of the San Francisco Bay Area*. Photographs by John Karachewski. Berkeley: University of California Press.

Smith, Gar. 2005. "Toxic Tour: Driving Through One of the West Coast's Deadliest Neighborhoods." *Earth Island Journal* 20, no. 3 (autumn): 29–33.

Smith, Henry Nash. 1950. *Virgin Land: The American West as Symbol and Myth*. Cambridge, Mass.: Harvard University Press.

Smith, Neil. 1984. *Uneven Development*. Oxford: Basil Blackwell.

Smith, Ted. 2003. "Pioneer Activist for Environmental Justice in Silicon Valley, 1967–2000." Interview by Carl Wilmsen, 2000. Tape recording and transcript. Regional Oral History Office, Bancroft Library, University of California, Berkeley.

———. 2004. Telephone interview with author. June 11. Founder and executive director (1982–2005), Silicon Valley Toxics Coalition, San Jose.

Soja, Edward. 1991. "Inside Exopolis: Scenes from Orange County." In *Variations on a Theme Park: The New American City and the End of Public Space*, 94–122, edited by Michael Sorkin. New York: Hill and Wang, Noonday Press.

Sokolow, Alvin. 1990. *The Williamson Act : Twenty-five Years of Land Conservation*. Sacramento : California Department of Conservation.

Solnit, Rebecca. 1994. *Savage Dreams: A Journey into the Hidden Wars of the American West*. San Francisco: Sierra Club Books.

———. 2000. *Wanderlust: A History of Walking*. New York: Viking.

—————. 2001. *Hollow City: Gentrification and the Eviction of Urban Culture.* London: Verso Press.

—————. 2003. *River of Shadows: Edweard Muybridge and the Technological Wild West.* San Francisco: City Lights Books.

Solomon, Daniel. 1992. *Re Build.* New York: Princeton Architectural Press.

Spangenberg, Ruth. 2004. "The Founding of the Committee for Green Foothills." Interview by Richard Walker. August 16, 2001. Tape recording and transcript. Regional Oral History Office, Bancroft Library, University of California, Berkeley.

Spencer-Wood, Suzanne. 1994. "Turn-of-the-Century Women's Organizations, Urban Design, and the Origin of the American Playground Movement." *Landscape Journal* 13, no. 2:125–37.

Spitz, Barry. 1997. *Mill Valley: The Early Years.* Mill Valley, Calif.: Potrero Meadow.

Stanbery, Van Beuren. 1954. *Regional Planning Needs of the San Francisco Bay Area.* San Francisco: Bay Area Council.

Stanford Environmental Law Society. 1971. *San Jose: Sprawling City.* Stanford: Stanford Environmental Law Society.

Stanger, Frank. 1967. *Sawmills in the Redwoods: Logging on the San Francisco Peninsula, 1849–1967.* San Mateo: San Mateo County Historical Association.

Starr, Kevin. 1973. *Americans and the California Dream, 1850–1915.* New York: Oxford University Press.

St. Clair, David. 1994–95. "New Almaden and California Quicksilver in the Pacific Rim Economy." *California History* 73, no. 4 (winter): 278–96.

Stein, Mimi. 1984. *A Vision Achieved: Fifty Years of the East Bay Regional Parks District.* Oakland: East Bay Regional Parks District.

Steinberg, Goodwin. 2002. *From the Ground Up: Building Silicon Valley.* Stanford: Stanford University Press.

Steingraber, Sandra. 1997. *Living Downstream: An Ecologist Looks at Cancer and the Environment.* Reading, Mass.: Addison-Wesley.

Stone, Clarence. 1989. *Regime Politics: Governing Atlanta, 1946–1988.* Lawrence: University of Kansas Press.

Storper, Michael, and Richard Walker. 1982. "The Expanding California Water System." In *Use and Protection of the San Francisco Bay System*, 171–90, edited by William Kockelman, Tom Conomos, and Alan Leviton. San Francisco: Pacific Division, American Association for the Advancement of Science.

Strasser, Susan. 1999. *Waste and Want: A Social History of Trash.* New York: Metropolitan Books .

Strickland, Eliza. 2005. "The New Face of Environmentalism." *East Bay Express*, November 2–8, 15–26.

Strong, Douglas. [1970] 1988. *Dreamers and Defenders: American Conservationists.* Lincoln: University of Nebraska Press.

Sturgeon, Timothy. 2000. "How Silicon Valley Came to Be." In *Understanding Silicon Valley: The Anatomy of an Entrepreneurial Region*, 15–47, edited by Martin Kenney. Stanford: Stanford University Press.

Sutter, Paul. 2002. *Driven Wild: How the Fight Against Automobiles Launched the Modern Wilderness Movement*. Seattle: University of Washington Press.

Svanevik, Michael, and Shirley Burgett. 2001. *San Mateo County Parks: A Remarkable Story of Extraordinary Places and the People Who Built Them*. Menlo Park, Calif.: San Mateo County Parks and Recreation Foundation.

Swain, Donald. 1970. *Wilderness Defender: Horace M. Albright and Conservation*. Chicago: University of Chicago Press.

Szasz, Andrew. 1994. *Ecopopulism: Toxic Waste and the Movement for Environmental Justice*. Minneapolis: University of Minnesota Press.

Teaford, Jon. 1984. *The Unheralded Triumph: City Government in America, 1870–1900*. Baltimore: Johns Hopkins University Press.

Thiessen, Janet. 2006. "Oh, That Vigor: A Life of Dorothy Erskine." Manuscript.

Thompson, A. C. 2002. "The Judi Bari Bombshell." *San Francisco Bay Guardian*, May 29, 21–26.

Thompson, E. P. 1964. *The Making of the English Working Class*. London: Penguin.

Thompson, John. 1957. "The Settlement Geography of the Sacramento–San Joaquin Delta, California." PhD diss., Stanford University.

Thorwaldson, Jay. 2001. "Gracious Activist: Lois Hogle's Fight for the Foothills." *Palo Alto Weekly* 94 (August 22): 3, 13.

Tilden, Freeman. 1962. *The State Parks: Their Meaning in American Life*. New York: Alfred Knopf.

Toogood, Anna. *See* U.S. Department of the Interior.

Treib, Marc, and Dorothée Imbert. 1997. *Garrett Eckbo: Modern Landscapes for Living*. Berkeley: University of California Press.

Treuttner, William. 1972. *National Parks and the American Landscape*. Washington, D.C.: Smithsonian Institution Press.

Trounstine, Philip, and Terry Christensen. 1982. *Movers and Shakers: The Study of Community Power*. New York: St. Martin's Press.

Trudeau, Richard. 2003. "From Skyline to Seashore: Twenty-Two Years of Leadership, Land Acquisition, and Lobbying at the East Bay Regional Parks District, 1964–1986." Interview by Ann Lage. Tape recording and transcript. Regional Oral History Office, Bancroft Library, University of California, Berkeley.

Tuan, Yi-Fu. 1977. *Space and Place: The Perspective of Experience*. Minneapolis: University of Minnesota Press.

Udall, Stuart. 1963. *The Quiet Crisis*. New York: Holt, Rinehart, and Winston.

U.S. Council on Environmental Quality. 1974. *The Costs of Sprawl*, by Real Estate Research Corporation. Washington, D.C.: U.S. Government Printing Office.

U.S. Department of Commerce. 1960. *Future Development of the San Francisco Bay*

Area, 1960–2020. Prepared for U.S. Army Corps of Engineers. Washington, D.C.: U.S. Government Printing Office.

U.S. Department of the Interior. National Park Service. 1980. *A Civil History of the Golden Gate National Recreation Area and Point Reyes National Seashore, California*, by Anna Toogood. Washington, D.C.: National Park Service.

———. 1990. *The Top of the Peninsula: A History of Sweeney Ridge and the San Francisco Watershed Lands, San Mateo County, California*, by Marianne Babal. San Francisco: National Park Service, Golden Gate National Recreation Area.

———. 1995. *A Good Life: Dairy Farming in the Olema Valley*, by Dewey Livingston. San Francisco: National Park Service.

———. Works Progress Administration. 1937. *History of the California State Parks*, survey of pamphlets by many authors, edited by Joseph Neasham. Sacramento: California Resources Agency, Division of Parks.

Vance, James. 1964. *Geography and Urban Evolution in the San Francisco Bay Area*. Berkeley: University of California, Berkeley, Institute of Governmental Studies.

———. 1970. *The Merchant's World*. Englewood Cliffs, N.J.: Prentice-Hall.

Van den Bosch, Robert. 1978. *The Pesticide Conspiracy*. Garden City, N.J.: Doubleday.

Varian, Dorothy. 1983. *The Inventor and the Pilot: Russell and Sigurd Varian*. Palo Alto: Pacific Books.

Violich, Francis. 1978. "The Planning Pioneers." *California Living,* magazine of the *Sunday Chronicle and Examiner*, February 26, 29–35.

Walker, Richard. 1978. "Erosion of the Clean Air Act of 1970: A Study in the Failure of Government Regulation and Planning." *Boston College Environmental Affairs Law Review* 7, no. 2:189–258.

———. 1981. "A Theory of Suburbanization: Capitalism and the Construction of Urban Space in the United States." In *Urbanization and Urban Planning in Capitalist Societies*, 383–430, edited by Michael Dear and Allen Scott. New York: Methuen.

———. 1995a. "California Rages Against the Dying of the Light." *New Left Review* 209:42–74.

———. 1995b. "Landscape and City Life: Four Ecologies of Residence in the San Francisco Bay Area." *Ecumene* 2, no. 1:33–64.

———. 1996. "Another Round of Globalization in San Francisco." *Urban Geography* 17, no. 1:60–94.

———. 2001. "California's Golden Road to Riches: Natural Resources and Regional Capitalism, 1848–1940." *Annals of the American Association of Geographers* 91, no. 1:167–99.

———. 2004. *The Conquest of Bread: 150 Years of Agribusiness in California*. New York: The New Press.

———. 2005. "A Hidden Geography." In *The Golden Gate*, 145–58, photographs by Richard Misrach. New York: Aperture Press.

———. 2006. "The Boom and the Bombshell: The New Economy Bubble and the

San Francisco Bay Area." In *The Changing Economic Geography of Globalization*, 121–47, edited by Giovanna Vertova. London: Routledge.

Walker, Richard, and Bay Area Study Group. 1990. "The Playground of U.S. Capitalism? The Political Economy of the San Francisco Bay Area in the 1980s." In *Fire in the Hearth: The Radical Politics of Place in America*, 3–82, edited by M. Davis, S. Hiatt, M. Kennedy, S. Ruddick, and M. Sprinker. London: Verso Press, Haymarket.

Walker, Richard, Michael Storper, and Ellen Widess. 1979. "The Limits of Environmental Control: The Saga of Dow in the Delta." *Antipode* 11, no. 2:1–16.

Walker, Richard, and Matthew Williams. 1982. "Water from Power: Water Supply and Regional Growth in the Santa Clara Valley." *Economic Geography* 58, no. 2:95–119.

Walker, Robert. 1941. *The Planning Function in Urban Government.* 2nd ed., 1950. Chicago: University of Chicago Press.

Wallace, David Rains. 2006. "Speak of the Devil: The Unexpected Landscapes of Mount Diablo." *Bay Nature* (July–September): 17–32.

Warf, Barney, and Joseph Cox. 1995. "U.S. Bank Failures and Regional Economic Structure." *Professional Geographer* 47, no. 1:3–16.

Warner, Kee, and Harvey Molotch. 1992. *Growth Control: Inner Workings and External Effects.* Berkeley: University of California, Berkeley, California Policy Seminar.

Washington Post. 1983. "San Jose's Choice: Tax Retrenchment Hurts Needy Most." September 18, A1.

Watt, Laura. 2001. "Managing Cultural Landscapes: Reconciling Local Ideology and Institutional Ideology in the National Park Service." PhD diss., Department of Environmental Science, Policy, and Management, University of California, Berkeley.

Wayburn, Edgar. 1985. "Sierra Club Statesman, Leader of the Parks and Wilderness Movement, etc." Interviews by Ann Lage and Susan Schrepfer. Tape recording and transcript. Regional Oral History Office, Bancroft Library, University of California, Berkeley.

Wayburn, Peggy. 1987. *Adventuring in the San Francisco Bay Area.* San Francisco: Sierra Club Books.

Weir, David, and Marc Shapiro. 1979. *The Circle of Poison.* San Francisco: Center for Investigative Reporting, Food First Institute.

Weiss, Marc. 1987. *The Rise of the Community Builders: The American Real Estate Industry and Urban Land Planning.* New York: Columbia University Press.

Wellman, Grace. 1982. Interviews by Carla Ehat and Genevieve Martinelli, February 19. California Room, Marin Public Library, San Rafael.

Wellock, Thomas. 1998. *Critical Masses: Opposition to Nuclear Power in California, 1958–1978.* Madison: University of Wisconsin Press.

Westman, Walt. 1985. *Ecology, Impact Assessment, and Environmental Planning.* New York : John Wiley and Sons.

White, Peter. 1995. *The Farallon Islands: Sentinels of the Golden Gate*. San Francisco: Scottwall Associates.

White, Richard. 1995. "Are You an Environmentalist, or Do You Work for a Living? Work and Nature." In *Uncommon Ground: Toward Reinventing Nature*, 171–85, edited by William Cronon. New York: W. W. Norton.

———. 2003. "Whole Earth." Paper presented at Berkeley Workshop on Environmental Politics, Institute of International Studies, University of California, Berkeley, February 21.

Williams, Matt. 2004. Telephone interview in Oakland with author. August 6. Board of directors (1994–2002), Alameda–Contra Costa Transit District; cofounder (1992–96), Regional Alliance for Transit; cofounder (1996), Transportation Choices Forum (now Transportation and Land-Use Coalition).

Williams, Raymond. 1973. *The Country and the City*. London: Chatto and Windus.

Wills, Jane, and Roger Lee, eds. 1997. *Geographies of Economies*. London: Edward Arnold.

Wise, Nancy. 1985. *Marin's Natural Assets: A Historic Look at Marin County*. San Rafael: Marin Conservation League.

Wolch, Jennifer, Steve Brachman, Jed Fehrenbach, and Jamin Johnson. 2001. "Parks and Park Funding in Los Angeles: An Equity Mapping Analysis." Working paper, Department of Geography, University of Southern California.

Wood, Samuel. 1983. "Administration, Research, and Analysis in Behalf of Environmental Quality." In *Statewide and Regional Land-Use Planning in California*, vol. 1, 105–236. Regional Oral History Office, Bancroft Library, University of California, Berkeley.

Wood, Samuel, and Alfred Heller. 1962. *California, Going, Going: Our State's Struggle to Remain Beautiful and Productive*. Sacramento: California Tomorrow.

———. 1963. *The Phantom Cities of California*. Sacramento: California Tomorrow.

Woodbury, John. 2004, 2005. Interview in Oakland with author. August 16, 2004; interview by telephone, December 16, 2005. Executive director (1993–2004), Bay Area Open Space Council. Oakland; staff director (2004–), Napa County Open Space Advisory Commission.

Workman, Bill. 2000. "Environmental Crusader for Mountain Carries On." *San Francisco Chronicle*, March 2, A19.

Wright, John. 1993. *Rocky Mountain Divide: Selling and Saving the West*. Austin: University of Texas Press.

Wurm, Ted, and Alvin Graves. 1983. *The Crookedest Railroad in the World: A History of the Mount Tamalpais and Muir Woods Railroad of California*. 3rd ed. Glendale, Calif.: Trans-Anglo Books.

Yaryan, Willie, Denzil Verardo, and Jennie Vararardo. 2000. *The Sempervirens Story: A Century of Preserving California's Ancient Redwood Forest, 1900–2000*. Los Altos: Sempervirens Fund.

Yee, Cameron. 1999. *Crash Course in Bay Area Transportation Investment.* San Francisco: Urban Habitat.

Yoshiyama, Ronald. 1999. "A History of Salmon and People in the Central Valley Region of California." *Reviews in Fisheries Science* 7, no. 3–4:197–239.

Young, Terence. 2004. *Building San Francisco's Parks, 1850–1930.* Baltimore: Johns Hopkins University Press.

———. "Heading Out: American Camping from 1869 to 1990." Manuscript.

Zaitzevsky, Cynthia. 1982. *Frederick Law Olmsted and the Boston Park System.* Cambridge, Mass: Harvard University Press.

Zakin, Susan. 1993. *Coyotes and Town Dogs: Earth First! and the Environmental Movement.* New York: Viking.

INDEX

Abascal, Ralph, 221
Abramowitz, Mark, 213
Adams, Ansel, 22, 100, 114, 125
Adams, Janet, 9, 104, 107, 115, 126
Adams, Seth, 50, 77, 147
Adirondack Forest Preserve, New York, 20, 28
agriculture, 3, 35, 37, 38–42; California Land Conservation Act (Williamson Act) (1965), 43, 189, 194; and canneries, 41; and capitalism, 37–39; decline of, 43–45; environmental effects of, 11, 28, 222; farmland loss, 43–44, 66, 86; farmland protection, 43–44, 189–91, 197, 199, 201; and food promotion, 41–42; and land trusts, 175–77; orchards, 38, 57; and processing, 40–41, 44, 45; Sonoma County Agricultural Marketing Program, 198; traces on landscape, 37, 39–40, 42; and transport networks, 40; and urbanization, 43–44, 146. *See also* land trusts; Napa; Sonoma; wine
Akono, Azibuke, 235
Alameda County, 112; agriculture, 38; growth controls, 148–49; mining, 54; parks, 72, 74, 75; protected land, 108; water supply, 47, 50. *See also* Berkeley; East Bay Municipal Utilities District; East Bay Regional Parks District; Oakland
Albany, 70, 113
Albright, Horace, 17, 65. *See also* National Park Service
Alcatraz, 94, 95, 97
Almaden, 41. *See also* New Almaden mines
Altamont Pass, 72, 167, 228
Angel, Bradley, 236, 241
Angel Island, 56, 84, 94, 97; saving of, 89; state park, 68, 90
Anthony, Carl, 232, 233, 237, 239, 243, 244, 245, 253
Antioch, 79, 124, 148
Appalachian Mountain Club, 20, 24
Arguimbau, Nick, 210, 211
Arrigoni, Peter, 92
Asian Pacific Islander Environmental Network (APEN), 231, 233, 246, 253

Association of Bay Area Governments (ABAG), 179, 190, 194, 247, 298n20; and growth controls, 148; origins of, 138; and regional shoreline trail, 157; and regional transportation, 140; and Smart Growth, 153

Audubon Society, 16, 20, 89, 92, 114, 120, 203, 208; Bay chapter, 217; Golden Gate chapter, 236; Marin chapter, 89, 91, 122; South Bay chapter, 218

Azevedo, Margaret, 91, 93

Bari, Judi, 14, 34, 232

Bauer, Catherine, 132

bay. *See* San Francisco Bay; *see also* Monterey Bay, Richardson Bay, San Pablo Bay, Suisun Bay

Bay Area, 3, 35, 42, 57, 108, 189; agriculture, 35, 37, 38–42; air pollution, 209–15, 227–28; antinuclear movement, 124–26, 293n34; and class, 6, 82, 163; as conservation hearth, 20, 22–23; and critiques of environmental regulation, 150–52; defined, 3; energy supply, 123–24; and environmental health, 226–27; and environmental justice, 13, 231, 232, 247, 252; and environmentalism, 17–18, 22, 34, 115, 130, 131, 132, 134, 209, 250; exceptionalism, 7; farmland loss, 43–44; greenbelt, 3, 26, 108, 138, 157–58, 180; and government reform, 136; growth of, 82, 131, 136, 146, 159, 160, 253; and growth controls, 130, 142–45, 147–50, 161; housing prices, 151–52, 162; as Left Coast, 17; and liberal environmentalism, 8–18, 134, 249–52; maps, iv, xxv; and militarization, 112–13; natural gifts, 7–8; open space, 3, 62, 83, 108, 180; parks, 59–79; and planning, 130–42; political culture, 13–15, 17, 130, 131, 132, 134, 159, 160, 168–69, 247, 249–50; population, 7, 19, 131, 159; refineries,

211–12; and second homes, 192, 194–95; and Smart Growth movement, 153–54; and toxics movement, 220–27, 228; urbanized area, 3, 36, 108, 253; wealth of, 10, 19, 38, 57, 162, 163, 181, 192; and wine, 187–88

Bay Area Air Quality Management District (BAAQMD), 210, 212, 214, 233, 235

Bay Area Council, 137, 141, 153, 247

Bay Area Open Space Council, 149, 153, 178–80, 191, 214

Bay Area Regional Water Quality Control Board, 144, 216

Bay Area Ridge Trail Council, 157

Bay Conservation and Development Commission (BCDC), 122, 126, 215; and Bay Shoreline Trail, 158; and Coastal Conservancy, 178; leaders of, 95, 127, 136, 141; and wetlands, 111, 115, 116, 119

Bay Institute, 52, 117, 217

Behr, Peter, 91, 92

Belliveau, Mike, 14, 211, 222, 223, 238, 241

Bennett, Marty, 204

Berkeley, 15, 17, 70, 154, 209, 253; and ABAG, 298n20; bay fill, 113; canneries, 45; creek restoration, 51–52; Hillside Club, 63; quarries, 55; parks, 63–64, 72, 79; and Save the Bay Association, 114; water supply, 48. *See also* East Bay Hills; University of California, Berkeley

Berkeley Creek Restoration Ordinance (1989), 52

Berryessa, Lake, 50, 190

Bicycle Coalitions, 215

Big Basin: state park, 28, 34, 58, 64, 68, 99, 100, 171; protection of, 29–30

Bodega Head, 124, 194–95

Bodovitz, Joe 115, 127, 141

Bolinas, 27

Bonwell, Art, 77

California State Parks Commission, 66, 99
California State Water Project, 50, 117
California State Water Resources Control Board, 118, 216, 218
California Tomorrow, 133, 138
Callenbach, Ernest, 133
Calthorpe, Peter, 153, 154, 203
camping, 24, 29. *See also* recreation
Carquinez Straits, 40, 45, 79, 148
Carson, Rachel, 9, 83, 132, 219
Center for Health, Environment, and Justice. *See* National Clearinghouse on Hazardous Waste
Central Park, New York, 20, 58, 59, 60
Central Valley, 43, 253; agriculture, 37, 38, 41, 44; water, 49–50, 117, 122. *See also* San Joaquin Valley
Central Valley Project, 117
Central Valley Project Improvement Act (1992), 118
Chain, David, 34
Chang, Vivian, 233
Chatfield, David, 221, 222
Cheasty, Ed, 70
Chevron Corporation, 211, 217, 227, 233
Chicago, 210; City Beautiful, 63; greenbelts, 72; playgrounds, 25, 61; pollution, 206, 207; and progressive urban reformers, 11, 25; urban parks, 59
city: and country, 3–6, 35, 37, 42, 46, 47, 57, 60, 83, 182, 193; development of, 35; green, utopia, 249; livable, 11; perceptions of, 6, 13, 240–41; pollution of, 205; reconciliation with country, 249–54. *See also* political culture
Citizens for a Better Environment (CBE): and air pollution, 17, 212–15, 227; and environmental justice, 233, 236, 237, 239, 241; origins of, 210–11; and water pollution, 17, 118, 216–17, 222, 223, 226, 227

Citizens for Regional Recreation and Parks, 135, 138
Citizens for West Oakland Revitalization, 234, 247
Clark, Henry, 223, 241, 242
class: in Bay Area, 6, 82, 163; and conservation, 20–32, 81, 82, 108, 109, 249; and consumption of nature, 11, 19; and environmentalism, 10–12, 16, 109, 152, 181, 229, 230, 242–43; elite opposition to development, 11; and environmental health, 222; and environmental organizations, 114, 208; and housing, 85, 86, 163; and large lot zoning, 152, 189; and open space protection, 31, 32, 81, 82, 98, 100, 108, 109, 167, 204; and parks, 61, 75, 76, 80; and perceptions of the environment, 240–41; and planning, 131–34; and pollution, 249; and race, 12, 16; and recreation, 80, 190; and suburbanization, 85, 86, 87; and toxics, 220, 240, 245; and wine consumption, 184–85, 186, 188; working class, 12, 29, 61, 63, 75, 80, 85, 95, 116, 157, 163, 190, 220, 229, 240–41. *See also* Bay Area: wealth of
Clinton, William Jefferson (Bill), 231
Coalition for Effluent Action Now (CLEAN), 218
coast, 27, 84, 94, 104, 105, 106, 107, 123, 126, 128, 129
coastal protection, 104–8, 126, 127, 195, 196, 207; California Coastal Initiative (1972), 195; Coastal Protection League, 107; Save Our Coast Initiative, 107. *See also* California Coastal Commission; propositions: 20
Coast Ranges, 7, 27, 37, 123, 157
Cohen, Andy, 50
Colby, William, 21, 66
Cole, Luke, 232
Collins, George, 90, 91, 100, 194

Contra Costa County, 148, 155; cost of, 151; counterattacks to, 150–52; and economy, 141; and environmental justice, 246; and environmentalism, 145; greenlines, 143, 144, 145, 146, 148, 148; and housing prices, 151–52; Knox-Nisbet Act (1963), 137; Livermore, 143, 149; Marin County, 93, 144, 148; and mass mobilization, 131, 134, 145, 158; and modernism, 132–35, 158; Napa County, 145, 148, 155, 189, 190; and New Deal ideology, 158; in Oregon, 148; and open space protection, 143; Petaluma, 143, 144, 145, 148, 194; Pleasanton, 149; recreation as, 72; San Jose, 144, 145, 147, 150, 161; Santa Clara County, 144–45, 149–51; and Save Mount Diablo, 146–57; Smart Growth, 153, 203, 204, 245, 246, 247; Solano County, 148; Sonoma County, 148, 155, 194, 196–97, 202; and social exclusion, 152; strategies of, 137; and water supply, 93, 144

growth, urban: community builders, 85; costs of, 103, 151; environmental damage, 112, 128, 215; and environmental justice, 245–46; and exurban development, 162; local government role in, 142–45; and the New Urbanism, 153; and suburbanization, 142. *See also* Bay Area; growth control; planning; suburbanization

Guadalupe River, 51, 52, 54

Gulick, Esther, 114, 135

habitat. *See* wildlife

Half Moon Bay, 28, 39, 106, 107, 166

Hall, Ansel, 73

Hall, William Hammond, 59, 60

Hamann, Dutch, 86

Hamilton, Mount, 39, 47, 71, 172

Hanko, Nonette, 103, 164

Haraszthy, Agoston, 41

Harrison, Marie, 235

Hart, Caryl, 201

Hawes, Amanda, 221

Hayes, Denis, 17

Hayward, 40, 55

Hedgpethe, Joel, 124

Hegemann, Werner, 63, 72

Heller, Alf, 133, 138, 173

Hewlett, Bill, 109

Heyman, Michael, 141

Hickel, Walter, 95

Hickey, Jim, 138, 190

highways, 137; Bayshore freeway, 113; Collier-Burns Act (1947), 84; Freeway and Expressway Plan (1959), 84; freeway revolts, 11; Highway 1, 84, 106–7, 194; Highway 101, 31, 35, 83, 84, 85, 86, 93, 126, 143, 164, 189, 193, 203; Highway 24, 35; and highway revolts, 82, 83, 84, 91, 93, 104, 106, 107, 108, 135; highway tax, 203; Interstate and Defense Highway Act (1956), 84; Interstate 80, 189; Interstate 280, 84, 103; Interstate 680, 146; Interstate 880, 213; and Marin, 83, 91, 93, 104; and San Francisco Bay, 113; and San Mateo, 84, 104, 106–7; Skyline Drive, 85, 104, 164; and suburbanization, 84, 85. *See also* California Department of Transportation, transportation

hiking, 29, 73, 79, 163, 167; Bay Area Ridge Trail, 157; Bay Shoreline Trail, 158; early Bay Area clubs, 24; National Trails Act, 157; and Sierra Club, 21, 24, 73; in watersheds, 49. *See also* recreation

Hill, Andrew, 29–30, 34

Hillsborough, 93, 98

Hinkle, Charles, 195, 196

Hogle, Lois Crozier, 101, 102

Holtzclaw, John, 214

Hopkins, Timothy, 28, 99

Hornbeck, Hulet, 76, 77

Lanza, Dana, 236
La Riviere, Florence and Phillip, 120
Larsen, Denny, 211, 223
LeConte, Joseph, 20
Lee, Charles, 231
Lee, Eugene, 133
Lee, Pam Tau, 9, 233, 237
LeFrance, Charles, 41
Leonard, Doris, 91, 100
Leonard, Richard, 22, 91, 125
Leopold, Aldo, 80, 241
Liberalism. *See* environmentalism;
   New Deal; planning; political
   culture
Linn, Karl, 232, 236
Litton, Martin, 16, 22, 33, 101, 102, 125
Livermore: city, 143, 149; valley, 37
Livermore, Caroline, 8, 14, 88–90, 91,
   92, 114
Livermore, Norman (Ike), 76, 88, 90
Livermore, Putnam, 173
Local Area Formation Commission
   (LAFCO), 137, 144, 148. *See also* gov-
   ernment; planning.
Look, Tony, 100
Los Altos Hills, 86, 87, 101, 102, 153
Los Angeles, 7–8, 42, 108, 110, 118, 126,
   179, 206, 217; air pollution, 138, 209,
   253; housing prices, 151, 163; planning
   failure, 89; and toxics, 211
Los Gatos, 56, 101

Mackensie, Andrea, 201
Mang, Bob, 141
Mar, Warren, 236
Marin Conservation League (MCL), 13,
   91, 92, 99, 114, 135, 176; and class, 204;
   and Golden Gate National Recreation
   Area, 97, and open space protection,
   88, 89; and wetlands, 114; and women,
   88, 109
Marin County, 23, 27, 94, 122, 134, 143,
   145, 193; and class, 109; early conser-

vationists, 23–26; general plans, 84,
   93; growth controls, 89, 93, 144, 148;
   highways, 83, 91, 93, 104; housing
   prices, 163; Marin Agricultural Land
   Trust (MALT), 87, 173, 176; Marin
   County Open Space District, 87, 92,
   122, 164, 176; Marin Municipal Water
   District, 25, 45, 47 49, 74, 87, 90, 93;
   open space, 82, 87–94, 98, 108–9;
   parks, 67, 71; planning, 84, 89, 93;
   population growth and stabilization,
   93; protected land, 87, 108; quarries,
   56; size, 108; suburbanization of, 83;
   water supply, 47–48, 197
marine sanctuaries: Channel Islands
   National Marine Sanctuary, 127;
   Cordell Bank National Marine Sanc-
   tuary, 128; Gulf of the Farallones
   National Marine Sanctuary, 127,
   128; Marine Protection, Research,
   and Sanctuaries Act (1972), 127, 207;
   Monterey Bay National Marine
   Sanctuary, 128
Marshall, Bob, 80, 134
marshes. *See* wetlands
Mather, Stephen, 17, 31, 32, 65–66, 75.
   See also National Park Service
Matsuoka, Marta, 233
May, Julia, 210, 211, 212
May, Samuel, 73, 132
Maybeck, Annie and Bernard, 63–64
Mazzocchi, Tony, 220
McAteer, Eugene, 115
McClaren, John, 60
McCloskey, Mike, 95
McCloskey, Pete, 102, 104
McCrackin, Josephine Clifford, 30
McDuffie, Duncan, 66, 74
McFarland, J. Horace, 65
McLaughlin, Sylvia, 16, 91, 114, 135
McPeak, Sunne, 247
Mendocino County, 32, 183, 185
Meral, Gerald, 118, 168

pollution, air, 206–15; Air Quality Act (1967), 206; in Bay Area, 209–15, 227–28; Bay Area Clean Air Coalition, 204; and class, 239; Clean Air Act (1963), 206; Clean Air Act (1970), 145, 207–8, 210, 213–14, 216, 253; Clean Air Coalition, 207; Motor Vehicle Pollution Control Act (1965), 206; diesel fumes, 235; and oil refineries, 211, 212, 217, 223, 225; and power plants, 212–13; public awareness of, 206; and race, 239; state regulation, 209–10, 215–16; and suburbanization, 83; and transit, 213–15. *See also* toxics

pollution, water, 215; Basin-Plan Coalition, 216; Clean Water Act (Federal Water Pollution Control Act Amendments) (1972), 145, 207–8, 216; Clean Water Act Amendments (1987), 218; Clean Water Action, 222, 226; Coalition for Effluent Action Now, 218; from electronics, 218; Marine Protection, Research, and Sanctuaries Act (Ocean Dumping Act), 207; McAteer-Petris Act (1965), 115; from mining, 53–54; Porter-Cologne Act (1969), 215–16; from refineries, 217; from San Francisco, 216, 218; in San Francisco Bay, 110, 112, 215, 218; from San Jose, 218; state regulation of, 209–10, 215–16; and suburbanization, 83; Water Pollution Control Act (1948), 206; Water Quality Act (1965), 206. *See also* toxics

Pomerance, Rafe, 209
Pomeroy, Hugh, 89
Pope, Carl, 213
Poras, Carlos, 238
Portola Valley, 27
postwar era: and agriculture, 39; development of farmland, 43–44; and parks, 75–79; and recreation, 80–81; and redwood cutting, 32–34; and

Sierra Club, 22–23; and urban growth, 82–83, 86, 113–14, 131
Potrero Point, 213
Potts, Jim, 108, 113
preservation, 21, 80. *See also* conservation; environmentalism; wilderness
Presidio, 94; and GGNRA, 97
progressives, 25, 26, 31, 32, 48, 65, 72; and city parks, 61; and greenbelt movement, 132–33; and planning, 136; women, 26, 29–30
propositions, state: Proposition 12, 168; Proposition 13, 43, 77, 145, 152, 153, 166, 167; Proposition 20, 127; Proposition 4, 167; Proposition 40, 169; Proposition 65, 209, 225, 227
public: access, 50, 80, 111, 126, 167, 190, 194, 195, 201, 202; commons, 21; good, 48, 115, 133; sphere, 13–14; space, 11, 61, 73, 82, 126, 160, 176
Putnam, Helen, 196

quarries. See *mining*

race: in cities, 240; and class, 12, 16; and environmentalism, 12, 70, 229–30, 240–41; and environmental justice, 229, 230, 232–33, 240– 41; and environmental movement, 16, 231, 241, 248; and open space, 167, 202; and parks, 168; and perceptions of environment, 240–41; and pollution, 239; and toxics, 223, 228–30, 240, 245
Ralston, Billy, 48
Ramo, Alan, 211
Ravenswood, 27
Reagan, Ronald, 76, 90, 167, 213, 218, 220, 252
recreation, outdoor, 21, 24, 29, 75; Bay Area Ridge Trail, 50, 157; California Public Outdoor Recreation Plan, 135; class, 80, 190; and coastal protection, 123; in conflict with preservation,

80, 167; and conservation organizations, 206, 209; as growth control, 72; Land and Water Conservation Fund, 68, 168; LandPaths, 202; and mass tourism, 80; in Napa County, 190; National Trails System Act, 157; and open space districts, 163; and open space protection, 61, 72, 81, 135; Outdoor Recreation Resources Review Commission, 135; and parks, 17, 58, 60, 61–62, 67, 72, 79–80, 163; postwar demand for, 67; and Sonoma County, 201–2; and watersheds, 47, 49–50. *See also* camping; fishing; hiking; parks

Redwood City, 27, 28, 37, 57, 114, 116, 120, 121

redwoods: in Bay Area, 27–28; Calaveras Big Trees, 26; California Redwood Park, 29; cutting, 25, 27, 28, 32, 33, 34; exhaustion, 28; extent, 27; Headwaters Forest, 33–34; on Mount Tamalpais, 25, 27; as national cause, 32; protection, 28–32; Redwood National Park, 25, 33, 97, 105; in state parks, 20, 28, 29, 32; and tourism, 30. *See also* Big Basin; forests; Muir Woods; parks; Save the Redwoods League; Sempervirens Club

Reed, Charles, 30

refineries, oil, 211, 212, 217, 223, 225. *See also* pollution: air and water

Register, Richard, 51, 52, 232, 253

Reid, David, 155

Reilly, William, 231

Rexroth, Kenneth, 16

Rhinehart, Chuck, 195

Richardson Bay, 89, 93

Richardson, Fred, 66

Richmond, 70, 77, 84; and agriculture, 41, 45; creek restoration, 51; and environmental racism, 231; parks, 70; quarries, 55; toxics, 223; water supply,

48. *See also* refineries; West County Toxics Coalition

Riley, Ann, 51, 52

rivers. *See* Guadalupe River; Napa River; Russian River; Sacramento River

Roberts, Lennie, 102, 107

Roberts, Tommy, 75

Roe, David, 209

Rohnert Park, 92, 193, 197

Roosevelt, Theodore, 25

Rosen, Marty, 92, 173, 241

Ross, Lorraine, 225

Rudee, Helen, 196

Russian River, 93, 195; dams, 49, 56, 197, 202; Friends of the Russian River, 202; Jenner Coalition, 195; protection of, 202–3; quarrying, 56

Rust, Audrey, 175

Sacramento, 44, 90

Sacramento River, 22, 54, 128, 211

Sacramento–San Joaquin delta, 116; agriculture, 39; Bay-Delta Accord (1994), 118; Bay-Delta hearings, 118–19; reclamation, 39, 117; size, 116–17; wetlands, 116–18. *See also* Central Valley; San Francisco Bay

Saika, Peggy, 233

Saint Helena: city, 192; Mount, 67, 90, 192

Salzman, Barbara, 122

San Andreas valley, 47; fault, 124; reservoir, 47

San Bruno, Mount, 55, 71, 105, 107, 114, 173; Friends of San Bruno Mountain, 105; San Bruno Mountain Watch, 105; Save San Bruno Mountain Committee, 105

Sand Hill Road, 101, 175

San Francisco, city and county, 27, 30, 31, 35, 61, 94, 97, 129, 131; airport, 112, 116; 113; and agriculture, 37, 38,

7; housing prices, 162; land-use battles, 106; and Midpeninsula Regional Open Space District, 164, 166; open space, 82, 98–99; parks, 67–68, 70–71; population, 161; protected lands, 108; quarries, 55; redwoods, 27–29, 99; San Mateo County Parks Division, 71; size, 108; water supply 47, 197

San Pablo Bay, 111, 116, 121; and Watershed Restoration Program, 52. *See also* wildlife refuges

San Ramon, 37

Santa Barbara, 126, 183

Santa Clara County, 145, 160; growth controls, 144–45, 149–51; housing prices, 162; and Local Area Formation Commission (LAFCO), 144, 149; and Midpeninsula Regional Open Space District, 164–65; mining, 54; open space, 82, 98, 103, 107–8, 164, 166; parks 67, 68, 71, 158; population, 160–61; protected land, 108; Santa Clara Committee on Occupational Health, 221; Santa Clara County Open Space Authority, 155, 166, 169; Santa Clara County Parks and Recreation, 71; Santa Clara County Planning Commission, 102; Santa Clara Valley Water District, 47, 49; Santa Clara Water Conservation District, 49; size, 108; toxics, 226; wealth, 162. *See also* Silicon Valley

Santa Clara Valley, 30, 31, 71; agriculture, 37, 39, 41–42, 44; farmland loss, 43; grains, 38, 57; orchards, 57; water supply, 49

Santa Cruz County, 27, 28, 29, 64, 123; highways, 84; and Midpeninsula Regional Open Space District, 164; open space, 100; parks, 67, 68, 71; redwoods, 27, 28. *See also* Big Basin

Santa Cruz Mountains, 71, 101; open

space, 99; quarries, 56; and recreation, 29; redwoods, 27–28, 29, 30

Santa Rosa, 49, 194, 196, 197

Sauer, Carl, 133

Save Mount Diablo, 13, 77, 146–47, 148, 155, 166, 169. *See also* Diablo, Mount

Save San Francisco Bay Association, 12, 17, 75, 76, 91, 119, 120, 135, 143; and Bay Area environmentalism, 115; and East Bay elite, 114; mass mobilization, 110, 111, 115, 120, 129; McAteer-Petris Act (1965), 115; origins, 114; and regulation of bay fill, 115; and Save Our Bay Action Committee, 115; and water inflow, 117, 119; and water quality, 222; and wetlands restoration, 121; and women, 114. *See also* Bay Conservation and Development Commission

Save the Bay. *See* Save San Francisco Bay Association

Save the Redwoods League, 25, 114, 171; character of, 31, 32; members, 31, 65, 76, 90, 100; and philanthropy, 32; protection of redwoods, 31, 32, |33; and Redwood National Park, 33; and state parks, 66, 99; and women, 30

Sayer, Jim, 142, 157

Schemmerling, Carol, 51, 52

Schooley, David, 105

Schoradt, Brent, 155

Schwarzenegger, Arnold, 169, 213

Scott, Mel, 62, 132, 133, 135

Sea Ranch, 126, 194–95; and Castle & Cooke (Oceanic Properties), 126

seashores, national: Cape Cod, 91; Point Reyes, 81, 90, 91, 94, 96 (*see also* Point Reyes)

Sempervirens Club, 26, 80, 100, 171; campaign for Big Basin, 29–31; founding, 29; and Sempervirens Fund, 100;

transportation *(continued)*
Area Rapid Transit (BART), 137,
194; Central Pacific Railroad, 28;
Metropolitan Transportation Commission (MTC), 140, 153, 214, 215;
North Pacific Railroad, 24; Sonoma-
Marin Rapid Transit (SMART), 203;
Southern Pacific Railroad, 21, 25, 30,
70, 124; Transportation and Environmental Justice Project, 245; Transportation Justice Coalition, 235;
Transportation and Land Use Coalition (TALC), 203, 214, 235, 247. *See
also* highways
Trudeau, Richard, 76
Trust for Public Land (TPL), 90, 170,
172–74, 199, 232, 241. *See also* land
trusts

Udall, Stuart, 33, 117, 124, 133
University of California, Berkeley, 62,
64; and agriculture, 42; and conserva-
tion leaders, 20, 31, 65, 73, 76, 77; and
planning, 132, 133, 134
U.S. Army Corps of Engineers, 51–52,
84–85, 113–15, 123–24, 131, 197
U.S. Bureau of Land Management,
80
U.S. Environmental Protection Agency
(EPA), 118–19, 202, 207–8, 210, 213,
219, 221, 225, 231; and Environmental
Justice Advisory Council, 233
U.S. Fish and Wildlife Service, 120,
121
U.S. Forest Service, 80
U.S. Geologic Survey, 222
U.S. Navy, 112–13
Urban Ecology, 153, 214
Urban Habitat: and air pollution, 214,
235; and environmental justice, 231,
232, 237, 242 243; financing, 244; and
regional policy, 246–47

Vacaville, 148
Vallejo, 35, 40, 55
Van Pelt, Helen, 88
Varian, Dorothy, 100, 102
Varian, Sigurd, 109
Vaux, Calvert, 58,
Vilms, Joan, 199, 201, 202
Vincent, Barbara and Jay, 76, 115
Violich, Francis, 132

Walker, Bob, 77
Walnut Creek, 146, 148, 153
Walter, Carrie Stevens, 30
Warren, Earl, 79
Washington D.C., 31, 208, 209, 231
Waste. *See* garbage; pollution; toxics
Water, ground, 42; pumping of, 48
watersheds, 45–50; map, 46; recre-
ational use, 49–50; watershed restora-
tion, 52. *See also* creeks; rivers; San
Francisco Bay
Water supply, 21, 30, 31, 216; for Bay
Area cities, 45, 47–49; Contra Costa
Water District, 47, 49, 50; East Bay
Water Company, 48; as growth con-
trol, 93, 144; Hetch Hetchy, 20, 21,
23, 34, 48, 49, 117; Marin Municipal
Water District, 25, 45, 47 49, 74, 87,
90, 93; private vs. public, 47–48; and
recreation, 49–50; San Francisco
Water Department, 45, 47, 104; San
Jose Water Works and Federal Water
Service, 48; Santa Clara County
Water Conservation District, 49;
Santa Clara Valley Water District, 47,
49; Sonoma County Water Agency,
47, 49, 197; Spring Valley Water Com-
pany, 30, 47; Tamalpais Land and
Water Company, 26. *See also* California State Water Project; Central Val-
ley Project; East Bay Municipal Utility
District

Windshield Wilderness: Cars, Roads, and Nature in Washington's National Parks
by David Louter

Native Seattle: Histories of the Crossing-Over Place by Coll Thrush

The Country in the City: The Greening of the San Francisco Bay Area
by Richard Walker

WEYERHAEUSER ENVIRONMENTAL CLASSICS

The Great Columbia Plain: A Historical Geography, 1805–1910 by D. W. Meinig

*Mountain Gloom and Mountain Glory: The Development of the Aesthetics
of the Infinite* by Marjorie Hope Nicolson

Tutira: The Story of a New Zealand Sheep Station by Herbert Guthrie-Smith

A Symbol of Wilderness: Echo Park and the American Conservation Movement
by Mark W. T. Harvey

Man and Nature: Or, Physical Geography as Modified by Human Action
by George Perkins Marsh; edited and annotated by David Lowenthal

Conservation in the Progressive Era: Classic Texts edited by David Stradling

CYCLE OF FIRE BY STEPHEN J. PYNE

Fire: A Brief History

World Fire: The Culture of Fire on Earth

*Vestal Fire: An Environmental History, Told through Fire, of Europe
and Europe's Encounter with the World*

Fire in America: A Cultural History of Wildland and Rural Fire

Burning Bush: A Fire History of Australia

The Ice: A Journey to Antarctica

RICHARD A. WALKER is Professor of Geography and Chair of the California Studies Center at the University of California, Berkeley. His publications include *The New Social Economy: Reworking the Division of Labor* and *The Conquest of Bread: 150 Years of California Agribusiness.* Photo by Zia Walker-Chinoy.